The Sustainability of Science
How Science Impacts the Environment, and What Can Be Done

The Sustainability of Science

How Science Impacts the Environment, and What Can Be Done

Edited by

Martin Farley
Email: martin@greenlabassociates.com

ROYAL SOCIETY OF CHEMISTRY

Hardback ISBN: 978-1-83767-068-0
PDF ISBN: 978-1-83767-220-2
EPUB ISBN: 978-1-83767-221-9

A catalogue record for this book is available from the British Library

The Royal Society of Chemistry is a charity, registered in England and Wales, Number 207890, and a company incorporated in England by Royal Charter (Registered No. RC000524), registered office: Burlington House, Piccadilly, London W1J 0BA, UK, Telephone: +44 (0)20 7437 8656.

For further information see our website at www.rsc.org

For general enquiries, please contact books@rsc.org

For EU product safety enquiries, please email books@rsc.org or contact Royal Society of Chemistry Worldwide (Germany) GmbH, Römischer Hof, Unter den Linden 10, 10117 Berlin.

Printed in the United Kingdom by CPI Group (UK) Ltd, Croydon, CR0 4YY, UK

Foreword

BEN L. FERINGA

University of Groningen, The Netherlands
Email: b.l.feringa@rug.nl

"Fascinated by the marvels of Nature" and "taken inspiration from living systems" are frequently used opening phrases of our manuscripts that present the amazing discoveries in the natural sciences. But let's ask ourselves the following question from a sustainability perspective: Do all these wonderful advances from scientific research and the way we perform research and arrive at these outcomes indeed match the beauty of natural systems that we admire? When I was confronted with the societal challenge of how to recycle our materials, the notion that our own body recycles nearly half its weight of the ATP fuel we use each day *via* tiny bio-nanomachines, was a pretty shocking experience. Modern society is built on the successes of the joined scientific research efforts of many disciplines. However, the rapidly growing scientific developments and their societal impact, while confronted with the many challenges in the world ranging from climate to resources, is also a wake-up call for all members of the scientific community. If we acknowledge that many facets of our society must adapt in the years ahead, scientists should take the lead towards a more sustainable way to perform their research. Shaping the path towards a green future based on scientific discoveries and innovations also demands a new way of training our young scientists, embedding the insights of environmental issues as an integral part of their education and scientific training. I hope this book is a source of inspiration to all of us and provides information and guidelines on how we can accomplish this important mission.

In this book the insights and knowledge of several experts not only confront us with the challenges ahead but also provide a perspective of what

The Sustainability of Science: How Science Impacts the Environment, and What Can Be Done
Edited by Martin Farley

Published by the Royal Society of Chemistry, www.rsc.org

research might look like in the future. For me personally it was an eye-opener to experience the broader environmental impact of our research aiming at sustainable chemical transformations. We are proud to have recently developed within our team a new catalytic process with a low environmental factor (E-factor) regarding atom-economy. However, it was the various insights regarding the total impact of the research process, as discussed in the chapters ahead, which dwarfed our wonderful results.

Obvious is the impact of our special research facilities, for instance high energy costs associated with ventilation in chemistry labs to create a safe working environment, but similarly the use of clean rooms in materials science. The scientific equipment and consumables like plastic gloves and pipettes, chemicals and solvents necessary to run even the simplest chemical experiments as well as the waste produced, confronts us with the real environmental impact of current scientific research. Beyond any doubt, the insights into these aspects of our scientific research will help to drastically improve and be a source of inspiration to change the face of scientific endeavors.

The authors of this book on the sustainability of science present fascinating insights into the environmental consequences of our contemporary scientific activities. With the era of digital science and the massive use of AI and large data processing at the doorsteps one should realize that it also comes with excessive energy demands. On the other hand, facing the challenges of more complex chemical systems with large parameter space, AI in combination with automation and robotics offers great opportunities to target more selected and environmentally benign experiments. New separation methodologies and micro-analytic techniques can drastically reduce the amounts of chemicals and solvents used. Recycling and re-use should be integrated into our daily efforts to establish the green chemistry of the future. Various impressive actions have already been taken to establish more sustainable laboratories and research protocols, including certification and policies, as discussed in this book.

Our notion of many other environmental aspects of scientific research might be much less obvious. Sharing the excitement of our discoveries with the scientific community and the benefits of international cooperation go hand in hand with the environmental costs of conferences, publishing, and the burden of traveling, among other factors. You will be amazed with the perhaps unnoticed impact of scientific research from a sustainability point of view while reading the various chapters. But far beyond creating awareness, the authors also provide perspectives on where and how we can make a difference and reshape our scientific research and adapt to a sustainable future.

If we scientists advocate to society at large to be the leaders in inventing our future, we should start in our own daily practice. Most important is our science education where sustainable aspects should be integrated from the start. Our young talents, who will be the leaders of our society and future industrial innovators, deserve to be trained in a system of "education for sustainable science". We will greatly depend on the creativity and perseverance of our students and young scientists to shape this new science. There are tremendous opportunities and let us not forget creativity and imagination are really sustainable in this world. I hope this book will give you lots of inspiration during your next adventure in the scientific world.

Ben L. Feringa

Preface

The writing of a book titled **The Sustainability of Science** could to some seem without frills. In reality though, I cannot stress enough the passion that has gone into this writing. The energy which has been invested reflects the excitement that is generated in the collective of chapter authors, which I hope as a reader you will sense. Each author is not a full-time author. These are individuals who have day jobs and are actively working to study and address the issues they investigate. The time these authors invested did not come from their paid working hours, it came from weekends, holidays, and during busy commute times in jammed trains. It came from early wake-ups to finish references, lunch breaks, and late evenings all piled up to represent a 3-year journey from beginning to end. And don't forget the emails, way too many emails. For this alone, I'm so incredibly thankful for their time, because while we all try our best to leave a better world, life can pass quickly.

With regard to why this book exists, I want to give a heartfelt thanks to the Royal Society of Chemistry, who for some reason trusted me to edit this book and showed incredible patience in the process. They not only were supportive in engaging authors and understanding in slipping deadlines, but have shown true commitment to the topic by making the content of this book open source. This keeps with the spirit of what many of the authors have highlighted is a crucial aspect of sustainable science, and will help with the recognition of those who provided their time as well as the dissemination of their knowledge.

While much has been done, the environmental impact of science is relatively still in its infancy. Many of the co-authors say this best, but the importance of this work will only grow. Science is growing, as are the benefits associated. But so are the challenges we face in terms of addressing and adapting to climate change. Science will be and already has been affected. To not address this would be to stagnate, something we cannot accept.

The Sustainability of Science: How Science Impacts the Environment, and What Can Be Done
Edited by Martin Farley

Published by the Royal Society of Chemistry, www.rsc.org

The title of this book is **The Sustainability of Science**, and as the word "sustainability" can mean many things to many people, I must emphasize that we are focusing on environmental sustainability. Addressing environmental sustainability does not happen in a silo. As science itself becomes increasingly interdisciplinary, so do the mechanisms of supporting effective scientific operations. That is why I'm particularly enthused that this book covers significant environmental aspects but also extends beyond to address research quality, ethics, reproducibility, growth, and much, much more.

While there is much to do in this area, significant work has been done. One particular challenge in editing this book was to cover all the appropriate topics, whilst ensuring a balance of voices are heard. The chapter authors in this book have a breadth of viewpoints, some of which you may not agree with. However, their involvement represents their experiences and wisdom, and I'm sure in sharing them we can help achieve progress and stimulate evolution. I should add that there are of course more experts who could have contributed, and this should not be considered a complete list of influential voices. Simply out of capacity and limits on time, not everyone I would have liked was engaged in writing this and I would encourage the reader to seek out the thousands of references provided. While this book took effort to produce, it represents an exponentially greater amount of work behind it from some incredible people around the world, some of which are sadly no longer here.

A Brief History of Sustainable Laboratories Beginnings

Regarding those who paved the way, I wanted to use my introduction to recognize the origins of this topic and provide a brief unofficial history of "green labs". The book covers more than this, but the concept of green labs is a core building block for many of us involved. To assist with this, I engaged two individuals who have shown particular brilliance over the years to pioneer this topic for the rest of us. The first is Kathy Ramirez-Aguilar, who founded the University of Colorado Boulder Green Labs Program in 2007. From its beginnings as a pilot relying on Kathy herself, it gained campus funding from the Environmental Center and Facilities Management in 2009 at first to cover 25% of her time. From there it's gone from strength to strength as an exemplary program for the sector, as Kathy herself is one of the most recognized names in green labs. At the time of this publication the team will have 3 full-time staff, and up to 17 student employees supporting efforts across the university.

The second individual I engaged is Allen Doyle, to whom I personally owe the opportunity I was given to work in this area (a longer separate story, but in short Allen thoughtfully responded when I reached out to connect me with the now defunct organization "S-Lab", and subsequently the University of Edinburgh, to as far as I know, create the first role in sustainable laboratories outside the US in 2013). Allen started green lab programs in the early 2000's with Katie Maynard at University of California (UC)

Santa Barbara and then later UC Davis. The effort at UC Santa Barbara was named LabRATS and produced publicly available content which for me and many others was an early inspiration in the area. He himself cites David Orr's book "The Earth in Mind" as his own inspiration for developing green lab efforts which extended beyond the benchtop or building to civic engagement.

In speaking with them both as well as early leaders around the sector, one name appears to be at the origins of this topic – Phil Wirdzek. Phil worked at the Environmental Protection Agency (EPA) as a microbiologist turned administrator and was the energy manager for EPA's laboratories, which use significantly more energy than its offices. He had the novel idea that the research buildings that support the Department of Energy (DOE) and EPA research operations should exemplify some of the efficiency and environmental concepts that they support. With that idea, Phil and DOE's Will Lintner started a conference in 1999 under the name Laboratories for the 21st Century (Labs21), which aimed to bring together engineers, architects, and others involved with public and private building operations to address this challenge. EPA and DPE ran Labs21 until Phil retired from EPA in 2006, when he created a nonprofit, the International Institute for Sustainable Laboratories (I2SL). To this day, I2SL hosts an international conference in the US each year which attracts professionals from around the world to share best practices and innovation on the subject. Phil sadly passed away in 2021, but his energy and passion for the subject lives on through those he inspired.

While not a comprehensive list of those who were early leaders in this area, I want to mention a few others who took the leap in those early years. There were early green lab programs at Harvard University and the Universities of Colorado (Boulder), California (Davis and Berkeley), Arizona State, and Michigan, the latter of which was an early effort led by Sudhakar Reddy. The imperative for these and the previously mentioned green lab programs extended consideration beyond building energy efficiency to overall resource efficiency in laboratories and paved the way for global green lab efforts in the future. Green labs moved internationally and by 2009, S-Lab was launched in the UK by Peter James. This organization provided a means for me to be funded in the aforementioned role (Sustainable Labs coordinator) at the University of Edinburgh. In Germany, an organization titled EGNATON led to the publication of a book titled "The Sustainability Laboratory Handbook", which while similarly titled, focuses predominantly on building operations and design.

I'll stop this brief history at around 2010, as the area started to blossom with green and sustainable lab efforts in many more organizations, more than can be listed within reason. Apologies to those who have not been named, and credit to all who took the time to invest in this area.

The growth of green science since its origins does not mean our job is done though, if anything it highlights the increased awareness of the task at hand. As such, this book represents a perspective in this moment in time, and we can hope that a decade from now, much of it might feel out of date with the progress achieved (if we do our job well). That being said, science doesn't take place in a vacuum and there will be many more impacts on it beyond the environment. We can't predict everything, but our effect on the planet appears clearer and clearer. If we want to continue to benefit from research, we must adapt it to the home we make for ourselves.

Martin Farley

Contents

The Sustainability of Science: How Science Impacts the Environment, and What Can Be Done
Edited by Martin Farley

Published by the Royal Society of Chemistry, www.rsc.org

CHAPTER 1

The Scientist's Guide to Designing a Sustainable Laboratory†

ALLEN DOYLE,*[a] JACOB WERNER[b] AND GEORGE KARIDIS[c]

[a] Albuquerque, New Mexico, USA; [b] Ellenzweig, Boston, Massachusetts, USA; [c] SmithGroup, Detroit, Michigan, USA
*Email: adoyle0128@gmail.com

> *"You never change things by fighting the existing reality. To change something, build a new model that makes the existing model obsolete."*
> *R. Buckminster Fuller*

1.1 Your Role in Sustainable Lab Design

Laboratory design starts with big dreams – followed inexorably by months of meetings, difficult decisions, and the minutiae of technical detail. From a sustainability perspective, good design is the process that gets a building off on the right foot, and makes essential selections that will determine both its functionality and its environmental footprint. Your voice and determination to reduce environmental impact is essential.

Take heart. Many low impact solutions have already been devised that can be integrated into this new building. This chapter is designed to give you the tools to communicate with and enrol all the design stakeholders, as each has important considerations.

Our intended audience is the Principal Investigator (PI) who has been asked by their dean or VP to serve on a design committee that will guide the

The Sustainability of Science: How Science Impacts the Environment, and What Can Be Done
Edited by Martin Farley

Published by the Royal Society of Chemistry, www.rsc.org

conceptualisation and delivery of an important new building or major renovation. While they may be new to design and construction terms, they grasp the essential components of delivering a high performing building and incorporating persistent features for reliability and efficiency. This "building committee member" has a natural desire and responsibility to do this with the least impact on the environment and resources. Researchers in this role may have to learn what seems like mundane principles and technologies, yet they are essential for long term productivity and rightly navigating complex and overlapping objectives. Their input will be essential early in the design process, and in following through to the end to make sure features are not removed shortsightedly (Figure 1.1.)

We also welcome readers that work with researchers to design laboratories. Design subject matter experts (SME) may already be familiar with these topics. Furthermore, they may find this chapter a useful reframing that will help to connect with researchers about sustainable strategies. Contractors, developers and operators of "Spec Labs", with unknown future tenants, may find that their target clients – start-ups and new investigators – have resource conservation on their mind, yet don't know where to turn. By demonstrating measurable sustainable design practices, developers may increase their reputation and market share of tenants.

Get it From the Top: Leadership, Teamwork, Life Cycle Performance and Low Impact Design

It cannot be overstated how important leadership buy-in is to sustainable design as well as life cycle performance of laboratories. The scientist member of a design committee may be able to nudge further understanding of lab infrastructure and technology to their chief business officer (CBO), and this will empower the whole design team. This leader should be versed in the fundamentals of building core and shell design, materials, safety measures, electrical and mechanical systems, and heating, ventilation and air conditioning (HVAC). Their focus on sustainability needs to be unwavering, and practical.

Wendell Brase, Associate Chancellor for Sustainability at the University of California, Irvine, has a rare mix of technical knowledge, financial vision and team building. His review of trouble points with staff reveals broad knowledge and action; "Sometimes sustainable design comes down to good gutters, and other times it comes down to advanced pre-qualifications of design firms". With this Smart Labs strategy, UCI has delivered or refurbished over one dozen laboratory buildings on time and on budget, and their energy use was cut by 50%.[1]

Read this chapter, then hand it to your dean, vice president or vice chancellor, and encourage them to adopt a similar approach and to learn about lab infrastructure. In Wendell's terms, "hire innovative, engineers-with-attitude who will question assumptions, then let them go to work".

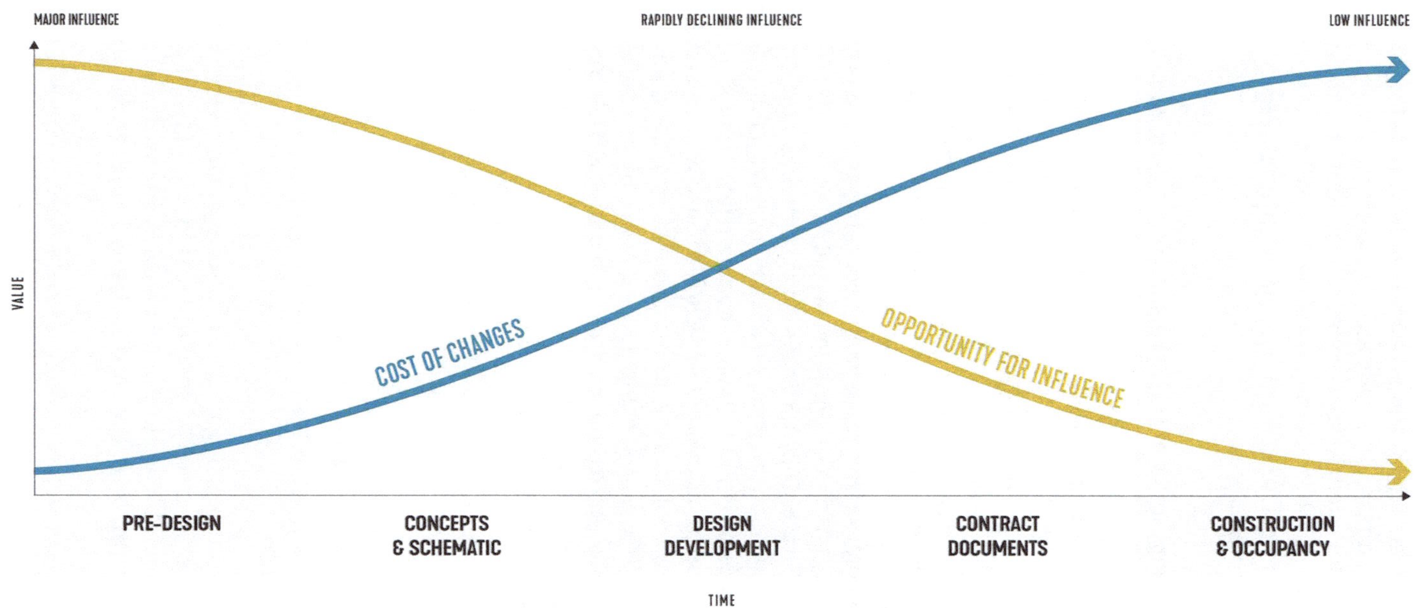

Figure 1.1 Input early in the design process has the most impact as changes become more difficult and expensive later on.

Design perspectives come from many angles, from research leadership, design departments, environment health & safety (EHS) staff, architects, engineers, contractors and technicians. Ultimately, these professionals are there to support the scientists occupying the building for the decades ahead. We hope to guide all these audiences toward sensible selection of building materials, mechanical, electrical and plumbing (MEP) strategies, automation and documentation.

Setting Your Energy and Carbon Targets: The Lab Benchmarking Tool and Labs2Zero

Labs have a dominant impact on the energy and greenhouse gas budgets of any enterprise, and energy and carbon conservation can slide off the table unless there is a challenging target for the design team. The International Institute for Sustainable Laboratories (I2SL) launched the Labs2Zero program in 2022 with a goal to help decarbonise the world's laboratories and create tools for owners and designers to reach this bold and imperative goal, which is consistent with the knowledge and values of science itself.

The I2SL Laboratory Benchmarking Tool (LBT) has the best public database of laboratory energy consumption available, and it is free. While historic data on the tool were from US institutions, I2SL has now opened it up to international institutions to broaden its database. With filters for percent lab floor area, science discipline, climate, commercial *vs.* public sector, you can find a comparable project cohort on which to base your sustainability goals. You can compare your design on a regression or receive an Energy Score between 1 and 100 (with 100 being the best performance against similar type labs). Adding additional details on your lab building's HVAC design, lighting, fume cupboards, and automation can also generate "actionable insights" to modify the design before it is too late in the process.

An operational emissions score, embodied carbon score, and "Design2Zero" module for scoring new construction projects are also in the works. Become a partner with I2SL and enter your data, then help optimise the tool.

How low can you go?? Dare to go lower!!

LBT.I2SL.org I2SL.org/Labs2Zero

1.1.1 Your Role

So, you were talked into being on a design team by your VP or your Dean, and it is going to take up a lot of your time. Why should you care, and what can you do? Well, a lot. Your influence on this team will have lasting impact, and significantly improve the environmental and social impact of your laboratory.

All stakeholders in the lab design process have the power to shape facilities toward sustainability – especially the owners and occupants. Project decisions are not just neutral choices. Seemingly small decisions can ripple through a lab's lifecycle, affecting energy use, carbon emissions, material consumption, water management, recycling, occupant comfort, resilience, and adaptability to a changing climate. Understanding these implications empowers you to weigh the cost-benefit of design choices and make informed decisions that lead to smarter, greener outcomes.

1.1.2 Putting Together Your Dream Team

You will need a strong team of systems thinkers and subject matter experts in design, engineering, construction, logistics, contracts, and finance to convert your dreams into a building. Whole-brained teams, representing a full cross section of your organisation, are more likely to make smart and well-considered decisions. Equally important, make time to have meetings not only with the designers and builders, but also internal discussions amongst your stakeholders. Consider "day one" factors, like providing the labs with the right number and size of fume cupboards and lab gasses, but also "future proofing" decisions, like how easy it will be to renovate for new research programs or HVAC modifications.

An early key decision is hiring an Architecture and Engineering (A/E) team. Select A/E firms with experience designing the kinds of laboratories that you are looking for. And make sure that they have a strong track record of progressive, sustainable design. Later, you may need to hire a builder, or the builder and architect will come as a unit. Select a builder with a strong track record of building sustainable laboratories.

Signs that your Capital Expenditures or design team may not be quite ready to sustainably build a lab:

a) Unrealistically low budget
b) Non-uniform enthusiasm about the design process and sustainability
c) Reliance on rules of thumb (proscriptive *vs.* performance-based specifications)
d) Lack of priority for sustainability as a driving motive for design
e) Your consultants suggest your expectations are out of alignment

1.1.3 Creating a Vision for Science and Sustainability

So how do you make sustainability a fundamental pillar of the project vision? Start by forgetting what is wrong with your current lab. Rather, imagine a new lab that works the way you want it to. Encourage your stakeholder team to "represent the future you": more productive, happier, and with a positive impact on the world and its environment.

A successful laboratory design matches the vision for the research program it is meant to house. Each research program has unique needs, ranging from specific spatial requirements, to equipment installations, to safety considerations. Your new lab can have a completely different configuration than your current lab. Yes, it can have a new and more efficient design with more sustainable mechanical and electrical equipment. AND it can also include flexible labs that adapt more easily to change over time. Most importantly, it can include more resource sharing, for example shared "core" laboratories, more efficient cold storage, more efficient office design, more collaborative spaces, *etc.* Work with your A/E team to explore smart lab design strategies to streamline both building systems AND modernise your laboratory workplace. Integrating these two powerful elements into your lab design will foster a productive, sustainable environment for scientific discovery.

Setting metrics for sustainability performance is critical and we hope to give you strategies and examples in the pages ahead. These targets could encompass space efficiency metrics measured as Net Square Metre per Full-Time Equivalent employee (NSM/FTE), water use intensity, energy use intensity (EUI), greenhouse gas emissions intensity, daylighting effectiveness, and more. Concrete metrics ensure that sustainability objectives are not just lofty aspirations, but tangible, trackable, and accountable outcomes.

Case Study: The UK's First Carbon Neutral Laboratory – GSK's Centre for Sustainable Chemistry

The Centre for Sustainable Chemistry (CSC) at the University of Nottingham, partially funded by GSK, is dedicated to pioneering research in sustainable chemistry. It aims to develop 'clean and green' standards to minimise environmental impact and ensure that the new chemistry developed is energy-efficient, resource-efficient, and sustainable (Figure 1.2).

The building embodies sustainability, utilising advanced technologies to achieve carbon neutrality over its lifespan. With a timber frame construction, it operates on renewable energy sources including solar power and sustainable biofuel. The building services throughout were designed to minimise energy use wherever possible, including an exemplar naturally ventilated laboratory. Excess energy generated by the building is intended to provide enough carbon credits over 25 years to offset the carbon used in its construction. It achieved both BREEAM 'Outstanding' and LEED 'Platinum' awards for its commitment to sustainability.

The main project drivers focused on:

- Minimising energy-in-use from environmental systems and user equipment
- Minimising day one embodied carbon content
- Not relying on carbon offsetting (*i.e.*, 'right-size' the systems)
- Using a fuel source that does not compete with food supply
- Regenerating its embodied and carbon-in-use over 25 years

Figure 1.2 Centre for Sustainable Chemistry (CSC) at the University of Nottingham, Reproduced with permission from Martine Hamilton Knight/Builtvision.

To meet the project's objective of repaying its embodied and carbon-in-use over 25 years, AECOM (the engineering firm tasked with its design) undertook an 'energy in use' simulation. This followed a methodology similar to that proposed by CIBSE TM54. The output data from this model assisted in refining operational schedules, informing the optimum level of renewables required, and evaluating the post-occupancy monitoring — enabling 37% less energy use than traditional chemistry laboratories.

1.1.4 The Lab Design Paradigm Shift

One key decision that profoundly influences a lab's sustainability is space planning. Traditionally, lab design has prioritised functionality, leaving sustainability as an afterthought. However, a paradigm shift is occurring, recognising that sustainable lab planning should be integrated from the outset (Figure 1.3).

Sustainable planning starts by considering various functional needs and grouping spaces based on energy use, ergonomic requirements, and implications for low-carbon materials. A hallmark of new laboratory designs is the smart separation between dry and wet lab spaces. This separation aligns with energy efficiency goals, ensuring that areas with different energy profiles are designed accordingly. For instance, dry office programs are excellent candidates for low-energy design, featuring elements like daylighting, natural ventilation, and low-carbon finishes. In contrast, wet lab spaces can be

Figure 1.3 Lab planning involves integrated information and stakeholders. Reproduced with permission from Ellenzweig.

optimised for energy efficiency through demand reduction, advanced automation, and flexible modular layouts. Mixing dry and wet spaces indiscriminately can short circuit the strategies for each, leading to suboptimal solutions for both.

1.1.5 If Sustainability is Tacked on, It Can Be Torn Off

Researchers have an essential perspective in experimental capacity and space planning. Experimental capacity may include built-in, large instrumentation such as MRI or cyclotrons, properties such as vibration control, electrical capacity or clean current, or shared equipment for washing, sterilising, cold storage, or drying. These functions are central to the research or testing mission, and must be met to have a high-performance building. We hope to provide examples of how this may be done in the most elegant, simple, and low impact manner. Similarly, space planning for the future is essential to meet the mission of an enterprise, and it also may be approached over time with focus and flexibility to avoid wasted resources and optimise interdependent functions for a net benefit.

All members of a design team may focus on the root issues, mission and necessary features of a building, and how they may be delivered in a simple and robust manner. This is exemplified from the capital planning perspective by a design overview from the University of California, Irvine,[1] where optimum width-to-length ratios and column spacing are proposed, for example, as well as details of cladding, and HVAC automation. These elements are fully integrated in a manner that is truly holistic, such that there are no superficial "sustainable features" that can be removed without making the building less efficient, uglier, and more expensive.

1.1.6 Support Spaces That Make a Difference

Even seemingly minor decisions about support spaces can have far-reaching consequences for building operations and sustainability. Paying attention to "back of house" areas like loading docks, trash zones, and storage can lead to a facility that is easier to maintain, clean, and repair. For example, providing adequate space for reuse, recycling, and donations can substantially reduce the volume of waste generated by the facility. Engaging custodial and facilities personnel early in design discussions and plans for support facilities will ensure that decisions align with the practical needs of those responsible for maintaining the lab.

1.1.7 Navigating Budget Constraints with Sustainability in Mind

Inevitably, construction projects encounter budget constraints. When faced with the challenge of realigning a project with budget limitations, a process called "value management" (or "value engineering") is often employed. This process involves brainstorming ways to provide the most of what you need and want for what you can afford, often rethinking how things may be accomplished to create dual benefits. But when poorly implemented, it can devolve into mere cost-cutting at the expense of future operating costs and functional utility.

It is essential to resist the temptation to disproportionately cut sustainability-related features to preserve laboratory functions and/or architectural gestures. Disproportionate cuts can result in a loss of space quality, reduced mechanical performance, compromised occupant wellness, and diminished sustainability, while investments during construction may result in savings over the future lifetime of the building. A more strategic approach is to make proportional cuts across all aspects of the project, ensuring that the building program and its features are balanced and logical. For example, a dramatic atrium may have an expensive first cost and add maintenance costs for the care of glazing and interior finishes and annual fire and smoke safety testing.

Watch how your colleagues on the design team are balancing these constraints. If there is some backsliding away from sustainable design, speak up to refocus and reinforce your commitment to high performance facilities, with low environmental impacts.

1.2 Sustainable Design Concepts for Laboratories

> *"When you change the way you look at things, the things you look at change." – Max Planck*

Traditionally, facility owners and designers have thought about the sustainability of laboratory buildings as the impact of facility energy use, water use, and land use planning. Sustainability programs like LEED and BREEAM have encouraged a comprehensive approach to sustainable design, but they tend to overlook the broader implications of development. These programmes limit their scope to the direct impact of the facility itself, neglecting the downstream effects of development on power infrastructure, supply chains, and manufacturing. While this focus on operational aspects like energy efficiency and water use remains crucial, it is no longer sufficient in the face of growing environmental concerns. Recent shifts toward a more holistic understanding of a development's impact, driven by movements like Environmental Social and Governance (ESG) investing, have prompted facility owners and designers to broaden their perspective.

1.2.1 The Three Scopes of Carbon Emissions

The emergence of ESG investing has pushed stakeholders to consider the broader context of sustainability, looking beyond the facility's direct impact. This includes understanding and mitigating greenhouse gas (GHG) emissions, classified as Scope 1, 2, and 3.

- **Scope 1 Emissions:** these are the direct emissions from a facility's operations, such as on-site burning of fossil fuels, methane releases, or refrigerant leaks.
- **Scope 2 Emissions:** this category includes indirect emissions associated with energy use, typically associated with the production of electricity from the grid. It urges us to consider the carbon emitted during the generation of the energy we consume, emphasising the importance of using renewable energy sources.
- **Scope 3 Emissions:** these emissions encompass the indirect emissions or "embodied carbon" associated with the entire lifecycle of a facility, including its supply chain, construction, and operation. They include the emissions from materials extraction, manufacturing, shipping, and the building process. The I2SL Laboratory Benchmarking Tool is developing an embodied carbon calculator for new construction of laboratories.

Embodied carbon, specifically in the context of laboratory facilities, is a significant concern. Lab buildings are characterised by intensive structural framing, robust finishes, and intricate mechanical systems. Lab building columns, beams, and foundations are designed for vibration performance, and thus are inherently heavier per square metre than those in office or residential buildings. Additionally, the finishes and built-in casework in lab facilities are far more substantial compared to their office counterparts.

1.2.2 Effective Design Choices for Reducing Embodied Carbon

When new laboratory construction is deemed necessary, it is essential to set ambitious low carbon design goals from the outset. This applies to both new ground-up buildings and to renovated areas within existing facilities. Key considerations for achieving low carbon and healthy design include:

- **Greenhouse Gas Reduction:** establish clear goals for reducing greenhouse gas emissions over the life of the facility. This should include not only operational energy use but also the carbon impact of construction materials. Prioritise low-carbon materials and construction processes.
- **Energy Efficiency:** prioritise energy efficiency measures, such as heat pump-based mechanical systems, all-electric designs, and efficient lighting. Incorporate renewable energy sources like solar panels to reduce reliance on fossil fuels.
- **Electric Vehicle Charging:** plan for electric vehicle (EV) charging infrastructure within the facility to support the transition to cleaner transportation options.
- **Mass Timber:** consider the use of mass timber for structural elements. Mass timber is a sustainable alternative to traditional building materials like concrete and steel. As trees grow, they pull carbon out of the atmosphere and store it in their wood. Responsible harvesting of timber captures the carbon stored in wood and machines it into low-carbon wood building materials. Using carbon storing materials in buildings diverts the stored carbon from the atmosphere for the life of the building. This is referred to as "sequestering" carbon. In addition to wood, many other biogenic, carbon-storing building materials are sustainable options for building projects, including: bamboo, fibre boards, straw bales, and wool insulation, among other emerging technologies. Many of these materials are also extracted and manufactured using low carbon processes, making them even lower carbon options.
- **Low Carbon Cladding and Glazing Systems:** opting for cladding materials with a lower carbon footprint, such as those made from recycled or sustainably sourced materials, will also contribute to reducing embodied carbon. Replacing metal elements in the building's exterior walls with comparable wood alternatives can further reduce embodied carbon, while adding aesthetically pleasing natural elements to the design. Wood cladding may be an option for some projects with compatible building height, site conditions, and climate. "Timber Curtainwall", which utilises wood framing on the interior with aluminium glazing on the exterior, is also an excellent option for compatible projects.
- **Demountable Partitions:** flexible and reusable partitions can enhance the adaptability of lab spaces, reducing the need for extensive renovations in the future and decreasing embodied carbon over the building's lifespan.

- **Wood Casework:** choosing wood casework over traditional materials not only reduces embodied carbon but also lends a warm, natural aesthetic to lab interiors.
- **Low-Carbon Concrete:** utilising concrete mix designs with lower carbon footprints is a promising avenue. Optimisation of concrete mixtures can significantly reduce the embodied carbon associated with structural elements. Better mix designs often have minimal impact on cost, performance, or schedule.
- **Low-Carbon Finishes:** choose finishing materials that have a lower carbon footprint. Low-carbon, lab grade finish options exist for many product categories, including flooring, ceilings, paints, and countertops.
- **Carbon-Smart and Water-Smart Landscaping:** integrate sustainability efforts to the exterior environment during construction. Implement carbon-smart landscaping practices, such as xeriscape, native plantings, salvaged materials, and bioswales to capture water, reduce runoff and recharge aquifers.

Case Study: A Timber Laboratory – Michigan State STEM Complex

Completed in 2021, the STEM Teaching and Learning Complex at Michigan State University is the campus' dynamic hub for science education. The Complex comprises two new teaching lab wings (11 270 m^2), an entry atrium/classroom addition (1235 m^2), and the adaptive reuse of a decommissioned Power Plant (4515 m^2) (Figure 1.4).

Figure 1.4 Michigan STEM Building. Reproduced with permission from Ellenzweig.

The STEM complex houses undergraduate teaching labs for chemistry, biology, physics, computer science, and materials science, as well as unassigned Project Labs that can be utilised for future courses and student projects. Prior to this facility, teaching labs for these disciplines were located in separate buildings across campus, with no linkage among them. In the new building, these teaching labs are co-located to enable students and faculty from different disciplines to intermingle. Consequently, students gain a greater awareness of all disciplines, which, in turn, fosters deeper interdisciplinary curiosity than was possible in the discipline-specific buildings.

The complex is the first large-scale mass timber project in Michigan. The new STEM wings use glue-laminated (GLT) and cross-laminated timber (CLT) for building structures, decks, and shear walls. The new floor area inside the Power Plant was created with structural CLT floor panels on new and existing steel framing. A conventional steel frame, for the STEM wings, would have required 24 times the embodied carbon needed for fabrication compared with this timber frame. In total, the mass timber components sequester 1856 metric tons of CO_2.

1.2.3 Adaptive Reuse: A Low Carbon Alternative

Before building a new laboratory, first consider whether an existing structure can be renovated to meet the research needs. Adaptive reuse of buildings not only helps preserve architectural heritage but also significantly reduces carbon emissions associated with new construction. Adaptive reuse has many advantages:

- **Carbon Efficiency:** existing buildings have already incurred carbon emissions from their construction. By repurposing them, we avoid the carbon-intensive processes of demolition and new construction, making renovation an inherently low-carbon option.
- **Urban Location:** many existing buildings are situated in urban areas with well-established infrastructure. They are often intricately connected to power and water grids, transportation networks, and amenities. This reduces the need for new infrastructure development and further lessens the project's carbon footprint.
- **Construction Impact:** demolition and new construction can have substantial impacts, from new materials manufacturing to increased waste generation. Renovation significantly reduces these impacts.
- **Heritage:** existing buildings often have cultural significance. Repurposing them allows us to reinvest in our cities, while transforming the building capabilities to meet our evolving needs.

Case Study: Renovating to Renew Science Facilities – Cornell University Plant Sciences

The phased renovation of the historic 15 330 square metre Plant Science Building provides the School of Integrative Plant Science (SIPS) at Cornell University with a revitalised home for its next century of research and teaching. The renovated spaces will flexibly support the five constituent plant science disciplines, create a unified identity for SIPS, enhance the student experience, enhance collaboration between students and faculty, and create a low-carbon environment for all of these activities (Figure 1.5).

In support of research synergies and energy conservation, the renovation creates ten lab suites where groups with complimentary research foci will be co-located, with four research suites renovated in Phase 1 and six suites in Phase 2. The research suites include flexible shared open labs and dedicated support labs. Shared support labs (environmental rooms, growth rooms and other specialised labs) are located outside of the suites for use by all of the research groups. Teaching labs, meeting and collaborative spaces, faculty offices, and student-focused spaces cluster about the rejuvenated historic corridors and central vertical circulation zone. The entire design incorporated the input of campus accessibility advisors. The designers worked with state code officials to implement state-of-the-art chemical management strategies into the design.

The renovation carefully integrates new infrastructure to support the phasing strategy and to minimise impacts on the historic structure, including custom pre-manufactured rooftop equipment and a machine-room-less elevator (state historic preservation authorities reviewed and approved all proposed alterations affecting the historic

Figure 1.5 Digital mock-up of the Cornell Plant Science building. Reproduced with permission from Ellenzweig.

character of the building). The design includes a variety of strategies to reduce operational carbon expenditures to a "net-zero-ready" energy use intensity of 319 $kW\,h\,m^{-2}$. These strategies include energy-zoning of the space program elements, ventilation systems that decouple thermal needs from air flow, heat-pipe energy recovery, cascading air flow from non-labs to labs, lake-water cooling and low temperature hot water heating, and an insulation and air barrier system design developed using hygrothermal modelling.

The renovation minimises embodied carbon expenditures through preservation and reuse of much of the original building (structure, masonry envelope and historic interiors), extensive recycling of construction waste, and specification of low-carbon interior materials (linoleum, rubber, and cork flooring, wood doors, "cradle-to-grave" carpet).

1.2.4 Wellness and Active Design

We spend the vast majority of our time indoors. Some statistics put indoor time spent at more than 90% of all hours. This has dramatic implications for human health and wellbeing. Buildings serve as filters for the environment, shaping the light we see, the air we breathe, the water we drink, and the food we eat. Thus, building designs have a direct impact on the health and wellbeing of future building occupants, whether we acknowledge it or not. Simple decisions, such as where to put stairways, or which type of window glass to use, can have significant impact on occupant steps walked, biorhythms, mood, and more.

- **Active Design:** research shows that occupants who have reasons to walk around during their work day, and especially those that use stairs, have both better overall health and better emotional wellbeing. Building designers can use this knowledge to design buildings that deliberately increase occupant activity to boost occupant health and wellbeing. This principle, and many related strategies, are called "Active Design". Reference Fitwel (https://www.fitwel.org/) for actionable strategies to incorporate active design into your facility.
- **Design for Wellness:** a host of design strategies have been shown to both improve occupant health and business outcomes. Specific design choices can help reduce sick days, increase attention and focus, and boost creativity. Reference the International Well Building Institute (https://www.wellcertified.com/) for a wealth of research on these topics, as well as direct actionable strategies for incorporating design for wellness into your facility. Basic ventilation recommendations may be found in ASHRAE Standards 62.1 and 241.
- **Healthy Materials:** during the 20th Century, changes in building technology have led to laboratories constructed with a wide array of materials, many with poorly understood chemical makeup. More research

and attention have been given to product durability, fire resistance, and stain resistance than to product impact on human health. Emerging research shows that many products in the built environment have an effect on human health, including impacts such as endocrine disruption and aggravating asthma. Reference the International Living Future Institute (https://living-future.org/) for information on selecting healthy building materials for your project.

1.2.5 Considering Future Reuse and Repurposing

Design choices have an impact far beyond the construction phase. They have a strong influence on the carbon footprint of the facility over its entire lifespan. Planning for cost-effective flexibility in lab design can significantly reduce the need for extensive renovations down the road, further decreasing embodied carbon.

A forward-thinking approach to design includes considering the potential for future reuse or repurposing of building materials and equipment. This can significantly reduce building waste at the end of the facility's life, aligning with sustainability goals.

1.2.6 Design for Climate Resilience

As our climate continues to change, laboratories must be designed with resilience in mind. This means planning for facilities that can withstand the challenges posed by a warming and more humid world in some climates, and hotter and drier world in other climates. Other disruptions may include utility interruptions or transportation infrastructure unreliability. Like many contingency investments, these added systems need to be designed with sound judgement, and use of available resources to balance likelihoods of occurrence with positive and negative consequences. Key considerations for designing resilient labs include:

- **Temperature Adaptation:** laboratories should be equipped to adapt to rising temperatures and humidity conditions, ensuring that research and experiments can continue in a changing climate. This might involve increased insulation, advanced window glazing, shading, and efficient HVAC systems. Some laboratories may be able to employ operable windows and advanced natural ventilation to both reduce energy use and provide alternative ventilation in case of power interruptions.
- **Extreme Weather Preparedness:** laboratories must be designed to withstand increasingly extreme weather events such as hurricanes, floods, and wildfires. Robust building envelopes, flood-resistant infrastructure, incoming air filtration, and fire-resistant materials are options to mitigate these hazards.

- **Utility Disruption Resilience:** labs may need backup systems in place to handle disruptions in power and water supply. This might include on-site power generation, energy and water storage solutions, on-site thermal storage (ice storage or chilled/hot water storage), and water recycling systems.
- **Transportation Contingencies:** plan for transportation disruptions that can affect staff access to the lab. Consider disaster plan scenarios, alternative means of getting to the lab, and remote work capabilities.
- **Data and Research Continuity:** ensure that critical data and research can be preserved and accessed even in the face of climate-related disruptions and/or data hacking. Implement data backup and recovery plans. Develop a strategy for sample preservation under different disaster scenarios.
- **Architectural and Engineering System Interdependence:** consider the mutual impact of the disruption of one system in a lab on other systems in the building. For more on this and other systemic considerations, with examples, see the *I2SL "Best Practices Guide: Laboratory Resilience"*.

1.3 Water Efficient Labs

"When the well's dry, we know the worth of water." Benjamin Franklin, Poor Richard's Almanack (1746)

1.3.1 Understanding Lab Water Use

A great resource for understanding lab water use is the United States Environmental Protection Agency's *WaterSense at Work*[2] publication. *WaterSense at Work* describes best practices for water management and water efficiency in a variety of commercial settings. Additionally, the *I2SL Best Practices Guide: Water Efficiency in Laboratories* contains specific recommendations for laboratory buildings.[3]

Laboratories, by nature, may consume a substantial amount of water especially if some simple oversights go unresolved, far exceeding the usage of office buildings of comparable size. The water flow within a typical laboratory, as depicted in Figure 1.6, follows a journey from incoming city water through three major water uses, and concludes with discharge into a sanitary sewer. The primary uses encompass domestic water, process water, and water for mechanical uses. The key to conservation is to use the "right" water quality, re-use it, reclaim it, and don't boil it if you don't have to.

- **Domestic Water:** while seen every day, this may be the smallest portion of water use. It includes water for drinking, handwashing, showers, and

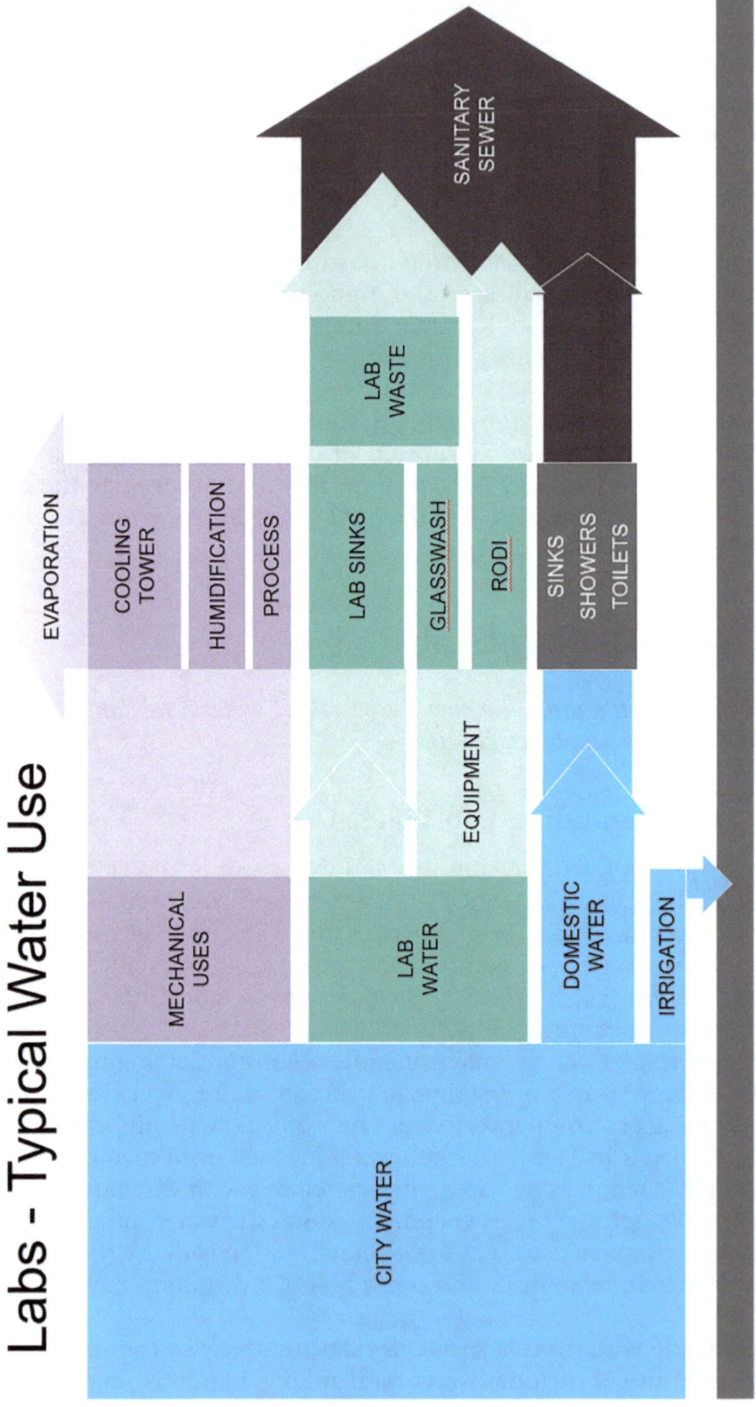

Figure 1.6 Water flow chart in a typical laboratory. Reproduced with permission from Ellenzweig.

toilets. An important water saving strategy is to provide high quality drinking water fixtures in convenient locations, to reduce the temptation to purchase bottled drinking water. Another good strategy is to reduce or eliminate potable water used for toilet flushing. It makes no sense to treat water to drinking water quality then use it to flush a toilet, however, that's what most commercial buildings do. On the other hand, waterless urinals can be a maintenance nightmare, so microflush appliances are recommended.

- **Lab Process Water:** lab sinks, glass washers, purified water, and various equipment utilise water in lab processes. Implementing water-saving technologies, such as aerators and flow restrictors, can significantly reduce water consumption in the laboratory. Replace single-pass cooling in labs with closed loop cooling, so that water is recirculated. This will likely require the use of a process cooling loop to district chilled water or a local chiller, which must be maintained. Consider adding a cistern to collect and store rainwater and cooling coil condensate for use in non-potable applications. Rainwater is remarkably high quality, and may be relatively easy to "polish" for laboratory use.
- **Mechanical Services:** lab buildings also use water for a variety of infrastructure needs, including for heating and cooling, humidifying the air and for moving heat around. Some of these water uses are "closed loop", meaning the water circulates around the building inside pipes many times with little waste, but some are "open loop", where the water evaporates to create a cooling effect or to add moisture into dry air. Strategies that replace open-loop systems with closed-loop alternatives generally cut both water and energy use substantially.

1.3.2 Savings Beyond the Sink: Lab Equipment and Infrastructure

Beyond sinks, significant water-saving opportunities lie in lab equipment. Glass washers, water purifiers, and other process equipment can be substantial water consumers, and jacketed autoclaves are the phantom guzzlers, discussed in the equipment section, so select jacketless autoclaves for occasional duty in labs. Upgrading to newer, more water-efficient models or exploring optional water-saving features from manufacturers can make a considerable difference. Note that the embodied carbon of water is typically low, and so when upgrading equipment, consider the embodied carbon of the new equipment itself with the impact of more efficient upgrades.

Old practices, like using single-pass water vacuum aspirators or condensers, have given way to inexpensive pumps, recirculating chillers, or even waterless condensers. Simple and cost-effective options exist to replace these outdated strategies, contributing to water conservation efforts in laboratories.

1.3.3 Don't Boil It; Don't Buy It

Water is often boiled to purify it, or to provide humidification or steam. Due to water's wonderful molecular attraction to itself, this phase change requires tremendous energy, and should be avoided where possible. Skip distillation for ion exchange; use sonication or spray nozzles for humidification, and implement heat exchangers instead of relying on steam heat. It is important to remember that ion exchange resins have "embodied water" consumption when they are returned to the supplier to be recharged with salt rinses that may contaminate aquifers, so check to make sure they are recharged appropriately. Bottled water can be expensive, and not necessary if adequate quality municipal water can be obtained.

1.3.4 A Shift in Perspective: Greywater

In the pursuit of water efficiency, labs are increasingly exploring alternative water sources, including captured rainwater and recycled water sources. For example, innovative labs are capturing air handling unit condensate as greywater for toilet flushing. These alternative water sources are often referred to as "greywater", to distinguish them as somewhere between drinking water and so-called "black water" (contaminated with human waste).

Some laboratories in hard water regions have realised that rainwater is already purified and ready to be polished. In addition, it is free of silica, a difficult solute to remove because it is neutral and not captured by ion exchange resins. Where feasible, rainwater may be captured in swales to drain back into the water table.

1.3.5 Water = Energy = Water

According to the US Department of Energy's report, the *Water Energy Nexus*, creating clean water uses a tremendous amount of energy. Energy for purification then disposal in municipal waste water treatment plants constitutes 80% of a typical water bill. And creating energy uses a tremendous amount of water. Power plants use as much as 50% of US freshwater consumption. So, by conserving water, you can conserve energy and *vice versa*.

1.3.6 Striving for Net Zero Water: A Vision for the Future

Climate change alters how we think about water consumption. Many hot, dry places, like the Mediterranean and desert climates are predicted to get hotter and drier in the coming decades. So future buildings, including labs, will grapple with increasing water scarcity. The idea that water is an infinite supply that just appears when you open the tap may feel like a forgotten luxury.

The concept of "Net Zero Water" is an ideal that we can imagine. Net Zero Water refers to treating the building as both a consumer and a harvester of water (see Figure 1.7). In this ideal, the building would, on balance, capture

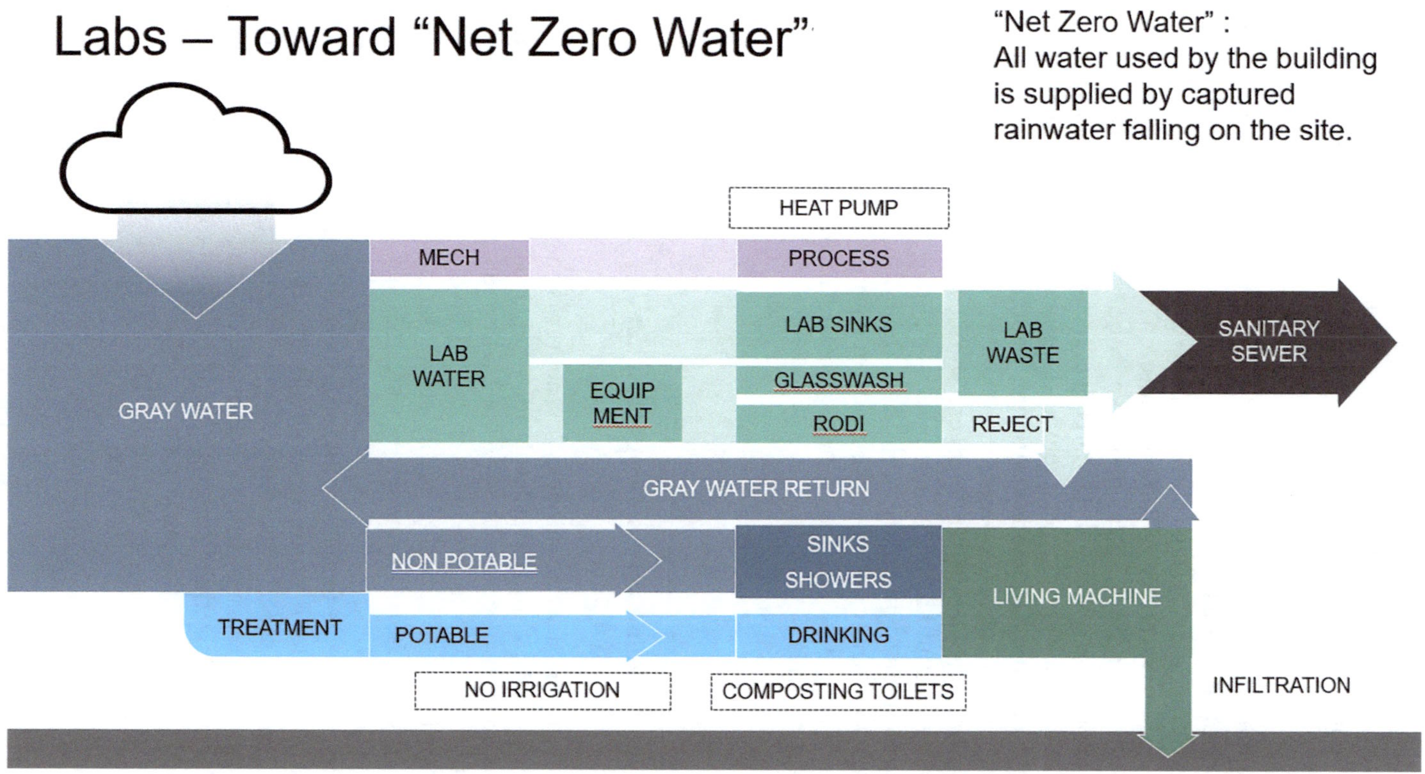

Figure 1.7 Water flow chart in a water-efficient laboratory. Reproduced with permission from Ellenzweig.

as much water as it uses each year, not taking any more water from the city than it could give back to the city through its own rainwater capturing system. Given how much water some labs use, this may be an unreachable goal, but there are some key Zero Water strategies that we may see in future lab designs:

1. **Eliminating Unnecessary Water Uses:** technologies like composting toilets and micro-flush urinals are mature, and they eliminate tremendous water waste (pun intended).
2. **Reusing Water:** we will focus more on reclaiming and reusing water within buildings. Strategies like condensate recovery and reverse osmosis and deionised (RO/DI) reject recovery will become mainstream. Contemporary examples include the Water Hubs at Emory University and Piedmont Hospital in Atlanta, Georgia, which go as far as on-site reclamation of black water.
3. **Embracing Greywater Sources:** sources of greywater will become more common. In some places we may see rainwater capture implemented in many buildings, including labs. Some cities may develop their own collective greywater systems and supply greywater as a service to high-demand facilities like factories and laboratories.

Case Study: Rainwater as Design Inspiration – Cuyahoga Community College STEM Building

An example of water efficiency in action is the Cuyahoga Community College STEM building in Cleveland, Ohio. This teaching lab achieves a 50% reduction in domestic water use and cooling tower water use. The facility not only prioritises efficiency but also serves as an educational tool. An interactive display dashboard within the building educates visitors on green features, including water efficiency strategies and stormwater management.

A key innovative feature of the project is the rain garden. During and immediately after a storm the area captures rainwater from the roof and transforms the space into a reflecting pool between storms. The space is an outdoor classroom and a social space. This feature in particular showcases the way that thinking differently about water can inspire better lab design in general (Figure 1.8).

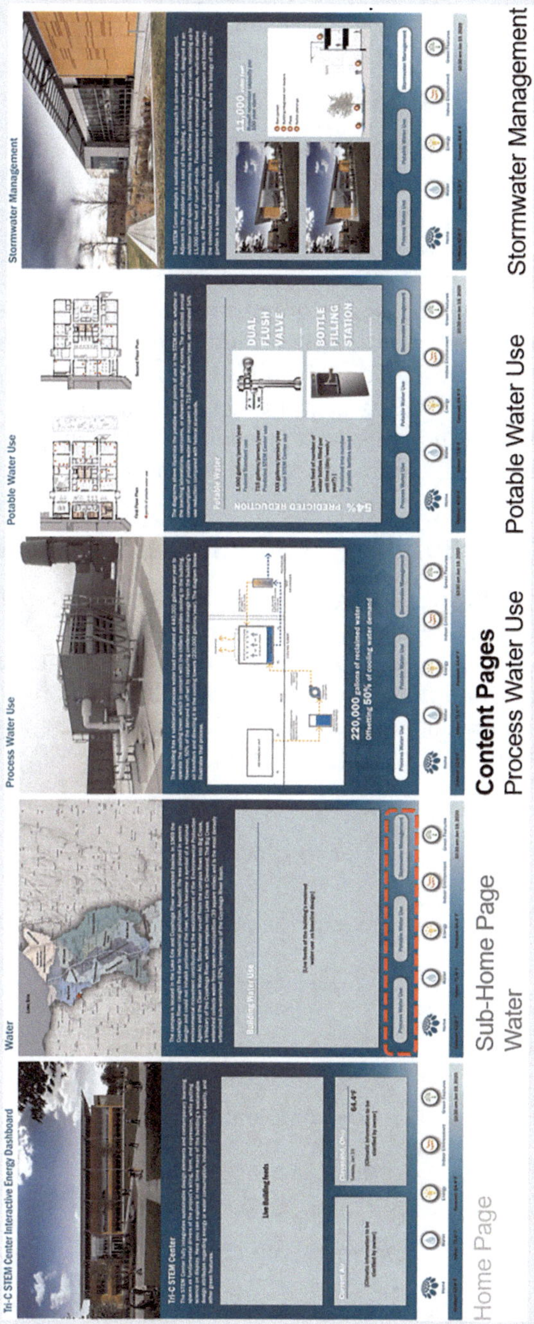

Figure 1.8 Cuyahoga STEM Building. Reproduced with permission from Ellenzweig.

1.4 Energy Efficiency and Carbon in Laboratory Systems

"Conserve first, then electrify." – *Wendell Brase*

1.4.1 Operational Carbon

Lab buildings are different from other buildings because their main goals must be to first provide lab safety, second create a controlled experimental environment and third provide comfort. With these priorities in mind, energy efficiency can be based on a sequence of strategies, starting with conservation (see Figure 1.9).

So, our high-level challenges are to: optimise end uses, increase efficiency, reuse energy where possible, and switch the rest to renewable energy. We will explore these at three scales: beyond the building, in the building, and at the lab/bench. While our perspective starts at the top, the design process for optimisation goes from the bottom-up. Once conservation has been optimised, electrification will be simpler and require less capital expense than an over-sized system.

1.4.2 Beyond the Building: Grid and Campus Scale Strategies

To meet the goal of reducing carbon emissions and resource use, first advocate for studying your local grid's emission characteristics in terms of carbon use and future outlook. For the UK grid, rapid growth in renewables and carbon taxing is curtailing fossil fuels, and nuclear energy provides base-loading. Grid-scale strategies for zero carbon include more renewables, more nuclear and other base-load sources, and deeply-integrated communication and energy storage from user scale to utility scale.

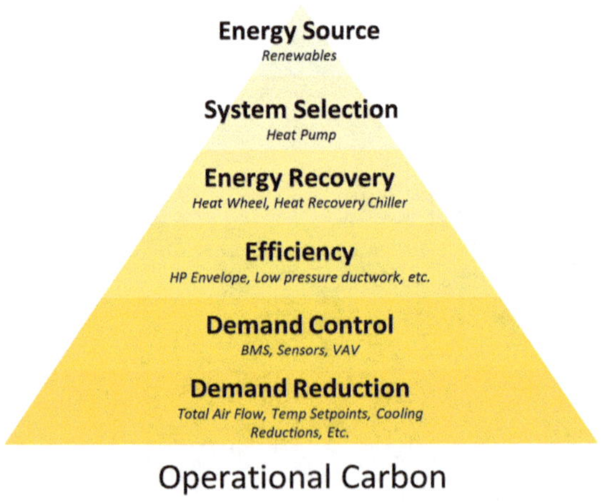

Figure 1.9 Energy & Carbon Efficiency Pyramid – Strategies at the bottom improve the effectiveness of strategies above. Reproduced with permission from Ellenzweig.

Case Study: Making the Grid Sustainable – Energy Systems Integration Facility (ESIF), National Renewable Energy Lab, Golden, Colorado. Photograph by Bill Timmerman, courtesy of SmithGroup (Figure 1.10).

This 16 900 m^2 facility—in a high desert, predominantly heating climate—consists of high-bay research laboratories, a high-performance computing data centre, and administrative office areas. Waste heat from the up to 10 MW data centre (currently 5 MW) captures data centre heat in 35 °C/95 °F water that is distributed to provide primary heating for administrative areas and preheating of laboratory outside air.

The high-performance computing data centre is cooled entirely by waste heat recovery and direct evaporative cooling processes (no compressors/refrigeration cycle required) and operates at a Power Usage Effectiveness (PUE) less than 1.04, meaning that less than 4% of computing energy is needed for operations, one of the lowest PUE's in the world. Waste heat from the data centre also provides campus heating in summer months, allowing campus boilers to be off for up to 4 months of the year.

The office wing (left) has an energy use effectiveness (EUI) of less than 80 kW h per m^2 per year utilising underfloor air, active chilled beams, and natural ventilation for space conditioning. The entire facility (besides the data centre) takes advantage of active daylighting from 10 AM to 2 PM each day. Two levels of high-bay laboratory (right) support renewable energy research. A research electrical distribution bus has the ability to move power from one laboratory to another, similar to an electric utility grid, allowing power generation from rooftop solar or research generators in one area to power research in others.

Figure 1.10 Energy Systems Integration Center at the National Renewable Energy Laboratory. Reproduced with permission from SmithGroup.

At the campus scale, advocate for an all-electric strategy with heat pumps, geothermal storage, ice/chilled water storage, and robust energy recovery. This may reduce or avoid large capital expenses to upgrade district energy boilers or chillers. Depending on the timeline of utility progress, this strategy *might* be supplemented with an interim, local, cogeneration-based micro-grid with lower distribution losses. Yet, the long view remains renewable grid electricity.

Renewable energy for buildings is primarily *electric* energy, notwithstanding biofuels—which make more sense for transportation than buildings. So, advocate for getting *away* from gas/oil boilers and siloed chillers/cooling towers that dump a similar-scale stream of kWh to the atmosphere. Promote moving *towards* renewable power for heat pumps, which can move 4–6 times their kWh input between cooling and heating needs and the ground, atmosphere, or other available sources. A significant data centre in a cold climate can be a great "campus heater".[4] The US District of Columbia Waterworks Department's office/lab building is heated and cooled from heat exchange with a sewer main. Ask your project engineers if they are modelling creative opportunities using REopt (Renewable Energy Integration & Optimisation) or HOMER (Hybrid Optimisation Model for Multiple Energy Resources)—freeware developed by the US Department of Energy's National Renewable Energy Laboratory, or ZENDEE applications for microgrid analysis.

1.4.3 In the Building: Energy Efficient Systems

At this scale, where efficient systems are paramount, your committee will likely have more direct influence. Dig into your lab operations and your group's culture to champion service demand reductions. Each building system is up for examination, and as we address below, each system is merely the sum of demands at the bench or open laboratory. Here are specific strategies you can encourage your team to consider:

1.4.3.1 HVAC is Where It's At

Lab ventilation consumes 60–80% of operational energy in most lab buildings, and modern systems are more complex than a laptop computer. On top of that, these systems are expected to last 25–50 years. This high energy use is because lab safety relies heavily on ventilation—a profound energy driver because the systems must constantly filter, heat, cool, dehumidify or humidify, and distribute large flows of conditioned supply air to "make up" what each lab must exhaust. While some of this makeup air may be recycled from non-lab areas, generally all the air *leaving* labs must be exhausted—either as general lab exhaust or hazardous exhaust (*e.g.*, from fume cupboards).

The best way to reduce HVAC energy use is to right-size the air flow, or total CFM going into and out of the building. The airflow needed may be optimised room-by-room (space) and hour-by-hour (time) based on a Laboratory Ventilation Risk Assessment (LVRA).[5] This determines conservative minimum air flows to dilute and remove airborne hazards based on factors including chemical hazard levels and volatility, and also includes volumes and operations such as heating or agitation that might generate vapour plumes. This risk profile can also empower the selection of unoccupied airflow setbacks for nights and weekends as determined either by scheduling or with occupancy sensors. For typical university, commercial and government labs, for most of the week the laboratory ventilation should be at its minimum.

The mechanical and automation capacities of the whole HVAC system above the ceiling need to be considered in determining the minimum flows and the setbacks or modulation when unoccupied. Recall that for most of the lifetime of the building it will be unoccupied, and its airflow needs to be optimised for that as well as during peak demands.

Advocate for the use of energy recovery[6] wherever feasible. In many US Building Codes, air-to-air heat recovery is now code-required and thus mandatory. For labs, this traditionally means a glycol "run-around loop" between heat recovery coils in the exhaust air stream and the intake of outside air stream, working both summer and winter. But this approach can be turbo-charged, in effect, by adding heat pumps between those airstreams for a greater percentage of energy recovery and a lower coil air-side pressure drop for systems running 24/7. Ask your project engineers what energy recovery opportunity they've found lately.

Encourage engineers to think about every service in terms of variable capacity and energy input. This includes variable-air-volume (VAV) HVAC systems with low turn-down, low pressure drop air valve capability, air handling units and exhaust fans with VAV controls, and fume cupboard sash monitoring and active controls. Performance based design and parametric design should replace rules of thumb.

Have your engineers scrutinise systems for opportunities to employ demand control technology. Because treating outside air is often the biggest energy demand, a sensor-based, demand control ventilation (DCV)[7] system can continuously sample air conditions in each lab to increase safety by detecting a concern and immediately ramping up ventilation. It may let the system "rest" when a lab is operating without incident and/or is unoccupied. Other avenues may focus on savings during out-of-hours operation.

Promote discussion of ground-source and/or air-source heat pumps. The earth's mass nearly begs to be a daily-to-yearly-cycle thermal battery, charged during days and summers and discharged during nights and winters. Closed-loop vertical bores are preferred and practical, with bore depths of 243 m/800 ft or more, consistent with geotechnical analysis and local drilling company rigs. If geoexchange is not available to the project, air source heat pumps can extract heat from outdoor air at temperatures of −23 °C/−10 °F or below. This entails some trade-off in capacity and performance at low temperatures. This lets more energy "get away" over the course of the year, but with less sitework cost. When used in tandem with geoexchange heat pumps, air-source heat pumps can offer a more cost-effective "hybrid" strategy.

It's never too early to start the building airflow management plan (B-AMP). This plan includes a written narrative of how the building should be ventilated and maintained, and includes, in particular, a comprehensive checklist of documents, testing methods, records of equipment, sequences of operation, duct schematics, ventilation risk assessment, automation intent with clear graphics, and maintenance procedures.

Each building's B-AMP is part of the campus wide Laboratory Ventilation Management Plan (LVMP) presently required by the US standard ANSI/ASSP Z9.5 "Laboratory Ventilation". The LVMP contains the standardised policies and procedures for managing lab HVAC such as testing procedures, maintenance cycles, standard and code references, responsible stakeholders and acceptable equipment specifications.

The "data layer" of a modern HVAC system is essential for long term performance and consistency. This includes having adequate sensors on fan speeds, temperatures, terminal unit valve status and pressure sensors to allow for both temperature and static pressure reset strategies. Actual measured airflows rather than calculated values can avoid faulty readings in future years. Furthermore, there are standardised conventions in the US for naming these data inputs, and standardised sequences of operation (SOO) so that automation is consistent from building to building. Building automation systems are best maintained by fault detection and diagnostics (FDD) platforms with careful scrutiny by trained experts to delegate work orders.

This level of complexity needs to be balanced with designing for simplicity and longevity of maintenance. For example, fine scale modulation of fume cupboard airflow may not be warranted using complex open area algorithms or flow sensors at their limit of detection under very low pressure drops. A more robust system with negligible airflow increases could be simple binary switches and coupled with BAS monitoring and automated emails when left open for more than 3 hours.

Challenge your project engineers to provide robust system flexibility for both planned and future lab needs and MEP technologies, and note this in the B-AMP. Perhaps the most essential future-proofing asset a lab can have is robust and integrated vertical and horizontal service corridors, utility chases, and interstitial spaces. This includes the "roofscape"/ penthouse to the lowest service entrances, the lab floor "cores" to the centre of open lab "monument free" zones, as well as generous space above the ceiling plane for accessible and adaptable services (whether or not a physical ceiling plane is installed). When today's technology wears out or becomes out of date, easy access will save considerable expense and effort.

What were you thinking and how did you do it? Be sure that the design team is required to give your estates team the information and tools they need during commissioning and building hand-off in a written Building Airflow Management Plan (B-AMP). This is similar to having a clear hypothesis and methods section in a research publication. This is often overlooked and becomes a migraine headache after the designers have gone. First, the airflow strategies, commonly called the sequences of operations (SOO), are buried within tables of hundreds of interacting valves and specifications. The design team should provide a text narrative of the design intent, assumptions of pressure and flow relationships, and strategy for HVAC sequences of operation. Similarly, the engineer of record and commissioning agent likely created a working spreadsheet of every airflow and water valve with their high and low setpoints and how they add up to total flow for the air handlers or exhaust fans. Be sure to require that they provide a working spreadsheet as part of the building handoff, not just static PDF tables and lists that have to be recreated. All of this goes into the Building Airflow Management Plan, and a detailed checklist of documents with their virtual or physical locations.

1.4.4 At the Lab/Bench: Judicious Use of Services

1.4.4.1 Creating Lab Services Uses Energy—and Water

Every service that you and your project committee ask for has an associated non-trivial operating cost, embodied carbon content, and a system capacity based on lab planner matrices developed with interviews of scientists. Incremental reductions in anticipated peak demands may represent one less piece of equipment, still allowing robust (*e.g.*, $N+1$ or other) reliability. For example, historically, many laboratories would pipe in gas for services like open flames at benches. Piping in gas could lead to increased use through ease of use, but also wastage through methane leaks, directly harming the environment with a greenhouse gas far more potent than CO_2. Engage your laboratories about what services they truly need moving forward to conduct research and associated impacts.

1.4.4.2 Lab Space is Also a Costly Service

The square metres of research space is an overarching driver of economic resources because of lab ventilation—with the other side of that equation being management of air change rates and demand reduction. Scrutinise your organisation's culture regarding space as prestige, fume cupboards devolving into storage, and the orderly transition from less to more to less space as projects and funding ebb and flow. Most organisation's goals demand efficient use of lab space. So, share more, have more open labs, and use less lab space if you can. Modern laboratories tend to favour larger shared laboratories, as opposed to smaller separated facilities, in attempts to make better use of precious laboratory space.

Fume cupboards are famously costly, especially if there is more than 1 cupboard per room. One open fume cupboard exhausts approximately the amount of air for a moderately-sized lab (or about 55 m^2/600 sf). Promote high-efficiency cupboards with your lab planner and engineers and discuss options such as presence detectors in front of a cupboard or sash position reminders. The installation of automatic sash closers may sound appealing, but they add cost, complexity and an eventual maintenance burden.

Other services also have costs. Reverse osmosis water is energy intensive, both to push domestic water through membranes and to back-flush and service those membranes. Another "service" that requires more attention than it normally gets is humidification, which is inherently energy-intensive due to its phase change. The colder the climate, the greater the need, and the greater the risk of mould and physical deterioration in exterior wall systems, especially in an "inherited" existing building. Humidification is also frequently used for animal facilities, which can require more precise environmental conditions. Providing such conditions to individually ventilated cages (IVCs) and the direct breathing zone of the animals inside can greatly improve their air quality while reducing the amount of energy that goes into maintaining the wider facility. Long range studies also show that washable cages will save money in the long term *versus* disposable cages.

Think critically about lab services and optimise to the real needs (don't ask for the moon). Discuss each system with your AE team's professionals to jointly understand and address what and how much of something is needed, and ways to manage services—so your project can be sustainable and so you can have more of what your capital and operating budgets *should* be enabling.

A controlled experimental environment includes air temperature and cleanliness and piped services like compressed lab air, vacuum, high-purity RO water, fluid-based process cooling, and specialty gases. The temptation is to think of these as "free" when you turn on the tap. But the reality is that, somewhere, a big machine is using energy (and water) to produce and deliver each of these to your lab.

This controlling of conditions introduces the maxim of *Put it in a Box*, akin to nested wooden dolls. Control of temperature, humidity, or inert atmosphere is best achieved at the smallest scale: a glove box, an incubator, a cold room, an IVC, or a vented enclosure rather than having an entire room or building as the controlled environment. Similarly, the generation of any hazardous fume or particulate is best contained at the smallest feasible scale, closest to its source: a bench-top apparatus, a glove box, a fume cupboard, or a walk-in fume cupboard.

Case Study: A Major Lab Achieving Net-Zero Energy - Electrical and Computer Engineering Building at the University of Illinois Urbana-Champaign (UIUC), neighbour to Beckman Institute—an earlier U of I Lab of the Year. Photograph by Liam Fredrick, courtesy of SmithGroup (Figure 1.11).

This 21 300 m^2 facility serves a world-renowned department with a tradition of innovative faculty, hands-on classroom experiences, and groundbreaking research in transistors, integrated circuits, light emitting diodes, and plasma displays. This ethos propelled energy efficiency as a driver for the project, while serving high energy research labs and an instructional clean room introducing nanofabrication technology to all undergraduates.

Notable features include a solar canopy shielding 80% of clear glazing from direct sun while promoting daylighting and views, and a high-performance terra cotta rainscreen over 70% of the wall surface. Chilled beams with heat wheel units serve classrooms, offices, and labs, and displacement ventilation serves the lobby and large auditorium. Net-metered heat recovery chillers reuse heating and cooling energy.

The lab received the 2016 Lab of the Year award for showcasing emerging thinking, sustainable practices, and creative responses to challenges in the design, construction and operation of laboratories.

Figure 1.11 Electrical and Computer Engineering Building. Reproduced with permission from SmithGroup.

In 2019, it received LEED Platinum certification from the U.S. Green Building Council. And in 2023 it became the university's first Zero Energy (ZE) Certification from the International Living Future Institute (ILFI). A 300-kW roof solar array provides 11% of its energy (and a basis for research), with the rest in solar renewable energy credits from the 12 MW Solar Farm 2.0 located south of the campus. More information about the ECE building is available at: https://ece.illinois.edu/about/buildings/energy-efficiency and https://ece.illinois.edu/newsroom/3534

1.5 Equipment: Safety and Exposure Control Devices

"A high-performance lab will be a safe lab, and a side benefit will be an energy efficient lab."

Wendell Brase

Introduction:

There are many devices in laboratories not found in office buildings, and they have very specialised designs that are essential for safety, accuracy, precision, stability and purified environments. While the initial impulse may be to require the most stringent criteria for all of these parameters, excessive overdesign comes at considerable cost and waste, and sustainable design provides strategies to comfortably exceed these needs without requiring an entire building to be a controlled environment. For each category of device or equipment, it is important to think about how to integrate it into building systems while right-sizing its scale and impact.

1.5.1 Protective Devices for Airborne Hazards

From a design perspective, the primary exposure control devices that may be part of the building mechanical system are fume cupboards, canopies, snorkels, and biosafety cabinets. Given the limits of this chapter, some basic guidance, systems approaches, and common pitfalls will be offered here. It is essential to remember that these devices are part of a system that needs to modulate and interact at both high *and* low flows, and the ideal lab building should spend most of its time at low flow at night, weekends and holidays. The building automation system and B-AMP needs to ensure this is the case.

An example of excessive overdesign is the specification of 2.4 m, floor mounted fume cupboards to contain 1 or 2 flash chromatography installations (Figure 1.12a). These instruments were installed with nearly sealed sources and waste bottles with vapour traps, thus reducing the potential for fugitive emissions. Enclosures of fraction collectors and flash chromatography are available that may be connected to exhausted fume ducts (Figure 1.12b). The airflow of a floor mounted fume hood may be 60–80 $m^3 m^{-1}$, while a ducted enclosure may have an air flow of under 3 $m^3 m^{-1}$.

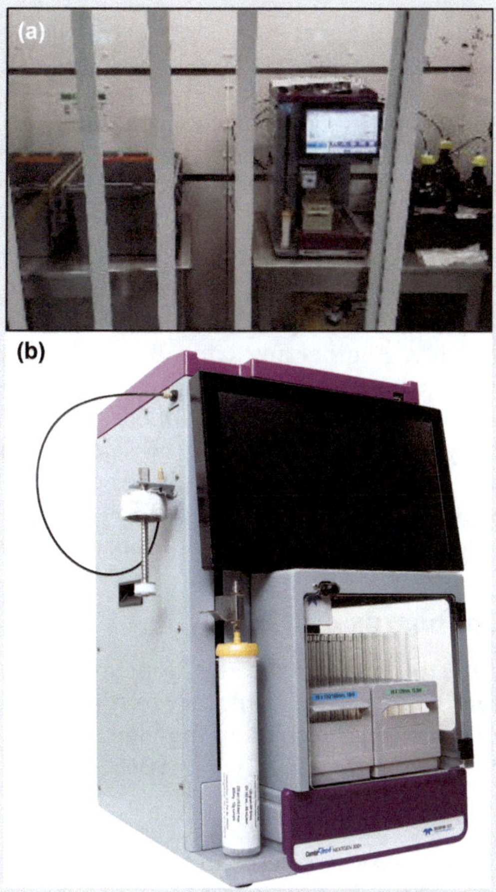

Figure 1.12 (a) Floor mounted fume cupboard with one flash chromatography system. Reproduced with permission of 3Flow. (b) Flash chromatography enclosure with rear exhaust port (not shown). Image was provided by Teledyne LABS for the purpose of this publication only.

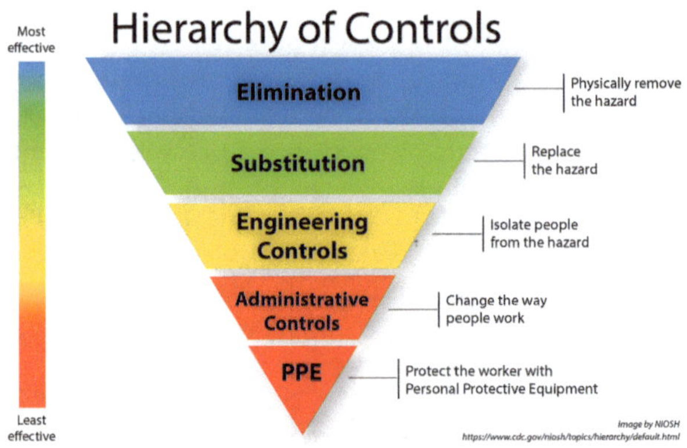

Figure 1.13 The Hierarchy of Controls can guide decisions for selection of source reduction, then exposure control devices and their projected use, then open room dilution. Reproduced from ref. 8, materials developed by CDC.

For breathing zone protection of airborne hazards, it is important to note that source reduction comes first, then containment, then dilution ventilation in the open lab, and finally personal protective equipment (PPE) when all else is not available. Engineering controls are third in priority in the hierarchy of controls (see Figure 1.13), and they can also be supported by administrative controls such as fume cupboard sash management and equipment sharing. As a representative of the science community, you can predict and attest to best practices, needs for devices, and the ability to flex among colleagues. Gaining the support of your colleagues will empower good selection of these devices and reduce their numbers to reasonable values.

1.5.2 Fume Cupboards

When specifying the number of Fume Cupboards in a building, asking for the moon can commit the building to high air flow for its life cycle. Understanding how to specify and manage cupboards can streamline the air flow and reduce this major driver of HVAC costs. If future needs are anticipated, provide stubs on the ducting that may be easily extended.

- **Limit the width**: keep the size of fume cupboards in small rooms as small as possible, and the number of fume cupboards in large rooms as low as possible. If cupboards less than 1.5 metres seem too small, then sliding face area reduction panels (see Figure 1.14) can provide a barrier that workers may reach around or move aside while permanently decreasing face area and full flow. These are especially appropriate for wide, constant flow fume cupboards.
- **Provide variable flow in rooms with highest abundance**: as a rule of thumb, rooms that have more than one cupboard per 60 m^2, each additional cupboard will "drive" the overall exhaust. In organic

Figure 1.14 Fume cupboard with reduced face area provided by sliding panels. Reproduced with permission from 3Flow.

chemistry or teaching labs with 4, 8 or 12 cupboards, these rooms may have very high exhaust flows and cost thousands of pounds in wasted conditioned air during most of the week.

- **Constant Volume may be OK:** in large rooms with a single cupboard, room air flow may be configured so that it provides the necessary overall, or "general" exhaust without adding much to total building exhaust.
- **Commit to Sash Management and Reasonable Average Openings:** design engineers often select a maximum average building-wide cupboard sash opening to be 100%, with a 50% average opening, and this can lead to an oversized system. These values are excessive and owners may consider a 50% average opening as the maximum, for the average opening over the work week may be 5–10% in molecular biology and low-solvent intensity research. In organic synthesis and teaching laboratories, the work-week averages for these rooms will likely be higher. Night time averages for teaching labs should be expected to be 0–5% open, and less than 10% for organic synthesis labs. These professional work practices render automated sash closers as redundant and needlessly expensive and complicated, adding to future maintenance burden. Sash position sensors that report to the BAS will greatly assist sash management.
- **Can Filtered cupboards work?** for R&D or educational work with minor volumes of hydrocarbons, filtered fume cupboards have the reliability and capacity for long term service with considerable savings.

Generally, one exhausted cupboard per room should be available for heavier service. Inappropriate use includes evaporative manufacturing or research that evaporates tens or hundreds of ml each day.

- **Canopies: Good for Heat and Steam, Generally Off:** open canopies are effective devices for removing heat and steam from ovens, furnaces, autoclaves and dishwashers, and are not generally used as safety devices. Because the equipment they support is off much of the time, they are prime candidates for demand control sensors or on/off switches and automated reductions in supply to maintain air balance.
- **Consider your face velocity:** historic face velocities for fume cupboards were higher than what modern "high performance" units may be designed and operated to. When setting up your fume cupboards, consider performing containment testing that reflects how the unit will actually be used, targeting lower (but safe) extraction rates to save energy. Older, constant flow fume hoods may be refurbished with aerodynamic upgrades and safety panels that may reduce their flow by 30–40% while maintaining containment (Figure 1.14). In the UK, face velocity rates in a standard laboratory can be dropped from the historic standard of 0.5 $m\,s^{-1}$ down to as low as 0.3 $m\,s^{-1}$ with containment testing to verify safety, though many will utilise 0.35 $m\,s^{-1}$ to compensate for potential losses in efficiency of extraction systems.

Snorkels: They Suck, and Badly

In some cases, snorkels can effectively collect heat and vapours above buoyant plumes or when directly connected to instrumentation such as ICP-MS or confocal microscopes. Snorkels appear to be targeted exhaust devices and give a nice "sciency" look to a lab, yet they may provide a false sense of protection except in rare cases, and they can be drivers of HVAC overdesign and wasted energy.

In a case of overdesign, 147 snorkels were installed in a laboratory because a design professional thought they would add flexibility to the building. When the building opened, the exhaust had over 15 cm of static pressure (a.k.a. suction) and the fume cupboards were in alarm or not working. Their flow had to be reduced for the building to open. Fifteen years later over 90% of them were unused, and they were still draining the exhaust system. A student summer project concluded that most of them could be capped, or "hibernated", saving over £50 000 per year.

Problems with snorkels include: (1) the intake velocity decays with the square of the distance from the opening, or 90% within one diameter of the tube opening. In order to have much collection capacity, they need to be within 10–15 cm of an emissions source; (2) they are untestable for safety; (3) an open exhaust is a wasted exhaust, and emission tubing from instrumentation or vacuum pumps should be plugged into caps with sealed connectors; (4) the suction (a.k.a. "static pressure") needed to provide 25–50 L s^{-1} at the end of a long duct can make the HVAC system work very hard; (5) occupants don't manage their snorkels very well and they can be left on 24/7, while dampers can be ineffective.

It is like trying to breathe through a straw. In the words of one commissioning agent, when considering snorkels, "Start with 'No'". Recommendation: If ducted exhaust needs are anticipated over the life cycle of a building, provide duct stubs that are capped.

1.5.3 Biosafety Cabinets: Recirculation is Just Alright with Me

General use biosafety cabinets (BSC) are designed to keep samples sterile or free of dust in the open lab, and to prevent low level hazards. These samples most often are biological in nature without hazardous solvents present, and recirculating the HEPA-filtered air into the lab is acceptable. Very rarely BSC's need to be exhausted because the samples or processes also include hazardous solvents or extremely infectious or toxic "potent compounds" of a pharmaceutical nature with unknown consequences if minute quantities bypass the filters.

The ducting of each BSC can add 30 $m^3 min^{-1}$ to exhaust demand, and generally should represent fewer than 10% of a building's BSC inventory. Because of the HVAC impact of ducted BSC's, these devices should be shared among groups needing them. Almost never should a building have 100% ducted BSC's, and rigorous review of the lab program should be double and triple checked before they are installed. Where ducted units are installed alongside recirculating BSC's, this should be communicated and advertised to users so they cannot overuse the more energy intensive exhausted ducted ones.

1.5.4 Benchtop Services, Equipment and Instrumentation

These items comprise the essential hardware of science, the electronic and mechanical tools that empower scientists to explore their realms of discovery. How can they be "sustainable?" Dare one intervene in tweaking a Stradivarius violin, or an electron microscope to save a few kW? Well, oftentimes yes; some simple distinctions and assumptions can save a lot of kW, cubic metres of water, money, and embodied carbon. Let's explore these ideas, and like good scientists, separate them into three components to help make decisions:

1. **Benchtop Services**: fundamental utilities such as electricity, water, pressurised air, vacuum, gases, or solvents. This infrastructure can be provided either by "house systems" or point of service apparatus. Making the right choice for what is needed *versus* the lifecycle of a building is important.

2. **Equipment**: apparatus that provides a service such as cold storage, centrifugation, computation, imaging, heating, drying, washing, or sterilising. Generally, these are larger items, and they usually stay with a building if a scientist moves away, except for centrifugation or freezers, for example. They may also be huge items like MRI imaging or cyclotrons, for example. These items can have a substantial impact on power, cooling, ventilation safety, and water needs.
3. **Instruments**: apparatus that analyses things to provide numbers, images or data. These are often benchtop items that are considered the property of the research group. Though smaller in size, building designers need to understand their demands on the building. Instruments that have flames, plasma generation or large vacuum pumps can be significant heat sources.

Of course, there will be overlap of impacts, yet these distinctions can help with building design, high performance science, and some basic strategies and decisions for procurement and even closed loop consumption. As with many decisions, request the operating conditions that are needed without excessive overkill, such as maximum power and variability, temperature and humidity, water consumption and temperature range, vibration, *etc.*

What is the typical electrical load and its associated waste heat in a laboratory? This is sometimes overestimated by a factor of 10 or more, as shown in a detailed study at the University of California, Irvine. They

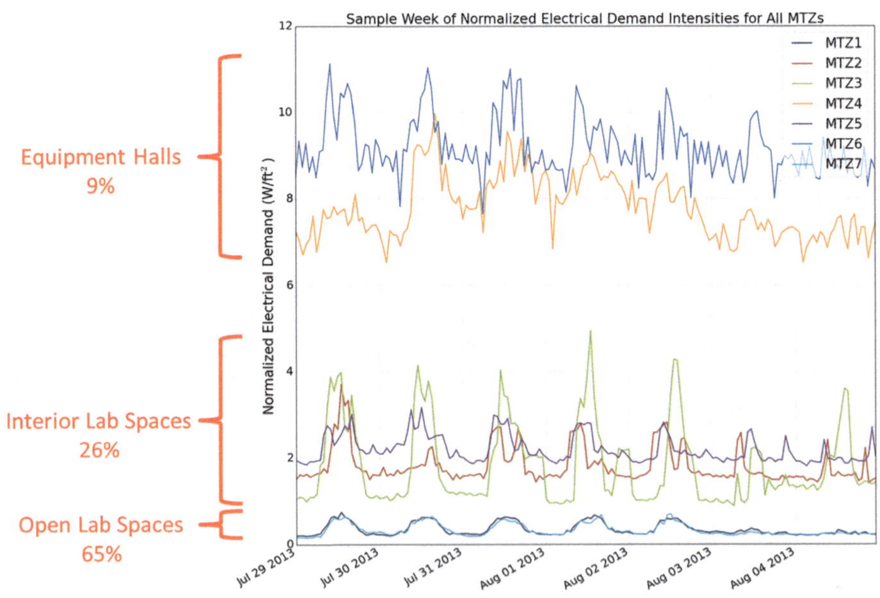

Figure 1.15 One week of plug load metered data for three types of lab zones in a molecular biology laboratory.

metered every circuit in a molecular biology building, and distinguished three typical zones: the open laboratory bench spaces (65% of space), support labs (26%) and equipment rooms (9%) (see Figure 1.15).

The data clearly show that most of the laboratory floorplan uses under 0.5 W ft^{-2} or 5 W m^{-2}. As a scientist this may make sense because you know that most research equipment is used as needed on a periodic basis, and instrumentation often does not use much power. There are exceptions of course when extensive heating is needed, and computer power, vacuum pumps, lasers or plasma flames can consume additional power and release considerable heat. Accommodating these sources and needs will be considered below.

1.5.5 Benchtop Services

The fundamental strategy to choose with each service is whether a central "house" system provides the correct stream (or suction), or whether a point-of-service approach provides the custom purity and intensity needed. First costs, equipment costs, maintenance, stability, accuracy, and refilling costs all need to be considered, along with flexibility for future remodelling. One underrated expense is maintenance or repair of failed central systems.

- **Domestic Water**: in laboratories, do not provide taps and sink wells for water aspirators that can consume thousands of litres of domestic water for a low-grade vacuum. Similarly, single-pass use of domestic water for cooling needs to be avoided both to conserve water and to avoid the risk of overnight flooding. Oh! The horror! Sometimes water isn't even needed for cooling if laboratories utilise air or waterless condensers.
- **Purified Water:** house systems that provide deionised water with 1–4 MegaOhm resistance are relatively easy to maintain, and users needing higher purity can polish on site. Purified water taps should have spring-loaded knobs to avoid endless usage.
- **Chilled Water:** chilled water can be an important way to cool environmental chambers, ultra-low freezers, vacuum pumps, servers, or even student distillations. Be sure that district chilled water is isolated from a building internal loop by a heat exchanger, and that the pumps for district side loop and the building loop are on emergency power.
- **Compressed Air:** this may seem like a simple system, but its need has been declining for many years, and the cost of upkeep and contaminated lines can be high. Point-of-service compressors or air tanks may be adequate for occasional use.
- **Vacuum**: house systems can be handy for general vacuum, but they can also be expensive to install with continuing maintenance expenses. Also, one bad decision can suck solvents into the system, and a £30 000 pump can fail two years after delivery. Vacuum needs come in many levels, and point of service systems can provide bespoke suction, either

low or high. If the first cost of a house system is avoided, the construction savings may include support for individual pumps where needed. Some high vacuum pumps generate a lot of heat, and it would be helpful if manufacturers offered enclosures and aerosol connections to remove the heat rather than release it into the open lab. Recommendations on required airflow ($m^3 \, min^{-1}$) for adequate cooling should be provided as well.
- **Nitrogen:** this gas may be well-served by point-of-service tanks for low to moderate use.
- **Flammable Gas:** this gas may be well-served for low to moderate use by point-of-service tanks for low to moderate use.
- **Solvents:** standalone cabinets serve this purpose well for scrupulously dry and oxygen free solvents. House systems are rarely necessary.
- **Liquid Nitrogen:** for applications where this is needed on a regular basis to fill smaller dewars, a central spigot of liquid nitrogen on each floor may avoid long term delivery fees to small dewars.

1.5.6 Process Cooling: Small Scale to Big Scale

Small scale process cooling may have substantial water use and high potential risks of flooding, ruining electronics and saturating interior walls and cabinets. To avoid this, University of California, Santa Barbara has installed many moderately-priced recirculation pumps on fume cupboards and provided air-cooled solvent condensers.

At the large scale, recirculated process cooling water can reduce the HVAC cooling load and be connected to district chilled water loops. Emergency power provisions are essential for these pumps.

1.5.7 Equipment

Some of this apparatus is highly specialised for scientific purposes, and with diligent shopping some of it can be provided by residential or commercial kitchen equipment without the "lab premium".

- **Incubators:** these are free-standing appliances and are meant to hold precise temperatures, and may include lighting or humidity control, similar to an environmental chamber. These should never be used as refrigerators because they generally have simultaneous heating and cooling, and waste a lot of energy for that purpose.

Can you have a Low Carbon Fruit Fly, *versus* a "Tail Wagger" that Drives Reheating?

Drosophila incubators may consume 600–800 Watts just to achieve room temperature while they are heating and cooling at the same time. A room full of inefficient incubators can get so warm that it becomes a "tail wagger" – the hot zone that drives the whole cooling system to a low supply air temperature, thus requiring the rest of the building to provide a lot of "reheat" energy for comfort. If the reheat system cannot keep up, then occupants may have to wear long underwear in a hot, California summer (true story). Efficient incubators will just use a little bit of energy to maintain consistent room temperatures. If precise drosophila maturation times are not a concern, colonies may be raised on a shelf at room temperature (also a true story).

1.5.8 Cold Storage *vs.* "Environmental Rooms"

The distinction between "environmental rooms" and walk-in refrigerators is the level of precision and fluctuation in temperature. Tight control generally takes more energy, and some systems may be toggled between tight and loose control, depending on needs. This is where your training as a scientist about accuracy *vs.* precision can help. Requiring short term precision makes a big difference, because cooling by compressors is usually "dumb" – either 100% on or off. This means that temperature control only by compressors will have relatively large temperature swings, three degrees Celsius, or more. For long term cold storage of samples, food or smelly cages, who cares if there are temperature swings over several minutes? In this case, cold rooms should have a setting for "low precision" where routine temperature swings are allowed and the heating measures described below for precision are turned off.

If temperature controlled biological experiments or sensitive chemical kinetics or chromatography are being analysed, then temperature swings of ±0.5 degrees may be tolerated *for the sample or vessel*. Remember that the intervals of the temperature swings *in the air* is 3–10 minutes, or less, and most vessels with water or dense substances will damp out short term temperature deviations with their thermal capacity, so ±0.1 degree precision in the air may be overkill, and a ±2.0 degree range may be just fine as long as those deviations in terms of short and long-term accuracy are sufficient.

The energy penalty for short-term precision in an environmental room (or incubator) may be severe. Some designs include simultaneous heating and cooling, with the compressor running continuously to fight the heating required to maintain a precise and unwavering air temperature. Old school compressors are designed with "hot gas bypass", a way to variably bleed off efficiency of the refrigeration loop while the compressor is 100% full capacity.

Recent technology adaptations that make cold rooms precise and energy efficient include variable frequency drive compressors, and intelligent controls that anticipate temperature swings, compressor overshoot, and fan requirements. Modern systems have advanced logic to turn off fans when evaporators are idle, and efficient fans have less reject heat, and then require less cooling.

Insulation, anyone? For freezer rooms, short term precision is rarely considered, so give this some attention if feasible. The insulation *R*-value of the walls is rarely mentioned, so you may ask the designers. Typically, these rooms only have 10 cm of insulation and a low *R*-value, while a house with a similarly large temperature gradient would have double that amount of insulation. See if there is a provider that offers more insulation. Designers and lab owners need to demand better insulation to reduce the cooling load and compressor size. For cold rooms with exterior roofs, white reflective roofs or shade structures may reduce the solar gain.

"Environmental Chambers" These are distinguished from environmental rooms because they may be growth chambers with lighting and humidity controls as well as temperature. With all of the lighting and compressors, they may have a high heat load and process cooling with water may improve efficiency.

Ultra-low Freezers: *Personal equipment or departmental service?* Proactive planning of laboratory floor space to accommodate and manage these freezers can alleviate user conflicts and provide appropriate short- and long-term cold storage. There is sometimes no plan on where to put them, how to remove the heat, and how to do this most efficiently. The function of cold storage has overlapping responsibilities among the scientists who select and operate them, and the building operators and designers who provide space, power and cooling for them. This overlap can lead to a lot of finger-pointing when freezers fail or hallways overheat while crowded with these 300 kilogram appliances.

In general, the larger the freezer, the lower the energy per cubic metre of storage. On the other hand, large shared freezers need collaborative management to minimise archival storage to valuable samples instead of forgotten test batches. One limiting factor is actually door size, in terms of moving them in and out of labs. Another limiting factor in "freezer farms" is the linear occupancy of wall space. The most limiting resource may be the time needed to maintain them, organise samples and remove expired ones.

Scientists prefer to have complete control of freezers as they hold decades of "priceless" samples, yet when they are queried about maintenance and sample disposal, most lab managers do not have a regular plan to defrost them or clean the condenser coils. Furthermore, when asked for an honest estimate of the percentage of samples that are out of date, mislabelled, or forgotten, the answer will often exceed 30 percent. There are many resources available for scientists to manage ultra-low freezers, such as My Green Lab and the International Freezer Challenge.

When constructing a new building or investing in a major HVAC upgrade, smart funders will also fund the turnover of legacy ultra-low freezers to Energy Star models so that cooling of the building can have low waste heat values and reduced first cost of the HVAC plant. With the breakthrough of Energy Star Competition among brands initiated by the United States' EPA and empowered by green lab champions, there are now several models that use under 10 kWh per day. The first model to get certification uses a remarkable "new" refrigeration mechanism, the Stirling Engine, which was theoretically invented 200 years ago. It is less than half the size of cascade compressor systems, and typically situated on top of the cabinet, where it feeds a small amount of low GHG-warming refrigerant into a sinuous tube by gravity instead of a pumped loop. Another advance for most models is that compressor-based freezers now have steady power consumption because variable frequency drives (VFD) have been fitted to the compressors. With the reduced load and absence of start-up electrical spikes, two freezers may now be put on one 20 Amp circuit with little risk of tripping the circuit breaker.

1.5.9 Sterilisers/Autoclaves

These items can be one of the biggest phantom energy and water consumers in a building, using 80–90% of their power and cooling while not being used, and these quantities are LARGE. A stand-alone model may consume 30–60 kW h per day, and hundreds of litres of water while not being used.

There is also a difference between medical grade and research grade autoclaves, and the latter may have more options for energy conservation as they do not require water jackets and are still capable of three to four cycles per day. To select the correct autoclave, you may wish to review user logs for 3–6 months, including the size of the load. In many academic departments the autoclaves are used only a few times per week, so a rapid cycling model is not required.

The ideal autoclave is as simple as possible and cold when not in use, and the most passive technologies are the most efficient. Air cooling after the end of the cycle takes longer, but does not consume water. Manual doors have only a few moving parts, while automated doors add a layer of expense and complexity that can lead to down time. Some manufacturers have passed rumours to users that the doors have to be left open while warm so the gaskets do not degrade. For small loads, a countertop autoclave should be available to sterilise a few instruments or small containers of media. Top loading autoclaves can be intermediate in size, air cooled, and quite efficient.

The procurement office should be wary of expensive service contracts offered by the manufacturer, as they have a conflict of interest in providing extra services, inadequate water monitoring, and expensive parts. The manufacturer should authorise contractor pricing for spare parts to any licensed service provider.

The trade-offs of cold autoclaves: they are not ready within 5 minutes of use and require the user to turn them on about 30 minutes ahead of time. This is usually convenient because the load or liquids need to be prepared. The payback can be a lot of energy and water conservation: the electricity of two to three houses and the water use of several houses.

1.5.10 Instruments

"Instruments" are different from "equipment" in that they analyse something and produce data or images. From a building design perspective, we

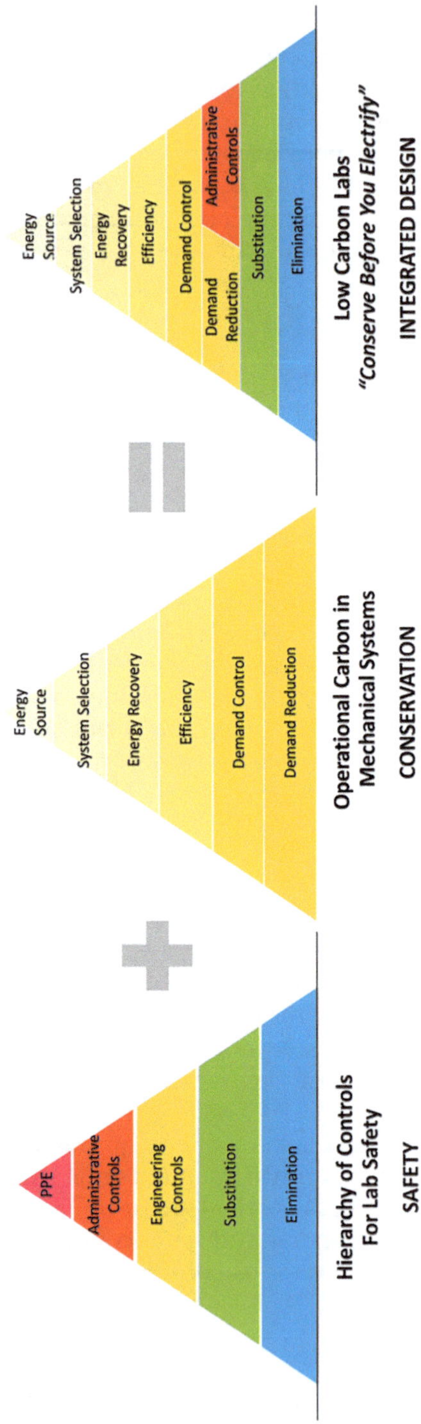

Figure 1.16 Sustainable design may be conceptualised with the combination of safety and conservation principles.

consider instruments that are built-in or demand considerable dedicated space, power, cooling or other features such as low vibration or noise constraints. These devices include MRI and NMR galleries, cyclotrons, and large mechanical devices for physical testing. These large systems require close integration with building design and operations, and it is important to maintain the same standards of right sizing and put-it-in-a-box strategy to maintain stability and precision, and at a larger scale than with bench-top instrumentation.

While benchtop instruments may be plentiful and require many outlets, their power draw and heat load may be relatively low because their constant power draw and frequency of use is low. Where possible, digital timers for outlets may be used where moderate loads are not needed at night and yet need to be on and stable during the day. This includes warming plates and ovens, for example.

1.6 A Call to Action

The hierarchy of safety and the principles of conservation may be combined into an integrated design approach (Figure 1.16). By addressing the fundamentals first, the final goals of reduced impact and effective worker protection are easier to attain than investing in later priorities first. The scientist advocating for sustainable design needs to keep these general principles in mind, as well as learn some of the important technical ways and products to reduce resource consumption during construction and for decades to come. Enrolling a wide range of colleagues, including technicians, labs managers, support services, vendors, scientists, and facility management in their support of sustainable operations will also give the design professionals breathing room to propose innovative solutions.

List of Acronyms Common in Ventilation

ACH	Air changes per hour
AFV	Average face velocity
AHU	Air handling unit
AIHA	American Industrial Hygiene Association
ANSI	American National Standards Institute
AMP	Airflow management plan
BAS	Building automation system
BSC	Bio safety cabinet
CAV	Constant air volume
CFM	Cubic feet per minute
CMH	Cubic metres per hour
DVA	Demand ventilation assessment
ECD	Exposure control device
ECM	Energy conservation measure
FPM	Feet per minute

GEX	General exhaust
HP	High performance
HVAC	Heating, ventilating and air conditioning
IWG	Inches of water gauge
I2SL	International Institute of Sustainable Laboratories
Labs2Zero	https://www.i2sl.org/labs2zero
LEL	Lower explosion limit
LET	Laboratory environment test
LVMP	Laboratory ventilation management program
LVRA	Laboratory ventilation risk assessment
LVS	Laboratory ventilation system
MPS	Metres per second
OAB	Outside air bypass
PA	Pascals
PEL	Permissible exposure level
PIB	Put it in a box
PIM	Performance improvement measure
QEX	Exhaust airflow
RAH	Return air handler
RCB	Risk control band
RH	Relative humidity
STE	Short term exposure
TAB	Test, adjust and balance
TMA	Too many acronyms
TME	Too much exhaust
TLV	Threshold limit value
VAV	Variable air volume
VEFF	Ventilation effectiveness

Further Reading

International Institute for Sustainable Laboratories, Labs2Zero, I2SL, 2000. Available from: https://www.i2sl.org/labs2zero [Accessed July 2025].

ANSI/ASSP. Z9.5: Laboratory Ventilation. 2022 Available from: https://webstore.ansi.org/standards/asse/ansiasspz92022 [Accessed December 2025].

ASHRAE Laboratory Ventilation Design Levels: https://www.ashrae.org/File%20Library/Technical%20Resources/Free%20Resources/Publications/ClassificationOfLabVentDesLevels.pdf.

University of Michigan, Design Guideline 230030 Laboratory Ventilation, Ann Arbor. Available from: http://umaec.umich.edu/desguide/tech/23/DG230030.pdf [Accessed July 2025].

EPA. WaterSense at Work, EPA 2017. Available from: https://www.epa.gov/sites/default/files/2017-01/documents/ws-commercial-water-sense-at-work-ci.pdf [Accessed July 2025].

Carbon Leadership Forum. Embodied Carbon – Overview and General Guidance for Owners of Buildings, CLF, Available from: https://carbonleadershipforum.org/clf-owner-toolkit/ [Accessed July 2025].

LEAF. Available from: https://www.ucl.ac.uk/sustainable/sustainability-contacts-ucl [Accessed December 2025].

My Green Lab. Available from: https://www.mygreenlab.org/ [Accessed July 2025].

References

1. B. Pratt and W. Brase, Life-Cycle and Sustainability Design Standards and Costs, University of California, Irvine, 2022. https://designandconstruction.uci.edu/_pdf/uci-life-cycle-design-standards-and-costs.pdf [Accessed July, 2022].
2. Office of Water, WaterSense at Work: Best Management Practices for Commercial and Institutional Facilities, Environmental Protection Agency. 2012. Available from: https://www.epa.gov/sites/default/files/2017-01/documents/ws-commercial-water-sense-at-work-ci.pdf [Accessed July, 2025].
3. S. Tanner, R. Pickering, K. Wagoner and B. Hoffman, Best Practices for Water Efficiency in Laboratories, International Institute for Sustainable Laboratories, 2022. Available from: https://www.i2sl.org/_files/ugd/59662e_464c82ed60d94cca9003e531101f08af.pdf [Accessed July 2025].
4. C. Pecchioli, J. Comella, D. Sickinger and O. VanGeet, Data Center Heat Reuse 101. Open Compute Project, Available from: https://www.opencompute.org/documents/20230623-data-centers-heatreuse-101-3-2-docx-pdf [Accessed July 2025].
5. International Institute for Sustainable Laboratories, Smart Labs Toolkit: Assess. I2SL.org. Available from: https://smartlabs.i2sl.org/assess.html#lvra [Accessed July 2025].
6. International Institute for Sustainable Laboratories, Energy Recovery in Laboratory Facilities, I2SL, 2023. Available from: https://www.i2sl.org/_files/ugd/59662e_6a20c95510c7468f88c5f6f820f41ff3.pdf [Accessed July 2025].
7. G. Sharp, Energy Savings with Demand Based Control, I2SL, 2021. Available from: URL: https://www.i2sl.org/_files/ugd/59662e_37d2dc9cf1c245bcbfe1ca24180e534e.pdf [Accessed July 2025].
8. CDC, About Hierarchy of Controls, Hierarchy of Controls, 2024. Available from: https://www.cdc.gov/niosh/hierarchy-of-controls/about/index.html [Accessed December 2025].

CHAPTER 2

The Environmental and Carbon Impact of Plastics in Scientific Research: Reporting, Accountability and Action[†]

ANNA LEWIS* AND LUCY WESTOVER

University of Bristol, England
*Email: anna.lewis@bristol.ac.uk

2.1 Introduction

When asked, "What do you consider to be the most significant challenge in your laboratory?" one of the most frequent responses is, "There is so much waste." This is especially true in medical, biological, or 'wet' labs, where people often add, "There is so much plastic waste, how on earth do we reduce this?" It's no surprise that plastic waste remains a frequent complaint; it is significant, ever-present, and highly visible. Since movements like Greta Thunberg's activism, the IPCC 1.5 °C report, and Extinction Rebellion, alongside universities declaring climate emergencies, there has been an increased awareness of the environmental impacts of plastics. Laboratory users and university staff are now more conscious of the carbon impact of plastics, including Scope 3 emissions and the embodied carbon within these materials.[1–4] Despite this growing awareness, plastic waste continues to dominate as a key concern.

Plastics tend to spark strong emotional reactions, whether frustration at their overuse or a desire to avoid them, but this concern can be disconnected

The Sustainability of Science: How Science Impacts the Environment, and What Can Be Done
Edited by Martin Farley

Published by the Royal Society of Chemistry, www.rsc.org

from the work done in research laboratories. Why? Firstly, because health and safety risks in labs are very real, particularly where hazardous substances and materials are involved, plastics are often deemed essential for personal protection. Secondly, although laboratories have long had the means to sterilise and reuse equipment, traditionally relying on glassware and flame or autoclave sterilisation, the rise of single-use plastics has led many researchers to perceive reuse as impractical or risky. Today, concerns about contamination, time constraints, and workload often deter researchers from moving away from disposables, even when alternatives exist. Most importantly, the concern stems from the research itself, a fear that moving away from virgin, sterile plastics might negatively affect the quality or reproducibility of results as reuse introduces a risk of contamination.

Yet, we are long past the point where these reasons can go unchallenged. Given the urgency of addressing climate change, it is essential to question every assumption, explore every option, and implement solutions wherever it is safe to do so without compromising research quality. The good news is that solutions often not only reduce environmental harm but also improve the operational efficiency and working conditions of laboratories. The challenge is clear, and so is the responsibility to act.

2.1.1 The Origins of Plastics

For centuries, humans have been using polymers found in the natural environment in their daily lives. For example, natural polymers like horn, hair, rubber, and cellulose were used to make items like combs, medallions, and cutlery.[5,6] Humans relied on these natural polymers until the 19th century when the development of man-made, synthetic polymers (also known as plastics) began. One of the earliest examples we have of a man-made polymer is the development of the thermoplastic, nitrocellulose (later patented as Celluloid) by Alexander Parkes in 1855. However, Celluloid still used cellulose (a natural polymer) meaning it was not an entirely synthetic plastic. This did not come until 1907 when Leo Baekeland created the first completely synthetic polymer, Bakelite.[6,7] The first half of the 20th century brought about a significant expansion of plastic development with at least 15 new classes of synthetic polymers being created.[5] Whilst we saw the development of synthetic plastics in the early 20th century, the large-scale production and use of plastics did not occur until around 1950.[8]

2.1.2 The Properties and Uses of Plastics

Since 1950, we have seen a significant increase in plastic production and use; this has grown from 1.5 million tonnes in the 1950s to now, where we are producing 400 million tonnes every year.[7,9] As of 2017, it was estimated that 8300 million metric tons of virgin plastics have been produced.[8] This rapid growth in plastics has surpassed the growth of most other man-made materials, excluding steel and cement, becoming an integral and transformative part of everyday life.[8]

The significant growth in plastics is because they are strong, durable, lightweight, inexpensive and have a range of mechanical properties, making them appropriate for a wide range of uses. Because of these properties, plastics are used in almost every sector and area of our lives including clothing, packaging, construction, transportation, medical applications, and many others.[9,10] Some of the most common plastics we use are polyethylene (PE), polypropylene (PP), polyvinyl chloride (PVC), polyethylene terephthalate (PET), polyamide (PA) and polystyrene (PS).[6,11] The typical timespan we use these plastics for is 1–50 years depending on what their use is; for example, plastics used in construction are likely to be in service for many decades, whilst plastics in fast fashion may be used for less than one year.[6,12] Whilst some of the benefits of plastic are that it is durable and long-lasting, approximately 50% of all the plastic that is produced is being used as single-use, rather than in applications where we continually re-use the product.[13,14] The service life of these single-use plastics can be as short as a few days to a few weeks.[6]

As a result of this growth in plastic production, we are also seeing an increase in plastic waste every year. This growth is in part due to an increasing demand for plastic, particularly for plastic with lifespans of less than five years which accounts for nearly two-thirds of plastic waste.[15] As well as an increasing demand for plastics that have a short lifespan, plastic waste is also growing dramatically due to poor waste management and recyclability. As a result of this, global plastic waste is predicted to almost triple by 2060 despite a growing movement that encourages manufacturers to use more recycled plastics in the manufacture of their goods.[16] We are also seeing an increasing amount of plastic waste because many plastics take a long time to degrade, with some taking more than 400 years to decompose.[17] Due to failing waste management systems, currently only 9% of plastic waste is recycled, 19% is incinerated, 50% goes to landfill and 22% evades the management systems and instead goes into unregulated dumpsites, is burned or ends up in the environment.[15] Whilst management of plastic waste is a global problem, the burden of plastic waste is disproportionally felt by low-income countries who pay a total lifetime plastic cost (in terms of the environment, health and a country's economy) that is 10 times higher than the cost high-income countries pay. This higher cost is despite these low-income countries consuming less plastic and the fact that these countries are the ones who are often expected to manage the end-of-life stage of these plastic products.[18]

2.1.3 The Ubiquitous Nature of Plastics

The amount of plastic being produced and consumed, as well as the lack of appropriate disposal for this waste, has led to plastics becoming ubiquitous in our environment. The plastics we find in our environment are commonly classified by their origin and size. The size categories that plastics are classified into are: macroplastics (plastics that are >20 mm in diameter), mesoplastics (5–20 mm), microplastics (<5 mm) and nanoplastics (1–1000 nm).[11] Whilst plastics are known for their longevity, they will slowly break down in the environment due to a range of factors including sunlight, wind, water and through biological processes. The breakdown of these plastics is what forms these micro- and nanoplastics.[19,20] Plastic waste has been found to pollute both terrestrial and aquatic ecosystems with microplastics even being found in Antarctic snow and the deep-sea.[9,21–24] Whilst plastic pollution is found in all environments, particular attention has been paid to the plastic pollution entering the marine environment. Since the 1980s scientists have been documenting the presence of plastic in the oceans and it is estimated that every year, approximately 6 million tonnes of plastics are leaked into rivers or coastlines of which 1.7 million tonnes is then transported to the ocean. There are estimates that 0.5% of the world's plastic waste ends up in the ocean.[11,25] Whilst much of the research has focused on plastic pollution of the marine environment, there is significant plastic pollution seen in terrestrial environments as well. It is estimated that 13 million tonnes of plastic are leaked into terrestrial environments.[11,25,26]

For years, the prolific presence of plastics in the environment has led to growing concerns of the negative impacts these plastics might be having on both human health and the environment.[10] The amount of plastic waste we are producing globally, combined with the fact that plastics are slow to break down, has led to plastic pollution becoming one of the most significant issues.

2.2 The Impacts of Plastics on Human Health and the Environment

2.2.1 Impacts of Plastic Pollution on Human Health

Despite the prevalence of plastics in our environment, relatively little is understood about how plastics could impact human health. Historically, plastics were considered a safe product due to their inert nature, but there is a major gap in our understanding on this topic.[27] Whilst some studies do indicate a potential threat to human health, there has not been enough research conducted to confirm the causality of these threats.[27]

Whilst we are not certain of the effects plastics are having on human health, we do know that humans are being exposed to plastics. These plastics are entering the body through ingestion, inhalation and direct skin contact.[27,28] The micro- and nanoplastics entering the body are being detected in a number of human tissues and biological samples including in the lungs, spleen, placenta, blood, urine and faeces.[20,28,29] Whilst plastics can enter the body through inhalation (for example, inhalation of aerosols containing plastics) and through direct skin contact (including through the use of skin care products that contain plastics and by touching contaminated water), ingestion is believed to be the main way plastics are entering the human body.[31] Plastics have already been found in some of the products we consume including seafood, salt, beer, and honey.[28,31,32] Microplastics have also been found in tap water; in 159 samples of tap water from around the world, 81% of these contained microplastics.[28,31] Plastics have also been found to be absorbed and accumulated in species and then transferred onto the next trophic level, leading to these plastics getting accumulated in the food chain.[33]

To modify and enhance the properties of plastic, most plastics contain high amounts of chemical additives. These chemical additives are added to the plastics during the manufacturing stage to give them properties like stability and flexibility.[6,31] Many of these chemical additives are classified as hazardous; for example, bisphenol A (BPA) is an endocrine disrupter which, if it were ingested or inhaled, could negatively affect human health. These endocrine-disrupting chemicals have been associated with several diseases including hormonal cancers, diabetes and asthma. However, this is not a topic that is well understood, and more research is needed to determine if there are any threats to human health by these chemical additives entering the body.[34] Not only do these plastics contain additive chemicals, but microplastics found in the ocean have also been found to accumulate persistent organic pollutants including polychlorinated biphenyls (PCBs) which may also lead to health impacts.[35] Whilst there is some evidence that plastics might have a negative effect on human health further research is needed to gain an accurate understanding of how much plastic we are exposed to and what effects this is having on our health.[26]

2.2.2 Impacts of Plastic Pollution on Wildlife, Ecosystems and the Climate

Whilst it is less clear the extent to which plastic pollution is harming human health, there is much clearer evidence about how plastic pollution is negatively impacting wildlife, ecosystems and the climate. Some of the ways in which plastic pollution can negatively impact these systems is through entanglement, ingestion, exposure to chemicals, facilitating transport of invasive species and increasing the likelihood of disease.[11] There has been extensive research conducted on the impacts of plastic pollution in the marine environment, but this is not the case for freshwater and terrestrial ecosystems which have not received the same level of attention. However, the impacts of plastic on these freshwater and terrestrial ecosystems are an area of research that has been growing in recent years.[11,36] Whilst it is challenging to know the exact number of species impacted by plastic pollution, it has been estimated that plastic pollution kills 100 000 marine mammals every year and 81 out of 123 marine mammal species have either eaten or been entangled in plastics.[37]

Entanglement is one of the ways species can be impacted by plastic pollution. This impact is well documented in marine organisms, but other species and ecosystems have received far less attention in the literature.[11,36,38]

Marine organisms can be entangled in ropes of ghost (lost) fishing equipment; it is estimated that approximately 640 000 tonnes of fishing gear is lost, left or dropped into the sea every year.[11,39,40] These ropes and fishing gear have entangled many marine species including all seven species of sea turtles, 67% of seal species, 31% of whale species and 25% of marine bird species.[40,41] Whilst there is more data available on entanglements in the marine environment, there is also evidence that entanglement does impact terrestrial and freshwater organisms as well. Freshwater fish, decapods, birds and reptiles have also been found entangled in plastics like bottle rings and plastic bags.[36,42] When organisms become entangled in plastics it reduces their ability to move, feed and breathe – often leading to lethal outcomes.[11]

Ingestion of plastics is also an increasing threat to wildlife, particularly marine species.[43,44] It has been reported that at least 1565 species across the terrestrial, freshwater and marine environments have been documented ingesting plastic.[45] The increase in the number of species ingesting plastic is likely a combination of a growing amount of plastic waste accumulating in the environment as well as an increasing interest in this field coupled with improved sampling and analytical tools.[45,46] There are several reasons why organisms might be ingesting these plastics. Organisms might mistake plastics for food and ingest it, but there are also examples of organisms that might be deliberately choosing to ingest plastic.[47] For example, one study showed that the odour coming from the plastic (specifically the odour of dimethyl sulphide) attracted some species of seabirds. This dimethyl sulphide is released when algae grow on these plastics and is then eaten by krill, attracting these birds and fish to the plastic which then eaten instead of the krill.[47,48] The ingestion of plastic has physical and chemical impacts on these organisms. Large pieces of plastic can physically damage or block the digestive tract. In marine animals, ingestion of plastic can also affect an animal's buoyancy, which can prevent some species from feeding and in turn lead to starvation.[11,42] The ingestion of plastic might also lead to false satiation, reducing an animal's appetite.[42] Plastics can contain and/or incorporate chemicals and additives either from their manufacturing process or these can be sorbed to the plastic from the environment.[11,49] Not as much is known about the effects of these chemicals, but some studies have suggested that these chemicals could have carcinogenic effects and act as reproductive toxins, for example.[11]

As well as effects on individual organisms, plastics are also influencing entire ecosystems and habitats. One way in which they do so is by increasing disease outbreaks. For example, in 124 000 reef-building corals from 159 reefs in the Asia-Pacific region, it was found that the disease risk increased from 4% to 89% when corals are in contact with plastic.[50] In the marine environment, plastic is also aiding the movement of invasive species to new habitats as these species can attach to it as it is carried to other habitats. Invasive species can have significant influence on the invaded environment and can alter the community structure and function through competition with the native species and the introduction of pathogens.[11,51]

As well as the effects seen with wildlife and ecosystems, plastics are also contributing to climate change. At all stages of its lifecycle, plastic produces greenhouse gas emissions from the production of plastics which relies on fossil fuels to the management of its waste which requires high amounts of energy and generates direct emissions.[52] In 2019, the production and incineration of plastics produced approximately 850 million metric tons of greenhouse gases.[53] The inappropriate disposal of plastics into the environment and non-sanitary landfilling is also contributing to climate change as it leads to methane emissions of approximate 2 million metric tonnes CO_2e every year.[52,53]

2.3 Plastic and Laboratories

2.3.1 The Use and Disposal of Plastics in Laboratories

One sector that is guilty of using significant amounts of single-use plastic is scientific laboratory-based research, particularly in the field of biomedical sciences. Whilst there is lots of *ad-hoc* evidence that these laboratories are consuming large amounts of plastics, just how much they are using is not clear, with few studies showing the true extent of the problem.[54] However, one study that has explored this estimated that medical, biological and agricultural laboratories dispose of more than 5.5 million tonnes of plastic every year.[55,56] Single-use plastics were brought into laboratories because they are cheap, lightweight, sterile and save laboratory users time compared to the glassware alternatives, which is what consumables were previously made of. These benefits made single-use plastics the preferred choice for laboratories.[55,57,58]

Laboratories have historically been limited in what they can do with plastic waste. What these laboratories can do depends on what local facilities and contractors are able to accept, with many being hesitant to take plastics coming from a laboratory due to fears of contamination. Because of this, currently less than 1% of plastic waste from laboratories is recycled.[59] Instead, laboratory plastic waste is often autoclaved (an energy and water intensive process used to sterilise the plastics), and then either incinerated or sterilised with chemicals and then sent to landfill.[56] However, not all plastic waste coming from laboratories is too contaminated to recycle and more contractors are now more willing to recycle these plastics than they historically have been. There are also now dedicated companies set up to recycle laboratory-based plastics. LabCycle is the first company on the UK market to claim to create a circular economy for single-use plastic waste from Safety Level 1 & 2 laboratories. The company is aiming to recycle 60% of laboratory waste in a recycling process that is much more sustainable than other recycling methods. LabCycle do not require the plastic waste to be sterilised by laboratory users before recycling, which is significantly reducing energy and water consumption.[60]

Whilst there have been steps taken to reduce the amount of single-use plastic consumers are using, for example by introducing a charge for single-use carrier bags and banning plastic straws, there has been far less action taken to encourage laboratory users to curb their plastic consumption. Not only has there not been the same pressure to reduce this plastic use, but it is also difficult to find suitable alternatives to the single-use plastics used in laboratories.

While limited, there still are some examples of reduction from within laboratory environments. For example, vials utilised for *Drosophila* (fruit flies) have been shown to be reusable through moving from a single-use plastic version to a reusable glass vial. A case study from the University of Sheffield showed great promise in reusing plastic and glass labware to reduce plastic waste.[61] Furthermore, the University of Manchester has been publishing work on how to reduce plastic for some time,[62] and recently produced a study on how to reduce plastic use in Western blotting and SDS-PAGE protocols.[63] At the University of Aberdeen, a recent assessment of preclinical oncology research estimated that each benchtop scientist produced 15 kg of single-use plastic per annum and provided recommendations on how to reduce this impact.[64] More and more publications are arising of new ways to reduce plastic use in laboratories, and we can all look forward to further innovation in this field.[65]

2.3.2 Reuse and Plastic Alternatives

Before single-use plastics, laboratories relied on glassware that was used and reused continuously. Whilst glassware can be found in some laboratories still, many common laboratory items have now been replaced with single-use plastic due to its high accuracy and sterility.[66,67] Some of the most common single-use plastics in laboratories are tissue culture flasks, syringe filters, pipette tips, multi-well plates, Pasteur pipettes, disposable gloves, weighing boats, petri dishes, vacuum filters and syringes. Whilst plastics are often favoured for lots of these consumables, glass substitutions have been found for the following consumables: falcon tubes, pipettes, filter bottles, petri dishes, bijou bottles, test tubes and weighing boats.[67,68] There are some negative aspects of reusing glassware which must be considered when balancing against the impacts of single-use including: decontamination of these glassware is required for their reuse which produces chemically contaminated wastewater, there is a need for technical assistance and time to process these consumables for reuse, glassware is often more expensive at the point of purchase and glass production itself is also carbon intensive. Glass petri dishes, for example, can cost around 30 times more than plastic ones and, in addition to this cost, require more time and resources to clean and sterilise this reusable alternative.[56] Whilst there are disadvantages to re-introducing glassware back into laboratories, the available data have shown that reusing glass will result in lower carbon emissions and can also be the same cost or cheaper in the long term than continually purchasing single-use items.[66,68]

As well as reuse of glassware, it is also possible to reuse some of these plastics that are marketed as single-use. For plastics to be reused they need to go through a process to sterilise them involving chemical decontamination and/or autoclaving. Whilst this process itself uses energy and has a carbon footprint, reuse of these consumables is still thought to have a lower carbon footprint than the use of these plastic consumables as single-use items.[66,68,69] By reusing plastic consumables, laboratories can see significant reductions in the amount of plastic they are consuming. For example, by reusing plastic Falcon tubes, one microbiology laboratory saved 1670 tubes over the 4-week testing period – this is a saving of 20 000 tubes and £1390 every year. This financial saving meant they had more money to purchase sustainable options for laboratory consumables including glass culture tubes.[69] Whilst not all plastics are suitable for reuse due to them not being able to withstand heat from autoclaves, there are some that have been found to be feasible if washed appropriately including reusing pipette tips (with specialist equipment), weigh boats, gloves (depending on application), tubes, cuvettes, and beakers.[68]

There are some circumstances in which it is not possible to reuse either glassware or plastics where we would still need single-use plastics, for example in some clinical settings or due to a risk of experimental cost.[66] Where this is the case, one solution that has been proposed is the use of bioplastics which are polymers that have been generated from renewable sources and biological systems that might be biodegradable.[70] These bioplastics could be made from renewable feedstocks like sugar cane, soybean and microalgae.[71] Whilst bioplastics would still be single-use, they should have less of an adverse effect on the environment. One study has shown that not only do these bioplastics have less of an impact than fossil fuel-based plastics in terms of a reduction in carbon emissions during the production stage, but they are also less damaging to the environment at their end-of-life stage.[70,72] Because of the potential of these bioplastics, there has been a huge growth in bioplastics production. In 2021 this production was around 2.4 million tonnes and is expected to increase to 7.5 million tonnes of bioplastics in 2026, accounting for 2% of the expected global production of plastics.[71,73] However, there are issues with bioplastics both in terms of their environmental impact and their suitability for use in laboratories. One issue is simply that while production is increasing, there is not currently enough generation of laboratory consumables from it to meet demand, and the ones that are generated can cost more. Furthermore the term bioplastics itself is vague; some bioplastics are not biodegradable and plastics that have petrochemical origin can be labelled as a bioplastic because they are biodegradable.[71] In addition, some of these bioplastics made from crops like corn or maize can create food conflict in some areas.[74,75] The main concern surrounding bioplastics is how well they degrade. Whilst there are bioplastics that are biodegradable, the majority of these can only do so in controlled conditions and if they are not disposed of under these conditions then they will likely take much longer to decompose; more information is needed to fully assess the end-of-life stage of bioplastics, however.[71,76]

In laboratories, there are currently very limited bioplastic options that could replace the single-use plastics used in laboratories. One reason for this is that even bioplastics that can be used in some laboratory processes (for example polylactic acid) are limited to what they can do because of their properties; for example, polylactic acid is currently used in some laboratory procedures but cannot be used in high temperature applications or in UV sterilisation.[77] Furthermore, new materials must be validated for existing experimental protocols to ensure that their introduction does not interfere with research outcomes, something that can take time. Whilst bioplastics have the potential to replace single-use plastics in the future, more research needs to be conducted both in terms of creating bioplastics that can be used in laboratories but also in continued exploration of the effects of bioplastics, particularly at the end of their life.[70,71]

2.4 The Carbon Impact of Plastics in Science

Plastics are essential in laboratory work, but their environmental impact, particularly their carbon footprint, is a growing concern. Laboratories consume vast amounts of single-use plastics for reasons such as hygiene, safety, and precision. The production, use, and disposal of these plastics all contribute to their carbon footprint.

2.4.1 Carbon Footprint Breakdown Across the Lifecycle of Laboratory Plastics

2.4.1.1 Production of Plastics

2.4.1.1.1 Material Extraction and Processing. Plastic production, largely derived from petrochemicals, is highly energy intensive. This includes the extraction and refining of fossil fuels like crude oil and natural gas, processes which emit substantial quantities of CO_2.

2.4.1.1.2 Manufacturing Process. The transformation of raw materials into various forms of plastic, such as polypropylene, polystyrene, or polyethylene, also releases greenhouse gases (GHGs). Manufacturing typically requires high heat and pressure, largely powered by fossil fuels.

2.4.1.1.3 Carbon Emissions. For every kilogram of plastic produced, around 1.7–3.5 kg of CO_2 equivalent is emitted,[78] depending on the type of plastic and efficiency of manufacturing processes. Recent studies have shown that emissions from raw material extraction and manufacturing account for the majority of the carbon footprint of laboratory plastics. This underscores the need for more sustainable production methods, such as incorporating recycled or bio-based feedstocks, to lower the emissions intensity of plastic production.[58]

2.4.1.2 Transportation and Distribution

Laboratory plastics are often transported over long distances, adding to their carbon footprint. The further the distance between manufacturing sites and laboratories, the greater the emissions from transport, which is typically reliant on fossil fuels.

2.4.1.3 Usage in Laboratories

2.4.1.3.1 Single-use Culture. Laboratories generally favour single-use plastics to ensure sterility, including pipette tips, Petri dishes, culture flasks, tubes, and vials. This results in a high rate of plastic consumption, with minimal opportunity for recycling or reuse, particularly when contamination from biological or chemical substances is involved. Recent data[79] show that a biosafety level 1 lab researcher produces an average of 116 kg of plastic waste per year.

2.4.1.4 Waste Disposal

2.4.1.4.1 Incineration. A common disposal method for contaminated lab plastics is incineration. While this neutralises hazardous materials, it also releases CO_2 and other harmful pollutants into the atmosphere. Plastics incinerated in this way significantly contribute to GHG emissions.[80]

2.4.1.4.2 Landfilling. If plastics are disposed of in landfills, they contribute to long-term environmental degradation. Although CO_2 emissions from landfilling may be lower than from incineration, plastics take hundreds to thousands of years to decompose, releasing methane—a potent GHG—in the process.

2.4.1.4.3 Recycling Challenges. Recycling lab plastics is difficult due to contamination from chemicals or biological agents. Contaminated plastics are often excluded from recycling programmes, limiting circularity and increasing their environmental footprint.[81]

2.4.1.5 Overall Carbon Emissions

A 2015 study estimated that plastic production alone accounts for 4–8% of the world's oil consumption and around 1.8 gigatonnes of CO_2 emissions annually (about 3.4% of global emissions).[8] Laboratory plastic consumption contributes to this broader statistic.

2.5 Understanding the Carbon Footprint: Data and Reporting

Sustainability professionals are acutely aware of the complexities associated with assessing the carbon footprint of products, particularly Scope 3 emissions. Universities, companies, and research institutions must address these complexities to establish a robust baseline for Scope 3 emissions within key expenditure categories, including the procurement and use of scientific research materials. Such baselines enable the setting of tangible reduction targets and the monitoring of progress toward sustainability goals. Critically, the estimation of product carbon footprints is indispensable for evaluating the feasibility of individual initiatives, such as the comparative benefits of reusing *versus* disposing of laboratory consumables.

For example, the decision to reuse glass or Pyrex laboratory materials instead of single-use plastics hinges on their respective life-cycle carbon footprints. The initial higher carbon cost of producing reusable materials, combined with the emissions associated with their cleaning and sterilisation processes, might offset the presumed environmental benefits. Conversely, the continued reliance on single-use plastics, despite their lower initial carbon impact, exacerbates long-term emissions due to manufacturing and disposal processes, often involving high-temperature incineration.

Recent studies have highlighted that reusing laboratory consumables—such as pipette tip boxes, reagent reservoirs, and glassware—can dramatically reduce plastic waste and associated emissions, with some interventions in clinical and microbiological settings cutting carbon impacts by up to 80%.[82,83] These findings emphasise the necessity of incorporating accurate product carbon footprints into sustainability assessments. Current methodologies, such as spend-based approaches, are inadequate for evaluating the nuanced impacts of laboratory consumables. Thus, establishing comprehensive carbon accounting frameworks, aligned with standards like ISO 14067, is essential for guiding sustainable procurement decisions and fostering carbon-neutral laboratory environments.

2.5.1 A Comparison and Critique of Current Carbon Accounting Methodologies

Assessing the carbon footprint of plastics is a critical element in managing Scope 3 emissions, particularly in sectors like scientific research where processes and operations are heavily reliant on plastic. This section evaluates three primary approaches to carbon accounting: the spend-based approach, bottom-up product weight and material estimation, and primary data-driven carbon assessments, including Product Carbon Footprints (PCFs). Each method has unique advantages and limitations, shaping its utility in accurately quantifying the carbon emissions of plastics.

2.5.1.1 Spend-based Analysis

The spend-based approach is currently the most widely adopted method and easiest for quantifying Scope 3 carbon emissions, including for plastics and other laboratory consumables. This method relies on secondary data to estimate carbon emissions based on the financial expenditure for a given product or category. Due to the widespread absence of Product Carbon Footprints (PCFs) for most items, including scientific consumables and plastics, this approach has become the default. While certain industries, such as information technology (IT), are making significant strides in providing product carbon footprints (PCFs) for items like laptops and servers, comparable data for science consumables remains sparse. However, some manufacturers—such as SHIELD Scientific, who now offer cradle-to-gate carbon footprint certificates for their nitrile gloves—are beginning to close this gap.[84]

In practice, spend-based carbon accounting assigns an emissions factor to each pound spent, typically using national datasets such as the UK Department for Environment, Food & Rural Affairs (DEFRA) carbon conversion factors or the U.S. Environmental Protection Agency's (EPA) Supply Chain Greenhouse Gas Emission Factors.[85] This approach provides an estimate of the carbon footprint associated with the product or expenditure category. However, it is highly prone to inaccuracies, with error margins reaching 40–50% in some cases.[86] For example, products with distinct manufacturing processes or material compositions may share the same conversion factor, leading to significant misrepresentation of their actual carbon impacts.

A fundamental limitation of this methodology lies in its failure to account for sustainable procurement decisions. For instance, a laboratory might choose to purchase centrifuge tubes made from recycled plastics or those with reduced weight and packaging. While these products demonstrably have lower carbon footprints, they may carry higher upfront costs due to the premium on sustainable production practices. In a spend-based analysis, such products are counterintuitively attributed a higher carbon footprint, undermining the validity of the approach and potentially disincentivising greener purchasing decisions.

2.5.2 Material Flow Analysis (MFA)

A more detailed approach involves estimating carbon emissions based on the weight and material composition of a product.[58] This methodology calculates the footprint using emissions factors assigned to specific materials, such as polyethylene or polypropylene, adjusted for their mass. For example, the carbon impact of a lightweight pipette could be calculated based on the weight of polypropylene used, providing a basic estimate of its emissions.

This method offers greater granularity than spend-based approaches, enabling differentiation between products based on their physical properties. It also highlights the benefits of using lightweight or recycled materials, which tend to have lower embodied carbon. However, it still falls short of capturing the nuances of production processes and supply chain logistics, as it generally ignores differences in manufacturing energy sources or transportation emissions. For instance, emissions from polypropylene production in a coal-powered facility differ significantly from those in a facility powered by renewables—data that this approach does not reflect.

2.5.3 Supplier-specific Carbon Data

The most precise carbon accounting method relies on primary data from suppliers, such as certified Product Carbon Footprints (PCFs). PCFs comprehensively assess emissions across all lifecycle stages, including raw material extraction, production, logistics, use, and disposal, using international standards like ISO 14067.

PCFs are particularly valuable for plastics, as they allow for detailed comparisons between single-use and reusable items. For example, PCF analysis might show that a single-use plastic pipette has a lower upfront carbon footprint, but a significantly higher lifecycle impact compared to a reusable alternative made from recycled materials. This level of specificity enables informed decision-making, encouraging sustainable procurement practices aligned with an organisation's carbon reduction goals.

However, the widespread adoption of PCFs is hindered by limited availability and the resource-intensive nature of their calculation. Few suppliers in the scientific consumables sector provide PCFs, making it difficult for organisations to incorporate them into their carbon accounting frameworks. Promoting standardised reporting across industries and incentivising suppliers to share this data will be critical to unlocking the full potential of this methodology.

Each of these carbon accounting methodologies has its own strengths and limitations. The spend-based approach offers accessibility but lacks precision and may misrepresent sustainable purchases. The product weight and material estimation method adds granularity but oversimplifies supply chain complexities. Primary data and PCFs provide the most accurate and comprehensive insights but face significant implementation challenges. A hybrid approach that leverages the simplicity of spend-based accounting while integrating detailed product and supplier data where available is likely the most practical path forward. Encouraging transparency and standardisation in PCF reporting will be essential to advancing carbon accounting practices in the plastics sector. However, producing full PCFs is resource-intensive and not yet feasible for many suppliers. While PCFs offer detailed insights, future frameworks may need to balance accuracy with practicality—for example, by requiring manufacturers to meet certain low-carbon criteria (*e.g.* renewable-powered production) rather than calculating PCFs for every product line. This would enable a more scalable yet still credible path forward in reducing emissions across supply chains.

2.5.4 UK Higher Education Scope 3 Emissions Reporting

UK higher education institutions currently use the Higher Education Supply Chain Emissions Tool (HESCET) to measure and report Scope 3 emissions, including those from laboratory plastics.[87] HESCET, recommended by the Department for Education (DfE) and funding bodies, forms part of a broader national strategy to standardise carbon reporting across the sector.[88] The tool is aligned with the Standardised Carbon Emissions Framework (SCEF), which is underpinned by the Greenhouse Gas (GHG) Protocol.[89] The SCEF establishes expectations for comprehensive Scope 3 reporting, covering supply chain activities such as the production and disposal of laboratory consumables.

The objectives of HESCET are both practical and strategic, encompassing the following:

- Standardisation of Reporting: establishing consistency and comparability in Scope 3 reporting across the UK higher education sector.[88]
- Providing Guidance: equipping institutions with a structured approach to assess and manage supply chain emissions, including those from laboratory plastics.[87]
- Ensuring Compliance: helping institutions align their reporting practices with established frameworks such as the GHG Protocol and SCEF.[88]
- Facilitating Insights: enabling institutions to pinpoint areas of high emissions, such as plastics, and focus on reduction strategies.[87]
- Informing Decisions: supporting sustainable procurement practices by integrating carbon impact considerations into purchasing and operational strategies.[90]
- Encouraging Collaboration: promoting shared benchmarks and practices across institutions to improve collective Scope 3 reporting.[87]
- Tracking Net-Zero Progress: providing tools to monitor emissions trends over time and measure progress toward long-term carbon neutrality targets.[90]

2.5.4.1 Limitations of HESCET for Laboratory Plastics

While HESCET provides a consistent framework for emissions reporting, its reliance on spend-based analysis introduces notable limitations, particularly for accurately assessing the carbon footprint of laboratory plastics. University expenditures are categorised using Higher Education Procurement Codes, which are aggregated into broad categories within the HESCET framework.[88] These categories lack the granularity needed to capture the wide variety of laboratory plastics and their differing carbon impacts.

For example, plastic items such as pipette tips, Petri dishes, and centrifuge tubes are often grouped under generalised categories like "STEM consumables." This aggregation obscures the lifecycle emissions differences between these products, which depend on factors such as material composition, manufacturing methods, and disposal practices.[66] As a result, HESCET may misrepresent or oversimplify the carbon impact of specific plastic products, limiting its effectiveness for targeted emissions reduction efforts.

2.5.4.2 Third-party Tools for Enhanced Granularity

To address these challenges, higher education institutions can complement HESCET with third-party carbon analysis tools that offer more detailed insights into the emissions profiles of laboratory plastics. Platforms such as Ecoinvent, CarbonScope, and OpenLCA provide advanced capabilities for lifecycle assessment (LCA), enabling institutions to evaluate the full environmental impact of specific plastic products.[91] These tools incorporate comprehensive datasets that account for raw material extraction, production energy, logistics, and disposal pathways, delivering far greater granularity than spend-based frameworks alone. However, these platforms often require costly licenses and are not open access, which can create barriers for less-resourced organisations.

By integrating such tools, universities can refine their analysis of plastic emissions within STEM categories. For instance, consumables like single-use plastics can be differentiated from reusable alternatives, providing a more nuanced understanding of emissions and supporting informed procurement decisions.[90]

The adoption of artificial intelligence (AI) offers additional opportunities to enhance Scope 3 reporting for laboratory plastics. AI can process vast datasets, identify patterns in procurement, and model the lifecycle emissions of various plastic products with precision.[92] By integrating AI-driven insights with tools like HESCET, institutions could significantly improve their capacity to identify high-impact reduction opportunities and develop tailored strategies for managing emissions from laboratory plastics.

2.6 Artificial Intelligence and the Digitisation of Science

The advent of artificial intelligence (AI) and computational modelling is revolutionising scientific methodologies, offering transformative approaches to mitigate the environmental impact of research. In particular, these advancements are enabling a significant reduction in the reliance on plastic laboratory consumables, addressing a critical aspect of sustainability in modern scientific practice.

2.6.1 Computational Modelling as an Alternative to Physical Experiments

Computational modelling, underpinned by AI, is increasingly replacing traditional experimental methodologies, offering a sustainable alternative to resource-intensive practices. By simulating chemical processes, material behaviours, and environmental interactions, researchers can conduct studies without the need for plastic-intensive setups. For example, in polymer research, computational tools now facilitate the design and evaluation of new materials, such as bioplastics, by predicting their properties and performance virtually, circumventing the need for iterative experimental testing.[78]

Environmental science has similarly embraced modelling techniques, with AI-driven systems used to predict the degradation pathways of plastics and their accumulation in ecosystems. These tools not only reduce plastic use in field and laboratory studies but also provide high-resolution insights into the global impact of plastic waste.[93]

2.6.2 AI in Laboratory Processes

AI technologies are reshaping laboratory operations by improving resource efficiency and minimising waste. Automated systems, such as AI-controlled pipetting robots, precisely measure reagents, significantly reducing the need for disposable plastics. In addition, advanced sterilisation technologies informed by AI can optimise cleaning cycles for reusable plastics, extending their usability while maintaining sterility standards. While detailed studies on AI's role in plastic reduction in laboratories are still emerging, its potential to enhance efficiency and enable reusability aligns with broader goals in sustainable lab practice.

These innovations demonstrate how AI can streamline laboratory workflows, not only reducing plastic waste but also contributing to cost savings and environmental benefits. This integration of AI into routine scientific practices represents a paradigm shift towards a circular approach to resource management.

2.6.3 Challenges and Future Directions

Despite its promise, the digitisation of science presents challenges that must be addressed to fully realise its potential. Energy consumption associated with computational modelling and AI infrastructure must be mitigated through the adoption of renewable energy sources to ensure net environmental benefits. Furthermore, rigorous validation of AI models is essential to establish their reliability as replacements for experimental data.[71] Should AI modelling be increasingly relied on, comparisons of environmental impacts from traditional experimentation *vs* AI energy consumption should be assessed.

2.7 How Laboratories Can Reduce Their Plastic Consumption and Carbon Footprint

The impacts of plastics, and in particular single-use plastics, on our environment show it is imperative that laboratories work to reduce their plastic consumption. Whilst this may seem like an impossible task, there are several steps laboratories can take to reduce their consumption. There is a hierarchy of actions that underpin sustainable waste management, known as the five Rs: refuse, reduce, reuse, repurpose, and recycle.[68] The idea of this hierarchy is that the initial four actions must be considered before reverting to recycling, as the other actions are more sustainable. Whilst not all these actions are applicable when a laboratory is considering how to reduce their plastic consumption, researchers can look to reduce, reuse, and recycle their plastic waste while engaging suppliers and minimising energy use to achieve meaningful reductions in their carbon footprint.[93]

2.7.1 Refuse

Avoiding Unnecessary Plastics: Refusing unnecessary plastics is the first step in reducing laboratory plastic consumption. This can include rejecting single-use plastics where practical alternatives exist or avoiding excessive packaging in delivered consumables. Laboratories can engage with suppliers to ensure deliveries are optimised with minimal packaging. Companies like NEB allow laboratories to return polystyrene packaging for reuse, reducing the need for new materials.[94] Collaboration with suppliers can ensure procurement processes favour low-waste packaging, preventing unnecessary plastics from entering laboratories in the first place. Notably, laboratories can also proactively better plan future experiments, or utilise local stores for bulk purchases of items to further decrease packaging.

2.7.2 Reduce

Minimising Plastic Use: Reducing plastic use is a critical component of sustainable laboratory practices. Laboratories can adopt smaller flasks or tubes and prioritise purchasing consumables with reduced plastic content.[94] Ordering chemicals and consumables in bulk and sharing them between laboratories can significantly reduce the number of plastic containers ordered.[94] Supplier engagement is also key; encouraging manufacturers to create products with lighter packaging or higher recycled content can further minimise plastic use. For example, some suppliers are developing consumables designed with less material, which reduces plastic consumption across their lifecycle.[95]

Minimising waste by carefully planning experiments, optimising designs to avoid unnecessary plastic use, and using materials efficiently can reduce both material and energy demands. These practices ensure that plastic use is only as high as necessary while maintaining sustainability goals.[96]

2.7.3 Reuse

Reusing plastics where possible can drastically reduce the environmental impact of laboratory operations. Many single-use plastic consumables can be decontaminated and sterilised for reuse. For instance, laboratories can invest in equipment to clean and sterilise pipette tips, significantly extending their usability. It is worth noting that such equipment must be well utilised to offset the emissions associated with its production and usage. Researcher groups should seek independent assessments of how much usage is required for this offsetting, as suppliers may use favourable figures.

Supplier engagement can also support reuse initiatives. Encouraging manufacturers to design products that are durable and reusable can help laboratories adopt sustainable practices more effectively. Additionally, revisiting historical practices such as centralised media preparation can reduce reliance on disposable plastics, enabling laboratories to reuse containers and equipment in a controlled manner.

2.7.4 Repurpose

Repurposing plastics offers laboratories an opportunity to innovate. For instance, companies like Polycarbin in the United States have developed closed-loop systems that collect used lab plastics, process them, and manufacture new consumables from the recycled material. They claim their products are safe for laboratory use and compatible with standard sterility and performance requirements.[97] While independent validation is still emerging, such initiatives demonstrate the potential for circularity in lab consumables beyond traditional recycling streams. Styrofoam boxes and other packaging materials can also be sent back to suppliers or creatively repurposed within laboratories. Supplier collaboration plays a vital role here, as companies can develop systems for reclaiming used materials and creating repurposed products that serve laboratory needs.

Beyond engaging suppliers, leftover materials may trigger artistic repurposing of plastics to push research groups to think critically about their practices. Several artists have taken steps to do just this, including tip-box art installations in the Netherlands,[98] and nitrile glove dresses developed by an independent artist, Saki the Artist.[99]

2.7.5 Recycle

Recycling should be considered once other options have been exhausted. Historically, recycling laboratory plastics has been challenging due to contamination issues, but advancements in recycling programmes have started to address this.

Engaging suppliers to accept and recycle specific plastic items—such as pipette tip boxes or polystyrene containers—can improve recycling outcomes. Specialised programmes, supported by supplier partnerships, ensure that recyclable plastics are diverted from landfill or incineration. Laboratories must also invest in infrastructure to separate uncontaminated plastics, enabling efficient recycling and reducing overall waste.[93]

2.8 Conclusion

Plastics have become indispensable to the modern laboratory, woven into the fabric of scientific progress with their unparalleled convenience, precision, and affordability. Yet, this reliance comes with a profound environmental cost. The story of laboratory plastics is one of duality: transformative in enabling research, yet deeply entangled with issues of waste, emissions, and resource inefficiency.

The path forward requires a reimagining (and sometimes remembering of old) of how laboratories operate, embracing not just the science of discovery but the science of sustainability. Reducing reliance on single-use plastics, rethinking workflows to prioritise reusable materials, and investing in innovative alternatives like bioplastics and circular economy initiatives are no longer just aspirational goals, they are imperatives. Beyond individual actions, laboratories must collaborate with suppliers, policymakers, and sustainability experts to forge systems that make sustainable practices the default rather than the exception.

This chapter has highlighted the environmental footprint of laboratory plastics and the immense potential for change. The solutions are within reach, but they demand a collective commitment to challenge norms, embrace innovation, and prioritise the planet alongside research. In this evolving narrative, laboratories have the opportunity not only to refine their own practices but to inspire broader change across the scientific community. Sustainability in science is no longer a choice between progress and preservation, it is the path to ensuring both.

References

1. IPCC, Global Warming of 1.5 °C, An IPCC Special Report on the impacts of global warming of 1.5 °C above pre-industrial levels, Intergovernmental Panel on Climate Change, 2018.
2. G. Thunberg, *No One is Too Small to Make a Difference*, Penguin Books, UK, 2019.
3. Extinction Rebellion, *This is Not a Drill: An Extinction Rebellion Handbook*, Penguin Books, UK, 2019.
4. Universities UK, Universities and Climate Change: Rising to the Challenge, 2019, https://www.universitiesuk.ac.uk/what-we-do/policy-and-research/publications/universities-and-climate-change (accessed May 2025).
5. A. L. Andrady and M. A. Neal, Applications and societal benefits of plastics, *Philos. Trans. R. Soc., B*, 2009, **364**, 1977–1984.
6. R. Geyer, A Brief History of Plastics, in *Mare Plasticum- The Plastic Sea*, Springer, 2020, 31–47.
7. Plastics Europe, Plastics- the facts 2012, 2012. Available at: https://plasticseurope.org/wp-content/uploads/2021/10/2012-Plastics-the-facts.pdf. Accessed: 30/04/2024.

8. R. Geyer, J. R. Jambeck and K. L. Law, Production, use, and fate of all plastics ever made, *Sci. Adv.*, 2017, **3**(7), e1700782.
9. Nature Sustainability, On the plastic crisis, *Nat. Sustainability*, 2023, **6**, 1137.
10. R. C. Thompson, C. J. Moore, F. S. vom Saal and S. H. Swan, Plastics, the environment and human health: current consensus and future trends, *Philos. Trans. R. Soc., B*, 2009, **364**, 2153–2166.
11. I. E. Napper and R. C. Thompson, Plastics and the environment, *Annu. Rev. Environ. Resour.*, 2023, **48**, 55–79.
12. B. C. Gibb, Plastics are forever, *Nat. Chem.*, 2019, **11**, 394–395.
13. J. Hopewell, R. Dvorak and E. Kosior, Plastics recycling: challenges and opportunities, *Philos. Trans. R. Soc., B*, 2009, **364**, 2115–2126.
14. K. L. E. Berry, N. Hall, K. Critchell, K. Chan, B. Bennett, M. Mortimer and P. J. Lewis, *Plastics. Marine Pollution – Monitoring, Management and Mitigation*, Springer, 2023, pp. 207–228.
15. OECD, Plastic pollution is growing relentlessly as waste management and recycling fall short, says OECD, 2022. Available at: https://www.oecd.org/environment/plastic-pollution-is-growing-relentlessly-as-waste-management-and-recycling-fall-short.htm#:~:text=Plastic%20consumption%20has%20quadrupled%20over,of%20global%20greenhouse%20gas%20emissions. Accessed: 30/04/2024.
16. OECD, Global plastic waste set to almost triple by 2060, says OECD, 2022. Available at: https://www.oecd.org/environment/global-plastic-waste-set-to-almost-triple-by-2060.htm#:~:text=Yearly%20value%20in%20million%20tonnes%20(Mt)&text=The%20report%20(available%20as%20a,rise%20than%20most%20raw%20materials. Accessed: 30/04/2024.
17. E. M. Veidis, A. D. LaBeaud, A. A. Phillips and M. Barry, Tackling the ubiquity of plastic waste for human and planetary health, *Am. J. Trop. Med. Hyg.*, 2022, **106**, 12–14.
18. WWF, Lifetime cost of plastic 10 times higher for low-income countries than rich ones, WWF report finds, 2023. Available at: https://www.wwf.org.uk/press-release/cost-plastic-10-times-more-low-income-countries. Accessed: 30/04/2024.
19. K. Zhang, A. H. Hamidian, A. Tubić, Y. Zhang, J. K. Fang, C. Wu and P. K. S. Lam, Understanding plastic degradation and microplastic formation in the environment: a review, *Environ. Pollut.*, 2021, **274**, 116554.
20. J. Barrett, Z. Chase, J. Zhang, M. M. B. Holl, K. Willis, A. Williams, B. D. Hardesty and C. Wilcox, Microplastic pollution in deep-sea sediments from the great Australian Bight, *Front. Mar. Sci.*, 2020, 7, 96–97.
21. R. Kumar, A. Verma, A. Shome, R. Sinha, S. Sinha, P. K. Jha, R. Kumar, P. Kumar, S. S. Das, P. Sharma and P. V. V. Prasad, Impacts of plastic pollution on ecosystem services, sustainable development goals, and need to focus on circular economy and policy interventions, *Sustainability*, 2021, **13**, 9963.

22. A. R. Aves, L. E. Revell, S. Gaw, H. Ruffell, A. Schuddeboom, N. E. Wotherspoon, M. LaRue and A. J. McDonald, First evidence of microplastics in Antarctic snow, *Eur. Geosci. Union*, 2022, **16**, 2127–2145.
23. H. A. Leslie, M. J. M. van Velzen, S. H. Brandsma, A. D. Vethaak, J. J. Garcia-Vallejo and M. H. Lamoree, Discovery and quantification of plastic particle pollution in human blood, *Environ. Int.*, 2022, **163**, 107199.
24. H. Ritchie, How much plastic waste ends up in the ocean? 2023. Available at: https://ourworldindata.org/how-much-plastic-waste-ends-up-in-the-ocean#article-citation. Accessed: 30/04/2024.
25. E. M. F. Kallenbach, E. S. Rødland, N. T. Buenaventura and R. Hurley, Microplastics in terrestrial and freshwater environments, in *Microplastic in the Environment: Pattern and Process*, Springer, 2021, pp. 87–130.
26. United Nations Development Programme, Microplastics on Human health: how much do they harm us? 2023. Available at: https://www.undp.org/kosovo/blog/microplastics-human-health-how-much-do-they-harm-us#:~:text=Different%20chemicals%20can%20leach%20from,decreased%20reproductive%20health%2C%20and%20cancer. Accessed: 30/04/2024.
27. Center for International Environmental Law, Plastic & Health, the hidden costs of a plastic planet, 2019. Available at: https://www.ciel.org/wp-content/uploads/2019/02/Plastic-and-Health-The-Hidden-Costs-of-a-Plastic-Planet-February-2019.pdf. Accessed: 30/04/2024.
28. E. Winiarska, M. Jutel and M. Zemelka-Wiacek, The potential impact of nano-and microplastics on human health: Understanding human health risks, *Environ. Res.*, 2024, **251**, 118535.
29. M. S. L. Yee, L. W. Hii, C. K. Looi, W. M. Lim, S. F. Wong, Y. Y. Kok, B. K. Tan, C. Y. Wong and C. O. Leong, Impact of microplastics and nanoplastics on human health, *Nanomaterials*, 2021, **11**, 496.
30. G. Kutralam-Muniasamy, V. C. Shruti, F. Pérez-Guevara and P. D. Roy, Microplastic diagnostics in humans: "The 3Ps" Progress, problems, and prospects, *Sci. Total Environ.*, 2023, **856**, 159164.
31. R. Lehner, C. Weder, A. Petri-Fink and B. Rothen-Rutishauser, Emergence of nanoplastic in the environment and possible impact on human health, *Environ. Sci. Technol.*, 2019, **53**, 1748–1765.
32. L. Alberghini, A. Truant, S. Santonicola, G. Colavita and V. Giaccone, Microplastics in fish and fishery products and risks for human health: a review, *Int. J. Environ. Res. Public Health*, 2023, **20**, 789.
33. M. Carbery, W. O'Connor and P. Thavamani, Trophic transfer of microplastics and mixed contaminants in the marine food web and implications for human health, *Environ. Int.*, 2018, **115**, 400–409.
34. C. Campanale, C. Massarelli, I. Savino, V. Locaputo and V. F. Uricchio, A detailed review study on potential effects of microplastics and additives of concern on human health, *Int. J. Environ. Res. Public Health*, 2020, **17**, 1212.

35. M. Smith, D. C. Love, C. M. Rochman and R. A. Neff, Microplastics in seafood and the implications for human health, *Curr. Environ. Health Rep.*, 2018, **5**, 375–386.
36. M. C. M. Blettler and C. Mitchell, Dangerous traps: macroplastic encounters affecting freshwater and terrestrial wildlife, *Sci. Total Environ.*, 2021, **798**, 149317.
37. WWF, Plastic in our oceans is killing marine mammals, 2023. Available at: https://wwf.org.au/blogs/plastic-in-our-oceans-is-killing-marine-mammals/. Accessed: 30/04/2024.
38. D. W. Laist, Overview of the biological effects of lost and discarded plastic debris in the marine environment, *Mar. Pollut. Bull.*, 1987, **18**, 319–326.
39. G. Macfadyen, T. Huntington and R. Cappell, Abandoned, lost or otherwise discarded fishing gear, UNEP Regional Seas Reports and Studies No. 185; FAO Fisheries and Aquaculture Technical Paper No. 523, Rome, 88, 2009.
40. A. Kurtela and N. Antolović, The problem of plastic waste and microplastics in the seas and oceans: impact on marine organisms, *Croat. J. Fish.*, 2019, 77, 51–56.
41. S. Kühn, E. L. B. Rebolledo and J. A. van Franeker, Deleterious Effects of Litter on Marine Life, *Mar. Anthropog. Litter*, 2015, 75–116.
42. P. G. Ryan, Entanglement of birds in plastics and other synthetic materials, *Mar. Pollut. Bull.*, 2018, **135**, 159–164.
43. J. F. Provencher, A. L. Bond, S. Avery-Gomm, S. B. Borrelle, E. L. B. Rebolledo, S. Hammer, S. Kühn, J. L. Lavers, M. L. Mallory, A. Trevail and J. A. van Franeker, Quantifying ingested debris in marine megafauna: a review and recommendations for standardization, *Anal. Methods*, 2017, **9**, 1454–1469.
44. S. Avery-Gomm, J. F. Provencher, M. Liboiron, F. E. Poon and P. A. Smith, Plastic pollution in the Labrador Sea: an assessment using the seabird northern fulmar Fulmarus glacialis as a biological monitoring species, *Mar. Pollut. Bull.*, 2018, **127**, 817–822.
45. R. G. Santos, G. E. Machovsky-Capuska and R. Andrades, Plastic ingestion as an evolutionary trap: Toward a holistic understanding, *Science*, 2021, **373**, 56–60.
46. M. Cole, P. Lindeque, C. Halsband and T. S. Galloway, Microplastics as contaminants in the marine environment: a review, *Mar. Pollut. Bull.*, 2011, **62**, 2588–2597.
47. J. Gabbatiss, Blue Planet II: Why marine animals can't stop eating plastic, 2017. Available at: https://www.bbcearth.com/news/why-marine-animals-cant-stop-eating-plastic. Accessed: 30/04/2024.
48. M. S. Savoca, M. E. Wohlfeil, S. E. Ebeler and G. A. Nevitt, Marine plastic debris emits a keystone infochemical for olfactory foraging seabirds, *Sci. Adv.*, 2016, **2**, 11.
49. E. L. Teuten, J. M. Saquing, D. R. U. Knappe, M. A. Barlaz, S. Jonsson, A. Björn, S. J. Rowland, R. C. Thompson, T. S. Galloway, R. Yamashita,

D. Ochi, Y. Watanuki, C. Moore, P. H. Viet, T. S. Tana, M. Prudente, R. Boonyatumanond, M. P. Zakaria, K. Akkhavong, Y. Ogata, H. Hirai, S. Iwasa, K. Mizukawa, Y. Hagino, A. Imamura, M. Saha and H. Takada, Transport and release of chemicals from plastics to the environment and to wildlife, *Philos. Trans. R. Soc., B*, 2009, **364**, 2027–2045.

50. J. B. Lamb, B. L. Willis, E. A. Fiorenza, C. S. Couch, R. Howard, D. N. Rader, J. D. True, L. A. Kelly, A. Ahmad, J. Jompa and C. D. Harvell, Plastic waste associated with disease on coral reefs, *Science*, 2018, **359**, 460–462.
51. J. García-Gómez, M. Garrigós and J. Garrigós, Plastic as a vector of dispersion for marine species with invasive potential. A review, *Front. Ecol. Evol.*, 2021, **9**, 770054.
52. OECD, Climate Change and plastics pollution: synergies between two crucial environmental challenges, 2023. Available at: https://www.oecd.org/environment/plastics/Policy-Highlights-Climate-change-and-plastics-pollution-Synergies-between-two-crucial-environmental-challenges.pdf. Accessed: 30/04/2024.
53. S. Sharma, V. Sharma and S. Chatterjee, Contribution of plastic and microplastic to global climate change and their conjoining impacts on the environment -A review, *Sci. Total Environ.*, 2023, **875**, 162627.
54. M. L. Tan, C. K. Ying and S. B. S. Hamid, Plastic pollution and sustainable managing of single-use laboratory plastic waste, *Sustainability Clim. Change*, 2022, **15**, 2001524.
55. M. A. Urbina, A. J. Watts and E. E. Reardon, Labs should cut plastic waste too, *Nature*, 2015, **528**(7583), 479.
56. L. Howes, Can laboratories move away from single-use plastic?, *ACS Cent. Sci.*, 2019, **5**, 1900–2009.
57. B. Joseph, J. James, N. Kalarikkal and S. Thomas, Recycling of medical plastics, *Adv. Ind. Eng. Polym. Res.*, 2021, **4**, 199–208.
58. I. Ragazzi, M. Farley, K. Jeffery and I. Butnar, Using life cycle assessments to guide reduction in the carbon footprint of single-use lab consumables, *PLOS Sustainability Transform.*, 2023, **2**, e0000080.
59. University of Bath, University of Bath is home to UK's first pilot plant for recycling plastic lab waste, 2023. Available at: https://www.bath.ac.uk/announcements/university-of-bath-is-home-to-uks-first-pilot-plant-for-recycling-plastic-lab-waste/. Accessed: 30/04/2024.
60. Industry & Energy, UK's first recycling plant for plastic lab consumables, 2023. Available at: https://www.industryandenergy.eu/circular-economy/addressing-the-plastic-problem/#:~:text=LabCycle's%20recycling%20process%20stands%20out,further%20decreasing%20its%20environmental%20impact. Accessed: 30/04/2024.
61. University of Sheffield, Case study: Washing and reusing labware for more sustainable research, 2024. https://www.sheffield.ac.uk/sustainability/news/case-study-washing-and-reusing-labware-more-sustainable-research. Accessed: 20/05/2025.
62. M. Fostier and R. Grady, Reducing single-use plastic in laboratories: lessons from Manchester, University of Manchester, 2020.

63. Y. C. Ow, R. Grady, M. Fostier, *et al.*, Strategies to reduce plastic in SDS-PAGE and Western blotting protocols, 2025.
64. A. Campion, L. Smith, M. Jones, *et al.*, Quantifying plastic use in preclinical oncology research and recommendations for reduction, University of Aberdeen, 2025.
65. R. Tay, Innovations in laboratory plastic reduction: case studies and approaches, 2024.
66. M. Farley and B. P. Nicolet, Re-use of laboratory utensils reduces CO_2 equivalent footprint and running costs, *PLoS One*, 2023, **18**, e0283697.
67. University of Leeds, Ashley talks sustainable labs and single use plastics, 2023. Available at: https://sustainability.leeds.ac.uk/ashley-talks-sustainable-labs-and-single-use-plastics/. Accessed: 02/05/2024.
68. University of Edinburgh, Reducing single-use laboratory plastics, n.d.a. Available at: https://www.ed.ac.uk/files/atoms/files/reducing_single-use_laboratory_plastics_0.pdf. Accessed: 02/05/2024.
69. J. Alves, F. A. Sargison, H. Stawarz, W. B. Fox, S. G. Huete, A. Hassan, B. McTeir and A. C. Pickering, A case report: insights into reducing plastic waste in a microbiology laboratory, *Access Microbiol.*, 2021, **3**, 000173.
70. A. Robertson, L. Vicenzotto and D. Costa, Reducing plastic waste in WPI laboratories, 2021. Thesis available at: https://digital.wpi.edu/concern/student_works/vh53wz83v?locale=en. Accessed: 02/05/2024.
71. A. Folino, D. Pangallo and P. S. Calabrò, Assessing bioplastics biodegradability by standard and research methods: Current trends and open issues, *J. Environ. Chem. Eng.*, 2023, **11**, 109424.
72. R. Halton, Barriers to laboratory sustainability, 2021. Available at: https://www.plantagbiosciences.org/people/ronan-halton/2021/06/01/barriers-to-laboratory-sustainability/. Accessed: 02/05/2024.
73. European Bioplastics, Global bioplastics production will more than triple within the next five years, in 16th European Bioplastics (EUBP) Conference ed. European Bioplastics, European Bioplastics, Berlin, Germany, 2021, pp. 1–2.
74. M. Carus and L. Dammer, Food or non-food: which agricultural feedstocks are best for industrial uses?, *Ind. Biotechnol.*, 2013, **9**, 171–176.
75. A. Nandakumar, J. A. Chuah and K. Sudesh, Bioplastics: A boon or bane?, *Renewable Sustainable Energy Rev.*, 2021, **147**, 111237.
76. A. Folino, A. Karageorgiou, P. S. Calabrò and D. Komilis, Biodegradation of wasted bioplastics in natural and industrial environments: a review, *Sustainability*, 2020, **12**, 6030.
77. B. Freeland, E. McCarthy, R. Balakrishnan, S. Fahy, A. Boland, K. D. Rochfort, M. Dabros, R. Marti, S. M. Kelleher and J. Gaughran, A review of polylactic acid as a replacement for single-use laboratory components, *Materials*, 2022, **15**, 2989.
78. J. Zheng and S. Suh, Strategies to reduce the global carbon footprint of plastics, *Nat. Clim. Change*, 2019, **9**(5), 374–378.
79. P. M. Weber, C. Michelsen and M. Kerou, What's in our bin?: Labs kick off and demand the transition towards a circular economy for lab plastics, *EMBO Rep.*, 2025, **26**, 297–302.

80. United Nations Environment Programme (UNEP), Single-use plastics: A roadmap for sustainability, 2018.
81. L. Shinn, Addressing the plastic problem in the lab, The Scientist, 2018. Available at: https://www.the-scientist.com/ (accessed May 2025).
82. L. Cerda, D. Gómez-Ríos and G. Torres, *et al.*, Environmental impact of reusable *versus* disposable lab consumables in clinical diagnostics: a comparative LCA, *J. Cleaner Prod.*, 20232023, **412**, 136622.
83. R. Schrijver, J. Alves and H. Stawarz, *et al.*, A case report: insights into reducing plastic waste in a microbiology laboratory, *Access Microbiol.*, 2021, **3**, 000173.
84. SHIELD Scientific, Responsible management for sustainable development: life-cycle assessment and carbon footprint certificate for nitrile and latex gloves, Wageningen, SHIELD Scientific, 2023.
85. W. Ingwersen and M. Li, *Supply Chain Greenhouse Gas Emission Factors for US Industries and Commodities*, U.S. Environmental Protection Agency, Washington, DC, 2020, Report No.: EPA/600/R-20/001.
86. Partnership for Carbon Accounting Financials, *The Carbon Accounting Financials Report*, 2019. Available at: https://carbonaccountingfinancials.com/files/2019-10/20191028-pcaf-report-2019.pdf (Accessed: 16 May 2025).
87. Universities UK, Universities and Climate Change: Delivering on Scope 3 Emissions – A Guide to HESCET and the Standardised Carbon Emissions Framework (SCEF). Universities UK, 2023. Available at: https://www.universitiesuk.ac.uk/ [Accessed May 2025].
88. DEFRA, Environmental Reporting Guidelines: Including Mandatory Greenhouse Gas Emissions Reporting Guidance, 2022. Available at: www.gov.uk.
89. Greenhouse Gas Protocol, A Corporate Accounting and Reporting Standard, World Resources Institute, 2021. Available at: www.ghgprotocol.org.
90. Green Alliance, *Plastic promises: Net zero, circular economy and materials policy*, Green Alliance, London, 2021.
91. ecoinvent Centre, ecoinvent database v3.11 (cut-off, consequential, APOS, undefined system models). St. Gallen, Switzerland: ecoinvent Centre, 2023. Available at: https://www.ecoinvent.org (Accessed: 20 May 2025).
92. Carbon Trust, The Carbon Trust Impact Report 2023: Accelerating Climate Action and Decarbonisation, Carbon Trust, London, 2023. Available at: https://www.carbontrust.com (Accessed: 20 May 2025).
93. R. Gopalan and R. Ramakrishnan, A study on AI solutions for plastic degradation, its benefits, and challenges, *J. Artif. Intell. Mach. Learn. Data Sci.*, 2024, **2**, 711–719.
94. C. Howes, Practical Steps for Sustainable Laboratories, 2019. Available at: www.sustainabilitylabs.org.
95. E. C. Trusler, M. Davies and B. Spurrier, Reusable Glassware for Routine Cell Culture, *Front. Sustainability*, 2024, **5**, 1447236.
96. M. Sethi, P. Sethi and R. Dhiman, Sustainable laboratory practices: Guidelines for waste reduction, energy conservation and product life cycle management, *Curr. Res. Green Sustainable Chem.*, 2020, **1**, 1–7.

97. Polycarbin, Lab Recycling Service & Sustainability Tracking Software, 2024. Available at: https://www.polycarbin.com (Accessed: 16 May 2025).
98. International Institute for Sustainable Laboratories (I2SL). *2024 I2SL Sustainable Laboratory Award Winners*, 2024. Available at: https://www.i2sl.org/2024-award-winners [Accessed 16 May 2025].
99. Saki the Artist, *Nitrile Glove Dress Project*, ISTA Sustainability Group and Green Labs Austria, 2024. Available at: https://sakitheartist.com/Nitrile-Glove-Dress#01-nitrile-glove-ball-gown (Accessed May 2025).

CHAPTER 3

Green and Sustainable Chemistry for a Circular Economy†

T. FREESE,*[a] J. L. VIDAL,[b] M. W. GEORGE,[c] J. C. WARNER,[b,d] M. POLIAKOFF[c] AND B. L. FERINGA[a]

[a] Stratingh Institute for Chemistry, University of Groningen, Nijenborgh 4, 9747 AG Groningen, The Netherlands; [b] Beyond Benign Inc., 18 Church Street, P.O. Box 1016, Wilmington, MA, 01887, USA; [c] School of Chemistry, University of Nottingham, University Park, Nottingham, NG72 RD, UK; [d] The Technology Greenhouse, 100 TradeCenter, Woburn, MA, 01801, USA
*Emails: t.freese@rug.nl; thom.freese@gmail.com

3.1 The Universal Presence of Chemistry

In societal discourse, chemistry frequently carries a negative connotation, often perceived as 'dirty' and 'toxic', being mentioned in the news due to well-publicized accidents in chemical factories or general waste pollution.[1] Consumer preferences mistakenly lean towards 'chemical-free' and 'bio' products during grocery shopping or product purchases.

Nevertheless, chemistry is omnipresent and impacts nearly all aspects of everyday life.[2] The fabric of existence itself consists of atoms and molecules engaging in constant chemical reactions, facilitating the essence of life. On Earth, life itself is based on chemical reactions and molecular interactions occurring constantly inside every living organism on the planet. The human

The Sustainability of Science: How Science Impacts the Environment, and What Can Be Done
Edited by Martin Farley

Published by the Royal Society of Chemistry, www.rsc.org

body, along with plants and animals, is a masterpiece of efficiency of chemical reactions, processes, and molecules, refined over billions of years through evolution.

From the compounds of our bodies to our smartphones, societal benefits and achievements are fundamentally grounded on chemicals and materials being tuned to their respective application (Figure 3.1): every shampoo bottle or plastic bottle for storage, the shampoo inside for washing our hair, every jacket that repels rain, our shoes, the pan and its non-stick coating for cooking, the cooking oil, the plastic foils and containers that keep groceries and food fresh, our toothbrushes, the toothpaste, all cosmetic products and creams, all coatings around everyone right now on the walls, the components of a car we use to get to work in the morning, the rubber wheels needed for mobility, our clothes and their color or properties, the coating and color on the chairs or furniture we sit on, all electronic devices, the chips and screens of our phones, and the materials of our laptops and computers are fundamentally composed of chemical substances. The growth of food we need to feed humanity and animals is accelerated through chemical reactions and fertilizers,[3] the gasoline or electric batteries in our cars, the laundry detergent in our washing machines to clean our clothes, alcohol and beverages, water bottles, toys for our children, and the pharmaceuticals and medicine we use to improve our health – life is chemistry, and we use chemical products at every single moment. Ranging from plastics to pharmaceuticals to cleaning, communication, and transport, modern life's comforts would be impossible without the chemical industry. Chemicals play a crucial role in the essence and existence of life.[1]

Figure 3.1 Examples of products relying on the chemical industry. Chemicals, materials, and chemical reactions are the key building blocks for our modern society.

3.2 Climate Crisis and Resource Depletion

Due to the pioneering efforts of scientists worldwide in understanding the dramatic extent and underlying mechanisms of climate change, both society and political stakeholders have become aware of the fact that the climate crisis represents one of the most pressing challenges ever faced by humanity.[4] Anthropogenic greenhouse gas emissions are the primary drivers behind extreme weather events, food shortages, biodiversity loss, and water insecurity.[5–10] The planetary boundaries concept identifies nine processes essential for maintaining the stability and resilience of the Earth's system (Figure 3.2).[4,5,11] Currently, of these nine boundaries, six have been exceeded, suggesting that Earth has moved well beyond the safe operating space for humanity.[11]

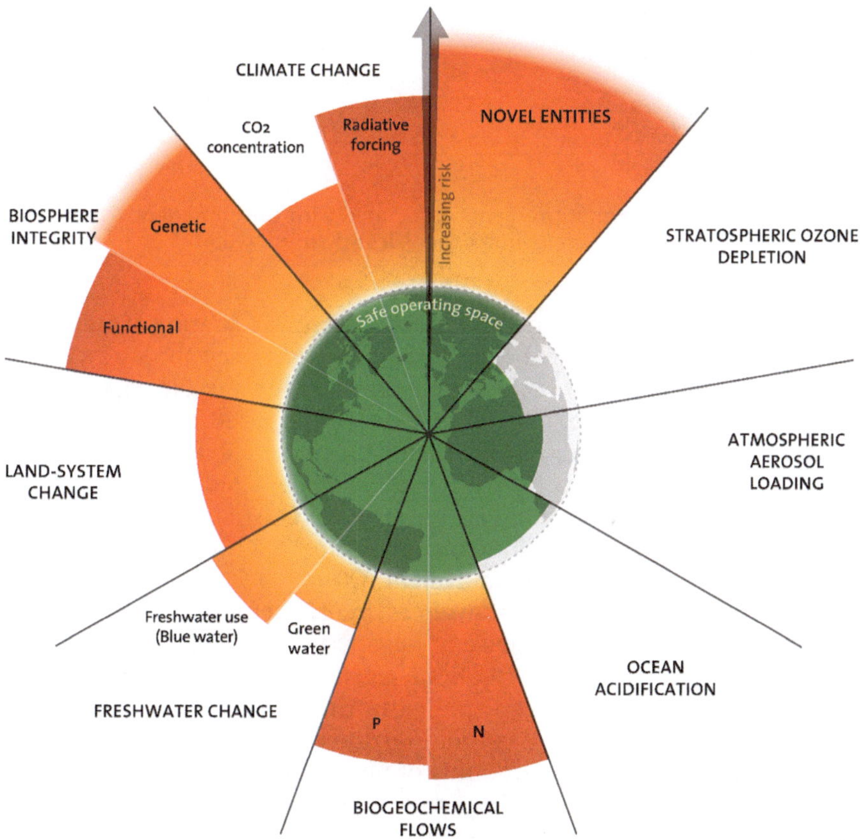

Figure 3.2 Status of control variables for all nine planetary boundaries. The green shading indicates the proposed safe operating space for these nine planetary systems. Reproduced from ref. 14 with permission from Stockholm Resilience Centre (Stockholm University) (based on analysis by Richardson *et al.*[11]).

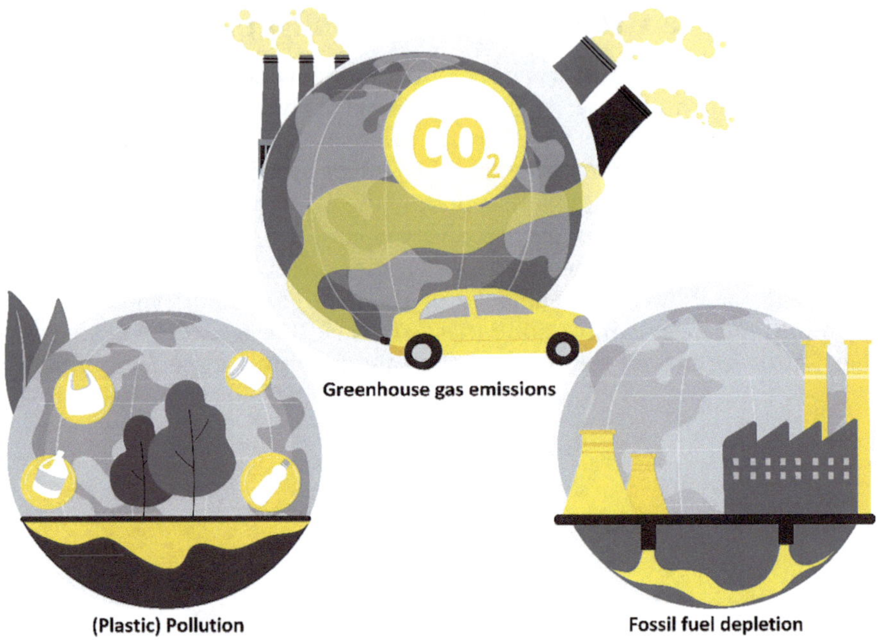

Figure 3.3 Environmental problems associated with the climate crisis. The chemical industry contributes significantly to its scale.

To achieve net zero by 2050 and limit further increased heating of our planet to 1.5 °C, these emissions must be cut by half every decade.[12,13]

These aspects become crucial when correlated to the fact that nearly all benefits of modern society and its products are based on fossil fuels: the products mentioned in Section 3.1 are mostly produced using petrochemical feedstocks, which are finite and at risk of being depleted in the future, while also embodying and releasing carbon dioxide (CO_2) and other greenhouse gases such as methane (CH_4) or nitrous oxide (N_2O) into the atmosphere (referred to as carbon dioxide equivalent, CO_2e).

Moreover, despite significant advances, many chemical industry processes still result in pollution and have a negative impact on the environment (Figure 3.3).[15] Several transformations rely on toxic or flammable organic solvents, which are not easily degraded in the environment.[1] Conventional plastics are built on fossil carbon resources and are usually non-biodegradable, leading to large amounts of plastic packaging being incinerated or finding their way into landfills or in our environment.[1]

Apart from petrochemical feedstocks for the production of materials, society's predominant energy production and consumption is primarily sustained by antiquated, carbon-intensive sources. Using oil, gas, or coal for heating and electricity significantly influences the environment, thereby impacting weather patterns and the temperature of our planet. Currently, both industry and society are in the process of transitioning from conventional energy production to renewable energy sources, such as wind and solar and a hydrogen-based economy among others.[16–20]

Because quality of life relies on the chemical industry, it is up to the chemical industry to drastically improve three major sectors: **Energy**, **Materials**, and **Feedstock**. We must now reinvent chemistry for a sustainable industry.

3.3 Sustainable Transitions

The chemical industry is crucial in supporting the global economy and addressing environmental challenges and transitions. Chemistry provides the foundation for green, sustainable, circular technologies, which can be divided into three key challenges (Figure 3.4).

3.3.1 Feedstock Transition

To promote the transition to a sustainable economy and society, products need to be manufactured from renewable rather than fossil and finite feedstocks, without adversely impacting food production and the further emission of CO_2. Employing biomass (wood waste, *i.e.* **lignocellulose**) and other waste streams (**plastics** and $\mathbf{CO_2}$) as raw materials for chemical production will play an important role in achieving full circularity in manufacturing everyday products and materials. However, it is crucial to implement such changes using a systems approach, as the removal of chemicals from supply chains can significantly impact the economics of other chemical processes. Highly efficient catalytic technologies for the transition need to be developed as well as energy efficient large-scale processing.

Figure 3.4 Key transitions necessary for a sustainable future, mostly developed through the chemical industry.

Specifically, CO_2 and (biobased) CH_4 need to be transformed into value-added products such as carbon materials and chemical building blocks for the manufacturing industry. Artificial photosynthesis and catalysis for the conversion of CO_2 to building blocks and fuels with high efficiencies need to be established. The conversion of biomass (wood waste) into new materials is required for fossil fuels to be phased out. However, wood- and biowaste are limited and therefore a focus on identifying and creating other sources for carbon (C) is pivotal. The recycling of plastic material is a promising solution for waste management and presents a potential new feedstock for sustainable fuels and platform chemicals for chemistry-using industries, allowing for circularity.

3.3.2 Materials Transition

The key to advancing the greening of both the chemical industry and society is the sustainable production of materials in the future. These products and materials need to have the same or better properties than current standards to be economically compatible. Plastics need to be durable, allowing for recycling in a circular economy, and (bio)degradable to prohibit pollution in landfills.

This material transition represents an important step to achieving a carbon-neutral future. The design, production, and recycling of future materials will play a crucial role in reducing greenhouse gas emissions and managing waste streams effectively. The inherent properties of molecules need to be considered in the design stage.[21] These aspects are necessary to determine whether compounds, processes, materials, and products are depleting or renewable, toxic or benign, and persistent or readily degradable and prone to recycling.[21] Envisioning materials derived from renewable sources and produced with minimal CO_2 emissions is of utmost importance. The development of innovative materials with augmented features 'smart materials', will significantly improve the sustainability of the materials sector. These aspects involve the design of new building blocks to facilitate the recycling of materials and the exploration of novel applications.

New catalyst technologies for converting greenhouse gas emissions and pollutants, producing biobased materials, and promoting 'greener' chemistries utilizing light and electricity as energy sources play a pivotal role. Shifting away from fossil fuels, chemical building blocks sourced from alternative, renewable sources need to be developed to reduce the carbon impact of materials. With a focus on recyclability, smart materials with groundbreaking properties that enhance durability while simultaneously exhibiting reversibility or self-healing attributes need to be developed. Dynamic, adaptive, and responsive features that provide unique optical properties, sensing, self-cleaning surfaces, self-repairing mechanisms, energy storage, and heat and light responsive coatings will be beneficial to future societies. The development of new chemical processes to convert plastic waste into valuable compounds will form the basis of a new materials and chemicals platform.

3.3.3 Energy Transition

All products, pharmaceuticals, and chemicals need to be produced utilizing green energy. This will require chemists to develop new efficient materials and chemicals for energy creation and storage (*e.g.* solar panels and batteries). The transition to an energy-efficient chemical industry necessitates the urgent development of secure and sustainable energy carriers, along with low-energy methods to produce chemicals. The development and enhancement of catalytic technologies that facilitate large-scale production of **energy carriers** such as hydrogen (H_2), hydrogen peroxide (H_2O_2), methanol (CH_3OH), formic acid (HCOOH), or ammonia (NH_3) is crucial. By switching to wind- and solar-energy, fluctuations in seasons and day-and-night cycles need to be stabilized through such long-term storage strategies. Through advancements in hydrogen technologies and ongoing investigations into electrochemical and photochemical processes, the way of producing chemical building blocks as well as the electrolysis of water into hydrogen (H_2) and oxygen (O_2) is enhanced.[22] Following subsequent storage, energy carriers can undergo a reverse reaction in fuel cells, yielding water and electrical energy.[23]

Energy-intensive processes need to be improved through innovative, earth-abundant, biobased, and more efficient catalysts to develop new sustainable chemical processes. Also, photochemistry harnessing light as an energy source for direct functionalization, inducing chemical transformations to produce bulk chemicals or building blocks will be a promising approach. Thereby reducing the steps needed in a chemical synthesis and thus enhancing its atom economy (which is a measure of how efficiently a chemical reaction achieves the desired product).[24] Electrochemistry and the application of electrons in redox processes to induce the conversion of chemicals into value-added products enable the circumvention of traditional energy-intensive processes such as heating.

3.4 Green Chemistry and Sustainability

The United Nations Sustainable Development Goals (UN SDGs) were agreed in 2015 as a global call for protecting our planet, ending poverty, and promoting peace by 2030. This framework of **17** goals recognizes that prosperity can only be achieved with approaches that promote economic growth while also addressing social needs and tackling environmental problems (Figure 3.5).

As chemistry is the molecular foundation of all products and processes involving materials, it has a fundamental role in addressing the UN SDGs. Chemistry has the capacity to reach and interconnect different aspects of our society, including the economy, technology, human health, and the environment. For these reasons, it is imperative that the chemical sector spearheads the development of solutions for a sustainable future.[25] Over recent decades, several approaches have been created to achieve more

Figure 3.5 The United Nations Sustainable Development Goals.[26] Adapted from https://www.un.org/sustainabledevelopment/sustainable-development-goals/ with permission from United Nations (UN). The content of this publication has not been approved by the United Nations and does not reflect the views of the United Nations or its officials or Member States.

sustainable chemical education, research, industry, and, ultimately, society. These diverse and valuable strategies are connected and complementary, utilizing the 12 Principles of Green Chemistry.

3.4.1 The 12 Principles of Green Chemistry

Green Chemistry is defined as the utilization of a set of principles that reduces or eliminates the use or generation of hazardous substances in the design, manufacture and application of chemical products.[27] The hazard of a chemical is its ability to cause adverse consequences to human health and/or the environment, being considered an important parameter in the control of accidents and damage risk. Risk is the probability that harm will occur, and it is expressed as a function of *exposure* and *hazard* (eqn (3.1)). The *exposure* can be minimized by using safety precautions, whereas the *hazard* can be reduced by designing and incorporating safer chemicals and processes. By applying Green Chemistry, the intrinsic hazard can be reduced, and the risk of damage remains low even in cases where exposure controls fail (*e.g.*, spills, accidents, leakages).[28]

$$\text{Risk} = \text{Hazard} \times \text{exposure} \quad (3.1)$$

To promote the application of Green Chemistry through the development of efficient, viable, and less hazardous processes and products, the **12 Principles of Green Chemistry** were developed by John C. Warner and Paul T. Anastas in 1998 (Figure 3.6).[27] This framework represents a holistic approach for

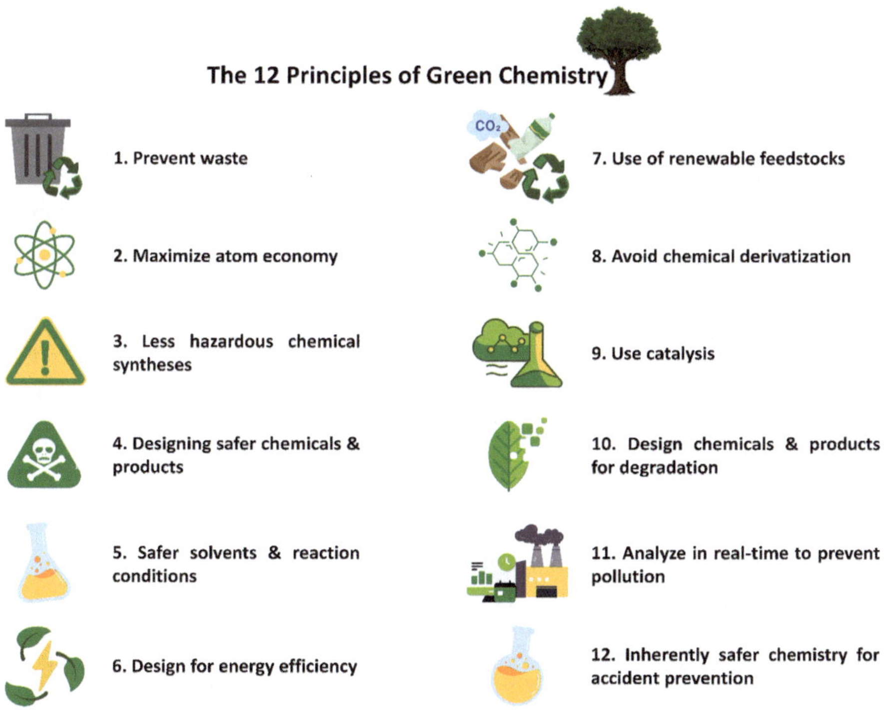

Figure 3.6 The 12 Principles of Green Chemistry created by John C. Warner and Paul T. Anastas in 1998.[27] This framework is a guideline for scientists and chemists to design processes and products that are less hazardous to human health and the environment, while still maintaining comparable performance and economics to the ones traditionally applied.

chemistry to provide benefits to our society without harming or negatively impacting human health and the environment.[29,30]

Principle #1: Waste prevention instead of remediation

It is better to prevent waste than to treat or clean up waste after it has been created. Metrics such as the environmental factor (E-factor) or Process Mass Intensity (PMI) can help assess and compare a specific chemical process regarding the amount of waste produced. While applicable to principle #1, it should be noted that E-factor, PMI, and atom economy are different metrics to also address principle #2. Whereas atom economy (principle #2) does not include yields, side products, and solvents, the E-factor offers a more flexible metric, which is defined as the mass ratio of waste to the desired product.[31] Within the E-factor, anything except the desired product is waste, ranging from side products, solvent losses, and additives. Generally, the E-factor increases with chemical complexity, as more synthetic and purification steps are required.[31–34]

Determination of E-factor, proposed by R. Sheldon:[31–33]

$$\text{E factor} = \frac{\text{Mass of waste (kg)}}{\text{Mass of desired product (kg)}} \tag{3.2}$$

Another environmentally relevant metric based on mass is the process mass intensity (PMI), which is defined as the total mass in kilograms of raw materials (reagents, solvents, *etc.*) required (input of materials) to produce one kilogram of the product (output of the synthesis). The ideal PMI value is 1 and when comparing two processes, the one with the lowest PMI is deemed the most environmentally sustainable.[35]

Definition of process mass intensity (PMI):[35]

$$\text{Process mass intensity (PMI)} = \frac{\text{Total mass used in the process (kg)}}{\text{Mass of product (kg)}} \tag{3.3}$$

Life Cycle Assessments (LCA): Crucially, not only for waste prevention but key to maximizing the potential environmental benefits of green chemistry research is to identify and fix so-called "hotspots".[36] A hotspot is a chemical, a process step, or a life stage that causes greater environmental impact than others. These can be identified through life cycle assessments (LCAs), where the environmental harm of a process is systematically calculated and attributed to the process inputs and outputs.[37–42] While LCAs are invaluable tools for quantifying environmental impacts, it should be noted that they were primarily developed with carbon accounting and climate policy in mind. Thus, their reliance on CO_2-equivalent metrics can sometimes downplay certain forms of waste. For instance, chemical by-products may have minimal carbon impact but pose significant threats to biodiversity or environmental health by being toxic, persistent, or ecologically disruptive — factors that are not fully captured through carbon-based assessments.

As such, more holistic environmental concerns often require advanced forms of LCA, such as *Life Cycle Impact Assessment (LCIA)*, attempting to incorporate additional categories beyond greenhouse gas emissions, including ecotoxicity, land use, and water consumption.[43–46] However, these extensions are not yet consistently applied and often face limitations related to data availability and methodological complexity.

Principle #2: Atom economy and efficiency

Processes and synthetic methods should be designed to maximize the incorporation of all materials used into the final product. Where yield and selectivity traditionally were the most important parameters for the development of new chemical processes, Green Chemistry established more criteria for the evaluation of reaction efficiency and sustainability. The ratio between the formula weight of desired products and the formula weight of all reactants, known as "percent atom economy", can help to assess how many atoms of the reactants have been incorporated into the desired product or wasted in by-products.[24,47,48] On a scale of 0 to 100, the higher the value, the better the AE of the process.

Determination of atom economy, proposed by B. Trost:[24,47,48]

$$\text{Atom economy (AE)} = \frac{\text{Formula weight of desired product (g mol}^{-1}\text{)}}{\text{Formula weight of all substrates used in reaction (g mol}^{-1}\text{)}} \times 100\% \quad (3.4)$$

The atom economy has significance in the initial stages of process design as it provides insights into the sustainability of a chemical transformation without the need for experimental results as it can be calculated *a priori*. Related to AE and to consider the reaction's yield and stoichiometry the Real Atom Economy (RAE) can be calculated *a posteriori* to allow for a more thorough sustainability assessment:[49]

Determination of real atom economy, including yield:[49]

$$\text{Real atom economy (RAE)} = \frac{\text{Actual weight of desired product (kg)}}{\text{Total weight of all substrates in process (kg)}} \times 100\% \quad (3.5)$$

Principle #3: Less hazardous/toxic chemicals and chemical synthesis

Whenever practicable, synthetic methodologies and processes should be designed to use and generate substances with no toxicity toward human health and the environment.

Principle #4: Safer chemicals and products by design

Chemical products should be designed to preserve the efficacy of function while reducing toxicity.

Principle #5: Safer solvents and reaction conditions

The use of auxiliary substances, including solvents, separation agents, and others, should be made unnecessary wherever possible and innocuous when used. Solvents usually facilitate chemical transformation but come with inherent properties and issues such as flammability, volatility, explosion risks, and greenhouse gas emissions. The amount of solvent should be reduced to a minimum (*e.g.* solvent-free) and biobased solvents prioritized.

Principle #6: Energy efficiency by design

Energy requirements should be recognized for their environmental and economic impacts and should be minimized. Synthetic methods should be conducted at ambient temperature and pressure. Most of the energy production in current society is based on fossil fuel energy consumption in the form of heat or electricity and should be minimized through renewable energy sources. Beyond energy sources, efforts should be made to reduce energy consumption wherever feasible.

Principle #7: Use of renewable feedstocks

A raw material or feedstock should be renewable rather than depleting whenever technically and economically practicable.

Principle #8: Reduce derivatives for shorter syntheses

Unnecessary derivatization (use of blocking groups, protection/deprotection, temporary modification of physical/chemical processes) should be minimized or avoided, if possible, since these steps require additional reagents and can generate waste. These steps improve the overall step economy of a chemical synthesis.[50]

Principle #9: Catalysis rather than stoichiometric transformations

Catalytic reagents (as selective as possible) are superior to stoichiometric reagents.

Principle #10: Design chemicals and products for degradation

Chemical products should be designed so that at the end of their function, they breakdown into innocuous degradation products and do not persist in the environment. New chemical designs and their products must undergo a thorough assessment to determine their potential environmental persistence and thus should be enhanced to facilitate degradability, recycling, and circularity.

Principle #11: Real-time analysis for pollution prevention

Analytical methodologies need to be further developed to allow for real-time, in-process monitoring and control prior to the formation of hazardous substances.

Principle #12: Inherently safer chemistry and processes for accident prevention

Substances and the form of a substance used in a chemical process should be chosen to minimize the potential for chemical accidents, including releases, explosions, and fires.

As ultimately, every interaction in our universe is based on atoms and molecules reacting with each other, chemistry expands into all other disciplines of science, technology, and engineering. Green Chemistry can be viewed as the foundational framework necessary for a sustainable future relying on the chemical sector to operate benignly.

There is thus an overarching applicability of Green Chemistry across the scientific spectrum ranging from physics through biology, pharmaceuticals and medicine, to engineering. Ultimately, the incorporation of circularity (the management of materials) and Green Chemistry (the intrinsic properties of materials) are unifying factors that bridge all scientific fields.

3.4.2 Current *vs.* Future Methodologies

Most of the current chemicals sector operates on a linear model (Figure 3.7, top), utilizing mostly fossil and finite feedstocks, where reagents with high reactivity but often unintended persistence and toxicity are involved.[21] The production processes frequently generate toxic, persistent, and bioaccumulating waste or greenhouse gases, often exceeding the intended product (5 to 50 times for specialty chemicals and 25 to 100 times for pharmaceuticals).[21,51] Pharmaceuticals enhancing human life or agricultural chemicals increasing crop yield must not persist in our water, as this results in the death of fish and the degradation of groundwater.[52–54] Optimizing chemicals for materials production with long-term durability and performance results in bioaccumulation in nature.[55]

In society and especially industry, current sustainability efforts are mainly directed towards the optimization of a linear model, where a reductionist approach focuses solely on either greenhouse gas emissions, energy or water consumption, or ecotoxicity to incrementally improve the sustainability aspects of an old system.[54,56]

However, for a truly circular and sustainable future, molecules and systems need to be considered and designed in their entirety, utilizing inherent molecule properties to not shift environmental impact from one problem to another. Traditional reductionist approaches optimizing outdated processes towards higher sustainability need to be replaced with holistic, systems thinking approaches to allow for truly circular and non-toxic designs for a sustainable future (Figure 3.7, bottom).[57–60] The chemical industry and academic research need to provide solutions that synergistically advance multiple sustainability challenges rather than optimizing just one. However, it is essential to recognize that incremental advancements, such as reducing toxicity and enhancing biodegradability, are valuable contributions, and the idea that one must achieve all-encompassing solutions should not discourage efforts toward partial progress.

A proactive design of chemicals and processes is required to not only optimize for risk reduction but rather reduce hazard.[61,62] By understanding molecular mechanisms and properties, a circular chemical sector allows for the utilization of renewable feedstocks with inherent functions. These

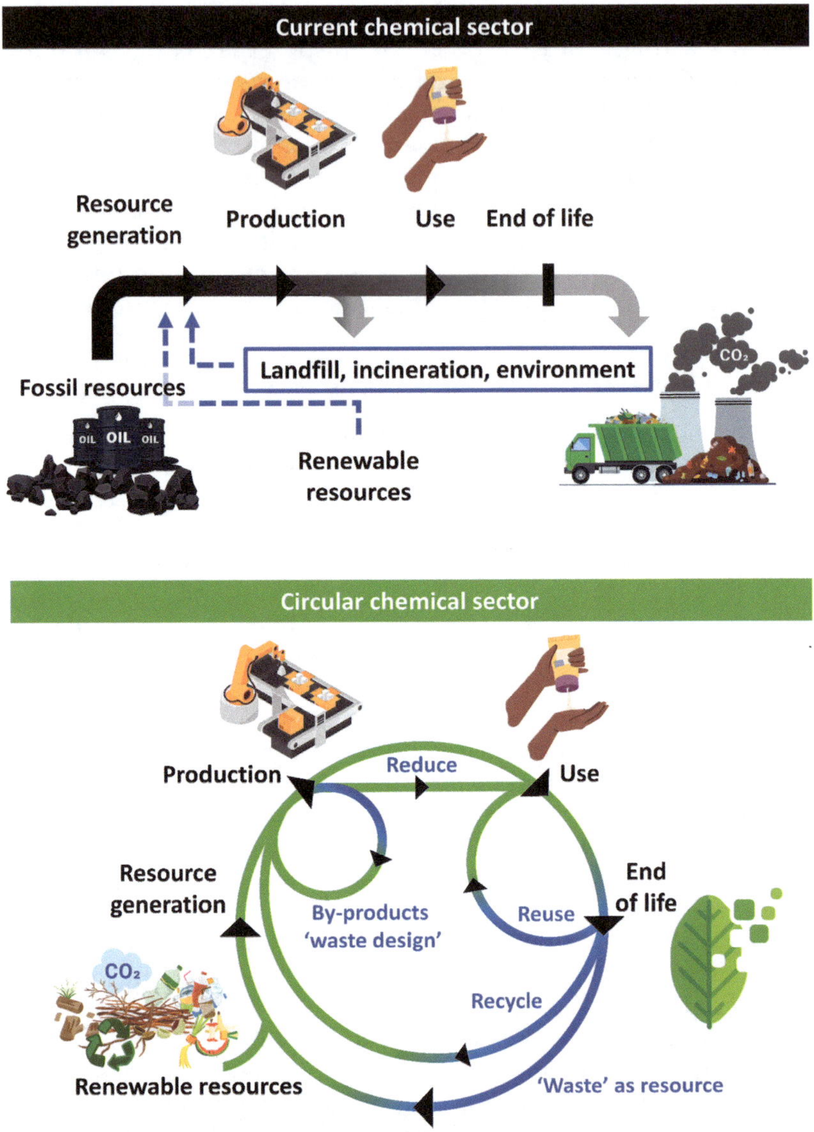

Figure 3.7 The current and future circular chemical sector to enable a sustainable society. Reproduced from ref. 21 with permission from the American Association for the Advancement of Science (AAAS), Copyright 2020.

feedstocks are transformed with renewable energy into products with high performance and durability but systematically allow for degradation, resulting in less pollution and a circular economy.

The complete life cycle of a product, spanning from its raw material acquisition through mining or agriculture to its end-of-life stage *via* recycling or disposal, involves multiple steps, each varying in environmental impact.

Whether evaluating global warming, ecotoxicity, or resource consumption, a predominant portion of harm often originates from one or two specific steps.[36] These critical steps, termed hotspots, warrant primary attention from green chemists and chemical engineers. Rather than solely dedicating resources to improving less impactful steps, the emphasis should be on utilizing life cycle assessments (LCAs) for existing processes to identify and prioritize such hotspots.[36,37] It should be noted that all steps in a process have the potential to cause harm and reducing harm is a key target of Green Chemistry.[63] While direct benefits to the environment may be enhanced by focusing on the most harmful steps first, creative solutions to improve non-hotspots may serve as inspiration for other researchers to fix problematic steps in other processes. Also, when primary hotspots are improved, secondary hotspots will become the primary ones, representing a greater fraction of the total harm of a process. After improving the known hotspots, past and future efforts to green non-hotspots will have greater value.[36] All these efforts are crucial to maximize the benefits of Green Chemistry, which will be further discussed in the following sections.

3.5 Minimizing Environmental Impact

By applying the *12 Principles of Green Chemistry* the chemical sector is aiming to develop and utilize platform chemicals from renewable feedstocks for products and consumer goods. Three key aspects are targeted for enhancement:[21,64]

1. Depletion of fossil fuel resources and its greenhouse gas emissions
2. Chemical toxicity
3. Chemical persistence in the environment (*i.e.* biodegradability).

Currently, most chemicals are based on finite fossil resources and thus depleting, while contributing to greenhouse gas emissions, which will be further elaborated in Section 3.6.

The **toxicity** of chemicals is another aspect that must be improved.[65] There is a need to explore the fundamental molecular mechanisms, not only for utilization of inherent functionalities and (bio)degradation, but also for the absorption, distribution, metabolism, and excretion of molecules within the body. As such, physiochemical properties such as solubility, reactivity, and cell permeability impact these processes.[21] Chemicals can have immediate and long-term (chronic) effects on generations, which need to be understood and minimized by applying Green Chemistry.[66] Green Chemistry aims to design safer, less toxic substances and the replacement of traditional chemicals with more sustainable alternatives.[67]

The last aspect, which is closely related to (bio)degradation is the **persistence of chemicals in the environment.** Therefore, it is imperative for chemists and chemical engineers to prioritize the mitigation of chemical release into the environment. If chemical processes and chemicals are not readily degrading and thus bioaccumulating, concerns arise for human and environmental health.[21] Notable examples are per- and polyfluoroalkyl substances (PFAS), which are concerning due to their high persistence (or that of their degradation products) and their known impacts on human and environmental health from well-studied PFAS.[68,69] Currently, thousands of different PFAS (>1400) are utilized in various applications, including textile impregnation, firefighting foam, and electroplating.[68] PFAS exhibit both hydrophobic and oleophobic properties, making them effective surfactants or surface protectors in several applications. The C–F bonds in PFAS lead to exceptional stability, a feature that also results in highly persistent terminal transformation products in the environment.[69,70] When molecules bioaccumulate, effects of low-dose and long-term exposure, reactions with other chemicals in the environment and high environmental mobility can have drastic consequences for our nature and society.[70] These aspects are further elaborated in Section 3.5.1.1.

3.5.1 Waste and Pollution Prevention

Waste prevention, or the minimization of waste, constitutes a fundamental principle that aims to diminish the quantity of unused material.[71,72] This concept is intricately connected with other principles, such as the reduction of surplus materials like (excess) solvents, stoichiometric reagents, or the generation of side products.[73] These reductions contribute to higher atom economy, increased energy efficiency, catalytic processes, and overall process safety (Principles 1, 4, 5, 6, 9, and 12, Figure 3.6). Furthermore, prioritizing waste prevention over waste remediation is preferable, as the latter entails additional energy input (such as solvent evaporation or purification, and isolation of products).

Generally, the waste and pollution associated with chemical processes is tracked through metrics such as atom economy or E-factor. Centered on the reduction of derivates and side products, reactions with full atom economy are realized through prime examples such as click chemistry,[74–76] which are often utilized in medicinal chemistry.[77–80] However, Green and Sustainable Chemistry also has significant impact in the pharmaceutical industry (Table 3.1).[15] Pharmaceutical companies typically manufacture a diverse range of chemicals in smaller quantities (500–5000 tons annually) and thus offer opportunities for incorporation of greener technologies.[15]

As discussed before, the waste generation and thus E-factor increases with chemical complexity, as more synthetic and purification steps are required. Thus, reaction pathways need to be optimized to avoid protection groups or tedious purification techniques to enhance minimum waste generation.

Table 3.1 Environmental factors of various chemical sectors.[15]

Industry	Product tons per year	Waste/product ratio (E-factor) by weight
Oil refining	10^6–10^8	<0.1
Bulk chemicals	10^4–10^6	<1–5
Fine chemicals	10^2–10^4	5–50
Pharmaceuticals	10^1–10^3	25–100

Some of these concepts are correlated to sustainable laboratory practices and Green Laboratories, which will be further explained in Section 3.9.

3.5.1.1 *Design for Degradation and Recycling*

The chemicals of the future should be designed for facile and rapid degradation into nonpersistent compounds that pose minimal adverse effects on the environment.[21,81] Novel chemical designs and their products necessitate systematic evaluation for their potential persistence in the environment and need to be advanced to allow for degradability, recycling, and circularity.[59,82] This includes the development of rapidly degradable pharmaceuticals,[83] fertilizers, insecticides, plasticizers,[84] polymers,[85] coatings, and plastics without compromising properties and performance.

Traditional polymers like polyethylene (PE) or polypropylene (PP) persist for several years after disposal. Tailored for durability, these polymers may not be suitable for applications where plastics are utilized briefly and then discarded. Furthermore, plastics are frequently contaminated by food and other biological substances, rendering physical recycling of these materials impractical.[86]

Plastics and other polymers used in our economy are currently not fully recycled. Often downcycling schemes are applied leading to environmental pollution downstream.

Avoiding bioaccumulation,[87] (bio)degradable polymers are specifically designed to undergo degradation through the influence of living organisms or chemical reactions when discarded (Figure 3.8). Remarkable advances have been achieved in creating products utilizing polymers like starch and cellulose,[88] polyesters,[89–92] or (poly)lactic acid.[86,93] Markets such as packaging materials (trash bags, wrappings, food containers, film wrapping), hygiene products (diaper back sheets, cotton swabs), consumer goods (fast-food tableware or containers) serve as optimal targets for (bio)degradable polymers.[86,94] Renewable and recyclable materials allow for circular processes, where chemicals taken from the nature are reused in the material over several decades and life cycles.[95] Due to their biodegradability properties those materials can be returned to nature without impacting the environment.[86]

Chemical recycling of synthetic polymers presents a viable solution for developing sustainable plastics and materials.[96] Considering technologies in the design stage to endow materials with dynamic and circular behavior has been key to the rational design of future smart materials.[97,98] Such dynamic chemistry, including supramolecular and dynamic covalent chemistry, enables sustainability by providing versatile molecular tools to design materials that can be adapted, repaired, and recycled.[99] These self-healing materials drastically reduce environmental impacts, as products are less prone to be discarded through fixing themselves (*i.e.* self-repair).[100,101] Furthermore, mechanically robust, tough, and processable crosslinked materials are being designed for chemical depolymerization under mild conditions, yielding virgin-quality monomers for future processing.[102]

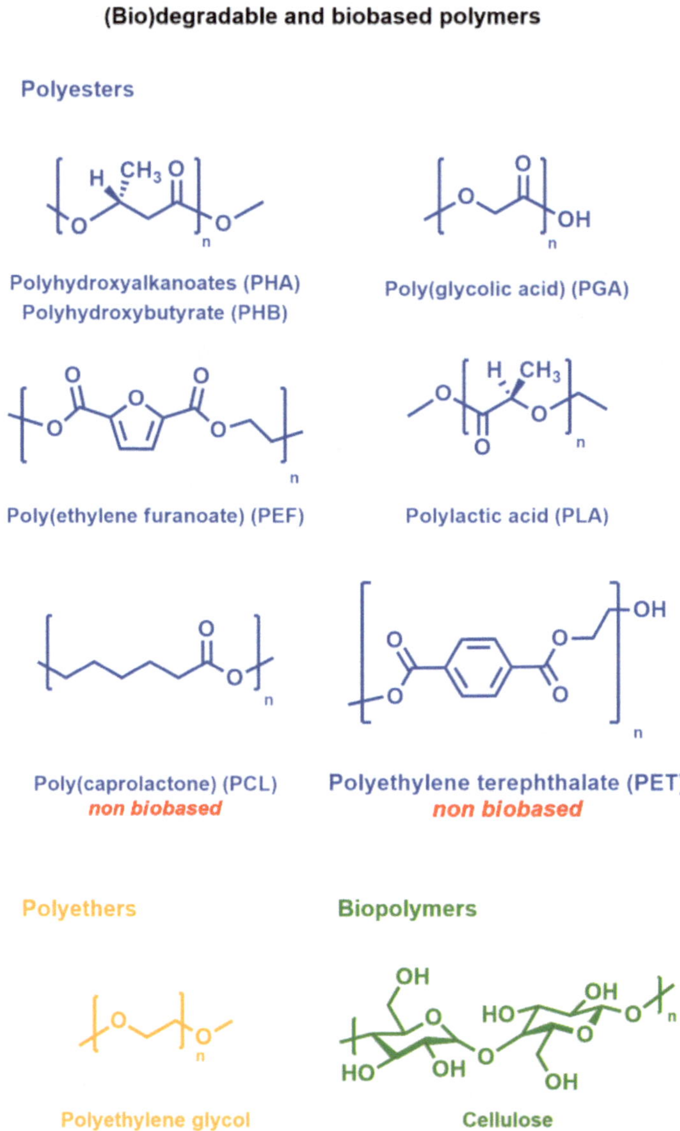

Figure 3.8 Structures of selected (bio)degradable and biobased polymers. PET and PCL are not biobased and plastics such as PP are not readily recyclable.

3.5.2 Sustainable Solvents and Solvent-free Reactions

Solvents find extensive use in various commercial manufacturing and service industries.[103] Organic solvents, as major sources of waste in the chemical industry, inevitably contaminate air, land, and water.[104] To address this issue, researchers must concentrate on minimizing solvent usage by developing solvent-free processes, improving recycling methods, and

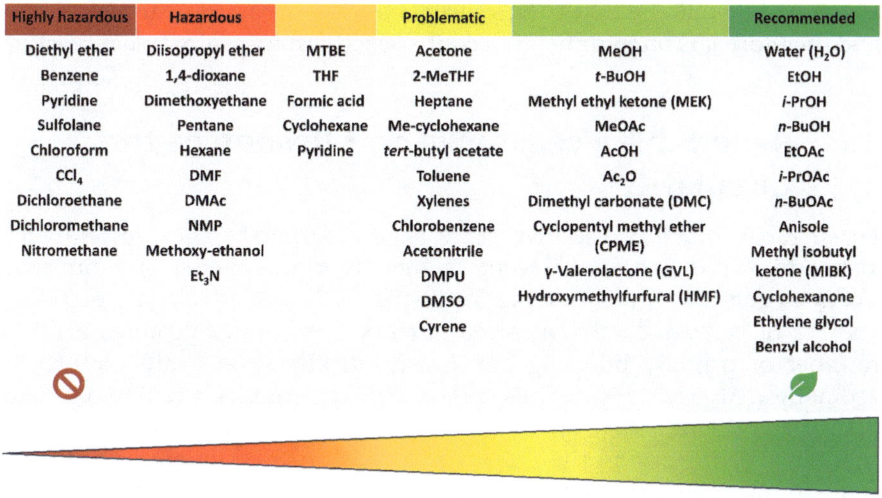

Figure 3.9 Ranking of solvents with respect to their sustainability.[111–116] Solvent guide for substitution of problematic solvents to enhance safety and reduce environmental impact.[117–121] Reproduced from ref. 122 with permission from ChemRxiv, Copyright 2023.

replacements through **greener alternatives** to drastically improve the environmental impact of a process (Figure 3.9).[105] There has been a plethora of research ongoing on improving the greenness of solvents, where new catalytic methodologies allowed for access to a variety of recommended and biobased solvents.[106] Also, classical organic reactions can be performed "on water",[107] at biphasic interphases between water and an immiscible solvent,[108] or in water itself through surfactants and the formation of micelles.[109,110]

Another class of solvents are **supercritical fluids**[123–125] and **ionic liquids.**[126–131] **Supercritical fluids** are gases at a temperature and pressure beyond their critical points, displaying densities and properties of both liquid and gases (*e.g.* CO_2).[132,133] **Ionic liquids** are salts (*e.g.* asymmetric organic ions) with melting points at relatively low temperatures (typically below 100 °C) with advantageous properties such as low volatility, recyclability, and solvating abilities.[15,134,135]

However, the most optimal way to further circumvent issues with solvents is to not use them, a method widely employed in the paints and coatings industries.[15,136] In recent years, there has been a significant shift toward solvent-free technologies such as powder coatings, high-solids formulations, and waterborne paints, which drastically reduce or eliminate the use of volatile organic compounds (VOCs) while maintaining performance and durability.[137–141]

This paradigm shift is also observed in the field of **mechanochemistry**[142–145] with the synthesis of a whole class of compounds, that were previously inaccessible or tedious to make.[146–153] Through such solvent-less reactions the environmental aspects of chemical reactions is reduced dramatically.

However, it should be noted that solvents play a big role in heat management and therefore it is often difficult to control exotherms in solvent-less reactions.

3.6 Renewable Feedstocks and Resources for Circularity

Fossil fuels, a finite resource, serve as the cornerstone of societal infrastructure, while also contributing to increased greenhouse gas emissions. Today's chemical production predominantly depends on oil, gas, and coal as sources of carbon.[21] Currently petrochemical refineries produce a limited number of primary building blocks, specifically short-chain olefins, and aromatics. Alongside synthesis gas, a combination of CO and H_2, these platform chemicals serve as the foundation of carbon based compounds and materials, giving rise to the extensive molecular diversity observed in more than 100 000 chemical substances in final products (Figure 3.10, right).[21]

Petroleum emerges as an unsustainable option due to the inherent challenges associated with the depletion of fossil resources, its connection to climate change, and inextricable links to toxicity. Consequently, the design of novel value chains based on non-fossil carbon sources and renewable energy will help establish a closed carbon cycle.[21] The chemical industry needs to shift from fossil- to renewable-based chemistries with a holistic approach to allow for a circular future. Molecules and products need to be designed to allow for (bio)degradability and recyclability, while utilizing the inherent properties of renewable feedstocks.[154] This also necessitates that their intrinsic properties and manufacturing methods be benign and includes a paradigm shift from linear to circular processes (Figure 3.10, left).[21]

Previously disregarded as low value 'waste' the future chemical sector is transforming these products and molecules as value-added resources to produce key building blocks for a circular economy.[155–157] Such carbon sources include the direct conversion of **CO_2** as valuable feedstock to replace petroleum-based carbon feedstocks[158–160] or **lignocellulosic biomass** from *e.g.* papermill waste to produce building blocks such as vanillin, catechol, or phenol.[161–167] In the framework of a circular economy **recycled plastic materials** are another major carbon source for the chemical sector.[168]

Platform chemicals obtained through renewable resources can be utilized in two strategies: the most desirable and perhaps challenging **Strategy (i)** is to develop an entirely new value chain starting from renewable resources (*i.e.* CO_2, lignocellulose, or plastic waste), through building blocks with improved functionalities to products with advanced and enhanced properties over existing fossil fuel derived ones.[169] Synthetic methods and industrial processes for achieving this kind of material and feedstock transition should be consistent with the 12 Principles of Green Chemistry (Figure 3.6). By taking advantage of the inherent higher complexity of renewable biobased feedstocks, shortcuts to desired functional groups and chemicals are opened. New commodity chemicals or new specialty

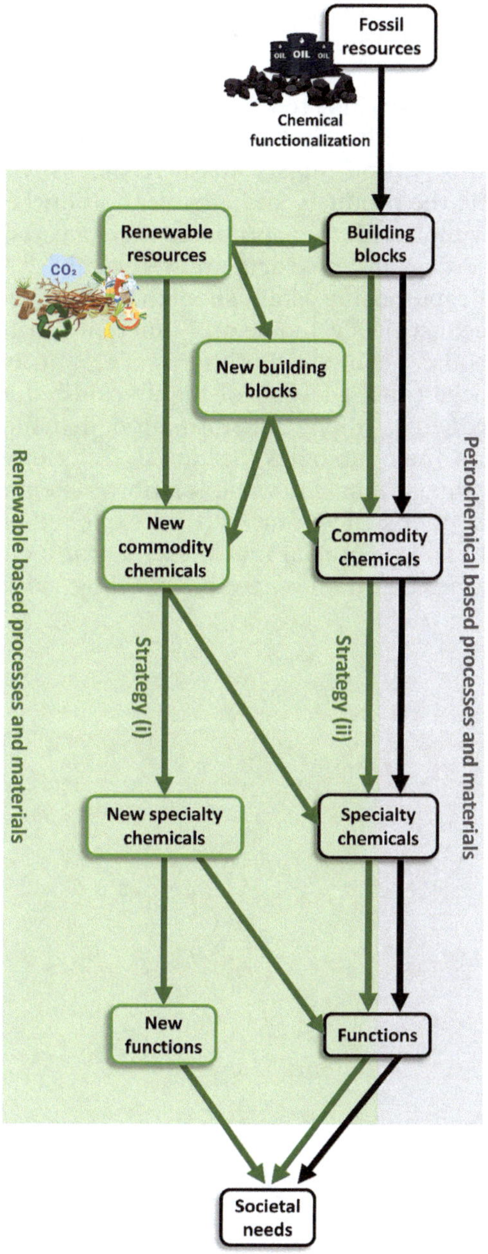

Figure 3.10 Overview of strategies for sustainable chemical processes towards societal needs. Petrochemical based pathway (black), and renewable based pathway (green). Strategy (i), development of new sustainable building blocks (*i.e.* platform chemicals) from renewable resources. Strategy (ii), development of sustainable shortcuts to existing key products from renewable resources. Adapted from ref. 21 with permission from the American Association for the Advancement of Science (AAAS), Copyright 2020.

chemicals obtained through this strategy allow for unprecedented reactivity and properties leading to new beneficial materials with improved functionality.[21]

The alternative **Strategy (ii)** is to take advantage of existing infrastructure, building blocks, commodity, and specialty chemicals, which are utilized in the current industry, and replace fossil resources with renewable ones (*i.e.* drop-in). While the products and chemical products would remain the same, but are now biobased, this approach offers a certain risk assessment as reactivity, properties, and products are well established.[21] Although this strategy allows for repurposing established knowledge and infrastructure, it dismantles and reconstructs valuable and inherent functionalities.

Unlike fossil resources composed primarily of carbon and hydrogen, these renewable feedstocks consist largely of highly oxidized and functionalized molecules. Consequently, their transformation demands innovative synthetic concepts and methodologies (Section 3.7).[21] However, these aspects also have the advantage that renewable platform chemicals have multiple functional groups preinstalled that can be exploited into new chemical families (Figure 3.11). It is crucial that these platform chemicals are established using strategies based on renewable building blocks outside the food chain.[170]

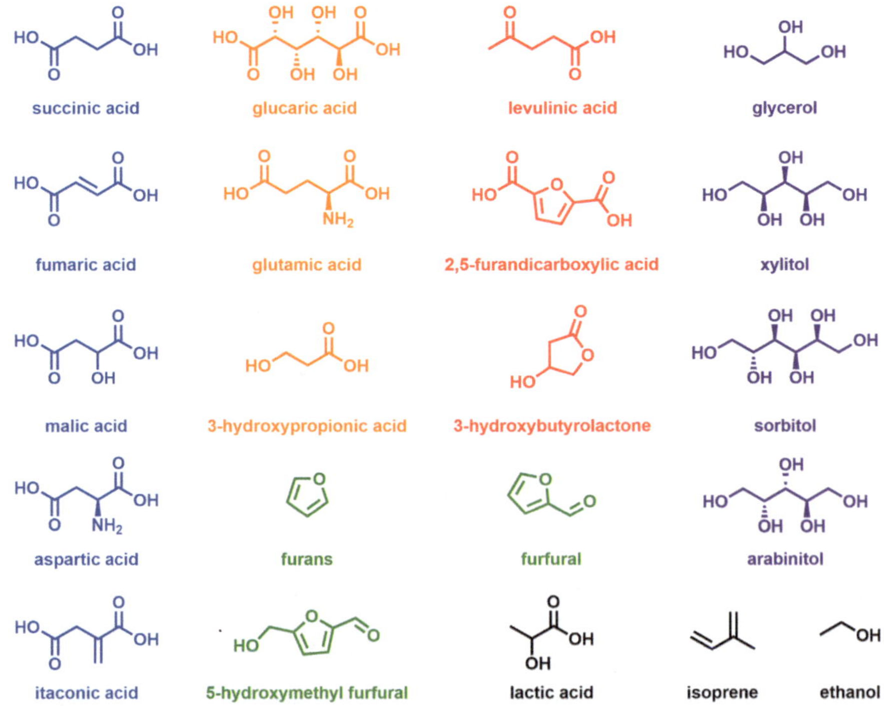

Figure 3.11 Top value-added platform chemicals derived from biomass.[171–173]

3.6.1 Biomass-derived Feedstocks

Utilizing agro-industrial waste streams and other biomass-derived feedstocks (Figure 3.12) allows for reduced emissions of fossil fuel-derived CO_2.[170] **Lignocellulose** consists mainly of cellulose (30–50% by weight), hemicellulose (20–40% by weight) and lignin (15–25% by weight).[174,175] Cellulose and hemicellulose are carbohydrate polymers, while lignin is an aromatic polymer. **Cellulose** is composed of glucose monomers linked by β-1,4-glycosidic bonds, creating high-molecular-weight linear polymers.[176–180] The structured arrangement results in high crystallinity, leading to poor solubility in water and thus making hydrolysis challenging.[176] **Hemicellulose,** on the other hand, is a smaller, amorphous, and branched polymer consisting of C_5-(xylose, arabinose and rhamnose) and C_6-sugars (glucose, mannose, and galactose).[181] Depending on the type of plant or material, different ratios of C_5- and C_6-sugars can be found. Hemicellulose is generally more soluble in water, allowing for easier hydrolysis into corresponding sugar monomers. **Starch** is also an inexpensive, renewable material obtained from the hemicellulose of corn and other crops. Serving as energy storage for plants, starch

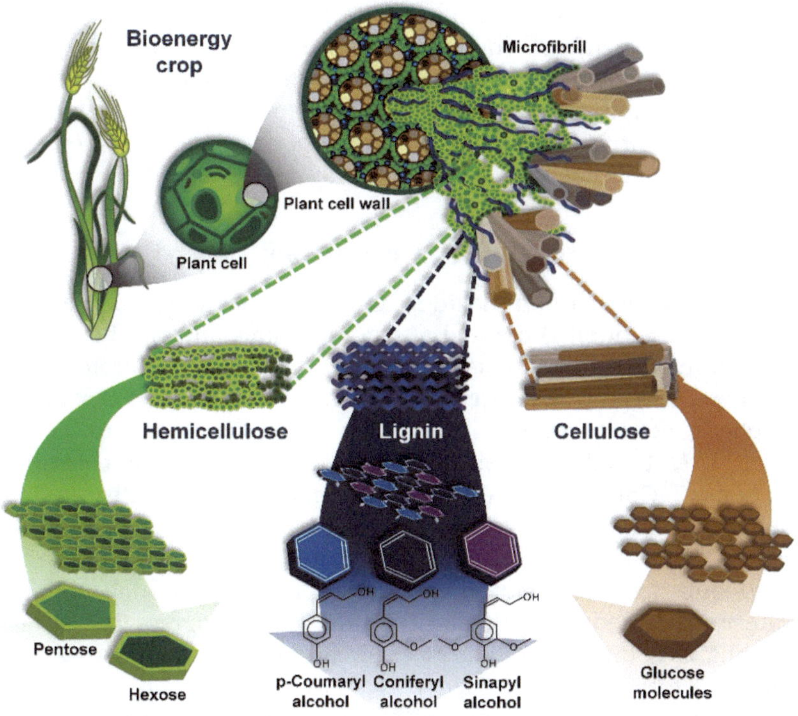

Figure 3.12 Structure of lignocellulosic biomass and its main biopolymers hemicellulose, cellulose, and lignin. Reproduced from ref. 174, [https://doi.org/10.3390/app9183721], under the terms of the CC BY 4.0 license, https://creativecommons.org/licenses/by/4.0/.

carbohydrates are generally biodegradable polymers used in eco-friendly materials and are key building blocks for the chemical industry.[86,182] The third major component, **lignin**, is a hydrophobic aromatic crosslinked polymer covalently bonded to hemicellulose.[183] It acts as a protective layer for the cellulose and hemicellulose network, providing strength and robustness while serving as a barrier preventing chemical and microbial threats.[184]

Lignocellulose as a feedstock obtained through wood waste offers pathways to molecules such as furane dicarboxylic acid (FDCA) derived from monosaccharides that serve as a potential building block for innovative polyester products.[185] Bio-derived lactic acid is also used as a feedstock for large-scale production of polymers.[86] The platform chemical furfural and its subsequent conversion into other building blocks for the replacement of *e.g.* acrylates and acrylic acid led to the development of high-performance polymers and coatings while adhering to the Green Chemistry Principles.[186–189]

Lignin, a major component of lignocellulose, is the largest source of aromatic building blocks on the planet and has great potential to serve as a starting material for value-added products such as phenol or benzene, toluene, and xylene (BTX).[190–195] Remarkable achievements in catalysis research allow for cleavage of C–O bonds of the most abundant β-O-4 moieties and of stronger moieties such as C–C linkages, to obtain well-defined compounds and platform chemicals from lignin.[196–202] At the interface of chemistry and chemical engineering, chemical processing techniques achieved advancements directed towards extracting lignin in biorefineries and subsequent transformation into high-performance plastics, biofuels, bulk- and fine chemicals.[203–207]

3.6.2 Utilization of CO_2 as a Feedstock

The greenhouse gas carbon dioxide (CO_2) stands out as a valuable renewable feedstock that is often generated at chemical sites and can be utilized as a carbon source directly after its production, prohibiting its release into the atmosphere. Direct air capture of CO_2 is often constrained through its low atmospheric concentration, a limitation which can be circumvented at chemical sites based on fossil fuels, where CO_2 is generated in high concentrations.[208] Consisting of only one carbon, CO_2 can be integrated into the current value chain through existing large-volume base chemicals such as methanol.[19,20,209] CO_2 conversion offers diverse applications, spanning from fuels to bulk and specialty chemicals, including pharmaceuticals.[210–212] Despite its thermodynamic stability quite some progress has been made integrating CO_2 into polymers (*e.g.* polycarbonates) and utilizing it in the synthesis of formic acid (HCOOH).[124,213] By utilizing CO_2 as a novel feedstock, not only are carbon emissions and fossil fuel dependency reduced, but a more sustainable chemical production can be achieved (Figure 3.13).[209,214]

At present, there are examples of CO_2-based processes that demonstrate decreased global warming impacts compared to fossil-based production routes. This may seem counterintuitive considering the thermodynamic

Figure 3.13 Utilization of carbon dioxide (CO_2) as a renewable feedstock to close the carbon cycle.[214] Through reduction with green hydrogen, platform chemicals and energy carriers such as methanol, formic acid and methane are generated.[19,215]

stability of the CO_2 molecule. The production of polyols *via* copolymerization of propylene oxide and CO_2 can replace the traditional homopolymerization of the epoxide to provide polyols. Traditional production of **formic acid** *via* CO or CO_2 starting from methane (CH_4) utilizes an energy-intensive separation step of the product mixture. The number of conversion steps can be reduced when directly starting from the carbon source CO_2.[215] While carbon dioxide itself does not constitute a carbon sink, the overall balance of the environmental impacts over the life cycle of products and processes can lead to improved and more sustainable production processes.[209]

CO_2 can also be applied as C1 building blocks for catalytic transformations to carboxylic acid functions and methyl groups.[216] This results in an entirely new family of synthetic methodologies (**Strategy (i)**, Figure 3.10). Utilizing CO_2 as a renewable feedstock has significant benefits for the replacement of hazardous reagents and atom economy.[217–219] Examples are new catalytic routes for the synthesis of carboxylic acids *via* direct catalytic carboxylation,[220–222] which traditionally start from chlorinated substrates *via* a Grignard reaction or the replacement of methyl iodide as a methylation reagent.[209]

The largest benefits however are achieved when coupling highly concentrated CO_2 sources from CO_2-emitting sectors with carbon-free hydrogen and/or electrons generated from renewable power such as wind and solar, resembling artificial photosynthesis.[223,224] Next to the production of formic acid, the production of **methanol** from CO_2 *via* low-carbon electricity offers great potential for power-to-methanol and thus energy storage. While methanol is already a key platform chemical in today's chemical sector, this method allows for harvesting renewable electricity as liquid fuel.[209] The direct incorporation of CO_2 into polymer chains for consumer products is already being industrialized.[158]

3.6.3 Upcycling and Plastic Waste Valorization

Plastics have surpassed most man-made materials and have long been scrutinized for their environmental impact.[225] Approximately >350 million metric tons (Mt) of plastics are produced annually. It is estimated that 8300 Mt of virgin plastics have been produced since 2017.[226] In 2015, about 6300 Mt of plastic waste had been generated, with only 9% being recycled, 12% being incinerated, and 79% accumulating in landfills or escaping into the natural environment. Current plastic pollution is barely addressed by **incineration** and **mechanical recycling.**[227] If these trends in production and waste management persist, an anticipated 12 000 Mt of plastic waste will be present in landfills or the natural environment by 2050.[226]

Most widely used plastics such as polyethylene (PE), polypropylene (PP), or polystyrene (PS) are not biodegradable and are challenging to recycle due to properties such as non-hydrolysable C–C backbones, high melting points, and high melting viscosities.[228–230] Also, per- and polyfluoroalkyl substances (PFAS) with their C–F bonds needed for exceptional stability result in highly persistent terminal transformation products harmful to human health and the environment.[69,70] Thus solutions and utilization of conventional plastics as feedstocks are of utmost importance to the rising environmental impacts of plastic waste production. Even with biobased and biodegradable options at hand, traditional plastics persist in nature. Technologies to utilize these wastes as valuable sources of carbon must be developed.

Pyrolysis presents a current commercially available solution, which suffers from high energy consumption and limited control over product selectivity.[228] However, the combination with several **chemocatalytic strategies** is giving life to fuels, chemicals, and monomers from plastic waste:[227,231,232]

- Pyrolysis (300–500 °C)
- Hydrocracking/hydrogenolysis (300–450 °C, H_2 at 20–150 bar)
- Gasification (>500 °C)
- Solvolysis (hydrolysis, methanolysis, glycolysis)
- Dehydrochlorination (mechanochemistry)[233–235]

Remarkable advancements into catalytic transformations are made in trying to overcome drawbacks and complement pyrolysis processes for chemical recycling and degradability (Figure 3.14).[236,237] Through such catalytic techniques and backbone scission, up to 80% conversion of PE into light alkanes at mild conditions can be achieved.[228,238,239]

Most notably in the context of pyrolysis and catalytic transformations, a discernible parallel exists with the catalytic cracking of hydrocarbons from fossil fuels, given the shared chemistry observed in PE, PP, PS, and oil components.[227] As such the knowledge accumulated in refining industrial catalysts can serve as guidance. Zeolites exhibit potential to mitigate deactivation pathways in the pyrolysis process allowing the possibility to simultaneously process biomass and plastics.[227,240,241]

In alignment with their large waste generation, PE, PP, and PET stand out as the polymers most widely tackled through catalytic conversion and utilization as feedstock for platform chemicals, energy carriers, and monomers.

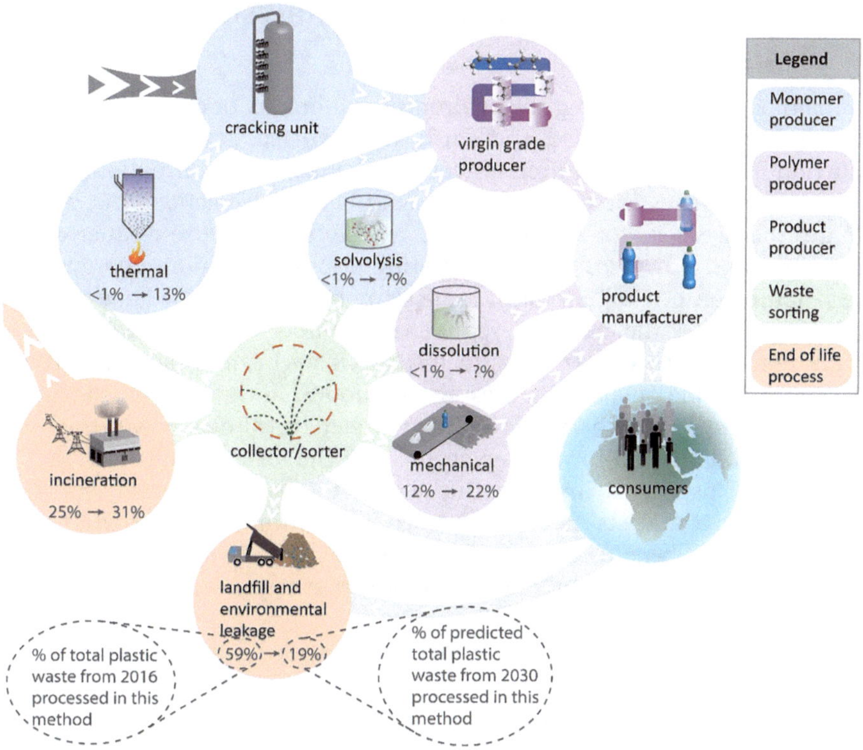

Figure 3.14 Illustration of an envisioned plastics value-chain that could enhance the transition to circularity. Currently most plastic is incinerated or landfilled (bottom left), because collection and sorting produce very contaminated and mixed plastic-waste streams. Better techniques for collection and sorting can lead to streams of plastic waste that can be recycled by various chemical recycling methods. These routes will still need to be complemented by traditional mechanical recycling for the purest steams of a single polymer. The share of each technique's contribution to waste processing in 2016 is displayed at the bottom left, while the projections for 2030 are shown at the bottom right. The plastic objects are then sold to the consumer and after its life cycle, collected again for sorting – to undergo another recycle. Reproduced from ref. 225, https://doi.org/10.1002/anie.201915651, under the terms of the CC BY 4.0 license, https://creativecommons.org/licenses/by/4.0/.

Further efforts are underway to enhance the degradation of PFAS into base chemicals such as fluorides, carbonates, or formates among others.[242,243] Presently, most attention is directed towards process engineering. Nevertheless, a more rational approach would be the prioritization of catalyst design and mechanistic understanding at the molecular level. Within the context of Green Chemistry and the utilization of renewable feedstocks, the realm of plastics depolymerization offers several opportunities to foster a sustainable future.

3.7 Energy Efficiency, Green Catalysis, and Upscaling

A major aspect in reducing the environmental impact of chemical reactions is the utilization of biobased and earth abundant catalysts to allow for a reduction in the energy required for certain chemical transformations. **Catalysis** not only enhances the energy efficiency of chemical reactions but also eliminates the necessity for stoichiometric reactants, leading to cleaner chemical reactions and processes.[244,245] By offering a lower activation energy pathway (*e.g.* temperature) for chemical transformation, catalysts, while remaining unchanged after the reaction, reduce the number of steps and energy consumption of a process, enabling it to participate in multiple reaction cycles. A prime example of such green catalysis is found in the manufacturing of the analgesic ibuprofen, where a novel catalytic step has almost halved the number of stages and allowed for the replacement of the toxic solvent CCl_4.[15,246,247]

A crucial aspect to improve the circularity aspects of catalysts is their recyclability: Catalysts need high turnover numbers and high turnover-frequencies to be effective on a large-scale. The **turnover number** is the number of moles of a substrate that a mole of catalyst can convert to product before becoming inactivated.[248]

Determination of turnover number:[248]

$$\text{Turnover number (TON)} = \frac{n_{\text{product}}}{n_{\text{cat}}} \tag{3.6}$$

The term **turnover frequency** (TOF) describes the turnover per unit time t (*e.g.* seconds). Where for the most relevant industrial applications the TOF is in the range of 10^{-2}–10^{2}.[248]

Determination of turnover frequency:[248]

$$\text{Turnover frequency (TOF)} = \frac{\text{TON}}{t} \tag{3.7}$$

Spent catalysts at the end of a reaction contribute to chemical waste. Hence catalysts need to be stable and recycled or reused. In particular, as many catalysts still contain toxic metals, new non-toxic catalysts must be invented.[15] Catalysts are only truly adhering to the 12 Principles of Green Chemistry when they are based on biobased and/or earth-abundant materials. If metals are necessary, iron (Fe) or titanium (Ti) should be the primary choices.

The following depiction of the periodic table (Figure 3.15), curated by the European Chemical Society (EuChemS), highlights **element scarcity** and underscores the rapid consumption of elemental resources. Certain elements face complete depletion unless effective recycling pathways are established. Additionally, it draws attention to element sources from countries where conflicts arise over mineral rights ownership. The table identifies 31 elements (with varying numbers up to 70) essential in smartphone production; devices often replaced more frequently than necessary. An element

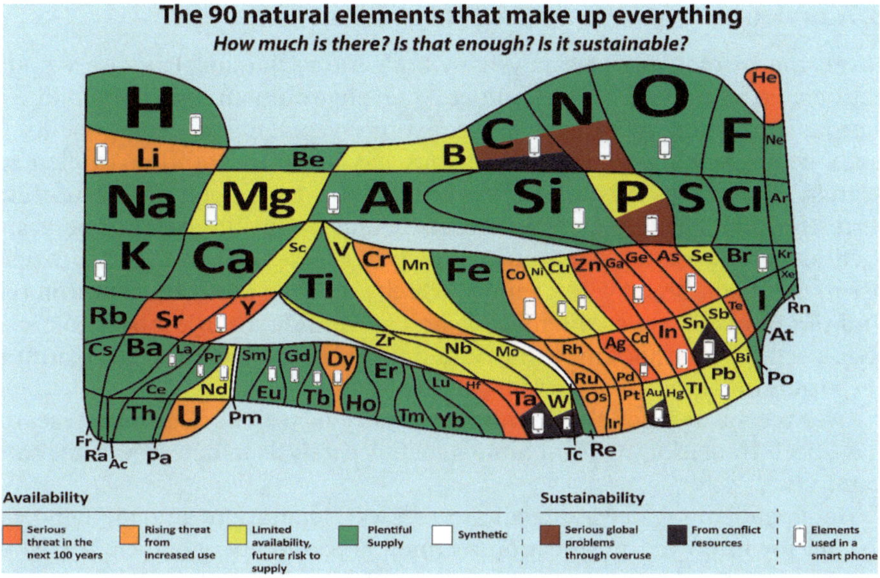

Figure 3.15 Element scarcity of the 90 natural elements that make up everything. Naturally occurring elements (except for rare ones beyond uranium) are illustrated through tiles. The size of these tiles serves as a logarithmic-scale indicator of an element's abundance in the Earth's crust and atmosphere. While these areas are approximations for all elements, the least abundant ones showcased are deliberately magnified to enhance visibility (technetium, promethium, polonium, astatine, radon, francium, radium, actinium, and protactinium). Reproduced from ref. 255, https://www.euchems.eu/euchems-periodic-table/, under the terms of the CC BY-ND 4.0 license, https://creativecommons.org/licenses/by-nd/4.0/deed.en

worth highlighting is lithium, as its extraction and use is projected to increase dramatically in the near future due to the production of lithium-ion batteries, particularly used in the automotive sector.[249] Efficient recovery processes and recycling in line with the Principles of Green Chemistry are paramount.[250–254] Another element is carbon, highlighted with three colors (green, dark red, and dark grey). Green is chosen as carbon is highly abundant through atmospheric CO_2 (too abundant), carbonate rocks, and vegetation. Red indicates that accessible reserves of fossil carbon are quite limited, and use should be restricted. Grey shows that those reserves are often located in conflict zones. The dark red color further highlights the overturning of the biogeochemical cycles of nitrogen, phosphorus and carbon, thereby severely affecting the biosphere. The massive release of CO_2 – as well as of nitrogen and phosphorus in soil and water bodies – has already largely overtaken the so-called planetary boundaries. The legend includes two color groups, corresponding to availability and sustainability. Catalytic materials often rely on metals, of which iron (Fe) or titanium (Ti) offer the greatest sustainability potential.[255,256]

3.7.1 Homogeneous *vs.* Heterogeneous Catalysis

Given the pivotal role of catalysis in both industrial and laboratory applications, it is imperative to evaluate the environmental sustainability of various transition metals, which constitute fundamental components of both heterogeneous and homogeneous catalysts. In addition to element scarcity, toxicity towards marine ecosystems (*e.g.* fish, algae, and plants) and terrestrial life, carcinogenicity, and elemental impurities need to be evaluated as parameters in a thorough life cycle assessment (LCA). The toxicity assessment combined with element scarcity analysis indicates that **iron** (Fe) and **titanium** (Ti) are as the most desirable candidates, in contrast to nickel (Ni), cobalt (Co), rhodium (Rh), ruthenium (Ru), palladium (Pd), platinum (Pt), iridium (Ir), and molybdenum (Mo).[257,258]

Over recent decades the chemical industry as well as academic research have been transitioning from **homogeneous catalysis** to **heterogeneous solid catalysts.**[259–261]

Heterogeneous catalysts facilitate catalyst separation and recycling, allowing for improved purification technologies and energy efficiency, when compared to homogeneous processes. While homogeneous (transition metal) catalysts were historically highly regarded for their superior activity and selectivity relative to their heterogeneous counterparts, research in heterogeneous catalysis, exemplified by the progress in zeolites, has continued to keep pace.[259,262]

Utilizing heterogeneous catalysts to produce fuels and chemicals reduces the energy demand and minimizes the formation of non-recyclable waste by increasing the activity and selectivity of renewable as well as fossil resource conversion.[263] Being an essential tool for the treatment of pollutants in air and water, especially *via* photocatalytic processes, demonstrates their major role for global environmental preservation.[264]

There is an extensive array of industrial processes and research endeavors that exemplify the advantages of heterogeneous catalysis.[265,266] In oxidation reactions, heterogeneous catalysts facilitate the replacement of toxic stoichiometric oxidizing agents with oxygen or hydrogen peroxide.[267–269] Heterogeneous catalysis contributes to reducing waste by enhancing reaction selectivity, saving energy, and increasing productivity through accelerated reaction rates.[264]

In view of the climate crisis catalytic processes need to be established, by moving away from precious metals or metal oxides.[270–272] Focus on earth-abundant metal sources as illustrated above is pivotal to reducing the environmental impact of chemical reaction. Biobased-, metal-free, or abundant (*e.g.* Fe, Ti) heterogeneous catalysts present an intriguing alternative for ptcertain industrial chemical processes.[273,274] Examples for the utilization of abundant resources include carbon[275–277] and carbon nitride,[278–283] iron[284–287] and iron oxide nanoparticles[288–291] or single-atom heterogeneous catalysts (SACs).[292–294] While still being heterogeneous solid catalysts, SACs inherently can mimic traditional homogeneous catalysis with all its benefits in activity and selectivity, while facilitating purification, separation, and catalyst recycling.[295,296]

Heterogeneous catalysis finds numerous applications featuring many key aspects related to Green Chemistry:

- Conversion of renewable feedstocks (CO_2 and plastics)[297–300]
- Biomass conversion (*e.g.* lignocellulose)[301–303]
- Ammonia synthesis from N_2 and H_2[304–306]
- Selective oxidations and cross-couplings[307–309]
- Fuel cells and electrochemistry[310–314]
- Renewable fuels[315–323]
- Renewable platform chemicals[324]

Heterogeneous mechanisms involve the **Langmuir–Hinshelwood mechanism** (two molecules adsorb on neighboring sites, and the adsorbed molecules undergo a bimolecular reaction), as well as the **Eley–Rideal mechanism** (one molecule adsorbs, and the other one reacts directly from the gas phase without adsorbing). Nonetheless, advanced *operando* spectroscopic and imaging techniques enable a comprehensive mechanistic understanding of heterogeneous catalyst materials.[325] Such surface, bulk, and ***in situ*** **characterization techniques** in combination with molecular modeling and sophisticated synthesis methods have revolutionized the preparation of solid catalysts and mechanistic insights in their functioning. Additionally, it has transformed the characterization of their physical and chemical (catalytic) properties from an art into systematic science.[262]

3.7.2 Photo- and Electrocatalysis

Following up on previously discussed heterogeneous catalysis and the energy transition, the direct conversion of solar photons into electricity and solar fuels stands as one of the most significant scientific and technological achievements of this century.[326–328] Carbon-free energy is required in the form of light, electricity,[329] liquid, and gaseous fuels to mitigate the severe consequences of global climate change and ensure a sufficient global energy supply.[262]

As such **photo- and electrocatalysis** are strong research fields enabling the (energy) transition towards a sustainable future.[330,331] Replacing the necessity of (harsh) thermal energy for chemical transformations, photo- and electrochemistry allow for mild, more energy efficient, and thus Green Chemistry processes.[159,332,333]

Cost-effective photovoltaic (PV) energy is based on chemistry, solar water splitting, redox catalysis for water oxidation and reduction, photoelectrochemical energy conversion,[334–336] and photo-induced electron transfer.[262,337] Utilizing sunlight directly (**photocatalysis**) or in the form of green electricity (**electrocatalysis**), water can be split to produce hydrogen – a clean and portable energy carrier that can subsequently be employed in fuel cells to generate electricity as needed (Figure 3.16).[338–340]

Figure 3.16 Direct application of photochemistry and electrochemistry for the conversion of solar photons into green electricity. Electrons and charge carriers generated can be utilized in *e.g.* electrolysers to oxidize water (H_2O) into oxygen (O_2) and hydrogen (H_2).[341] Green hydrogen can be further reacted into liquid fuels such as H_2O_2,[289] CH_3OH,[19] NH_3,[342] or HCOOH.[343–345]

Sophisticated electrode materials in **electrochemistry** not only allow for water oxidation (*i.e.* water splitting or water electrolysis),[346–349] proton reduction and oxygen reduction,[350] but also for chemical conversion of biomass,[351] transformations of organic platform chemicals, production of energy carriers,[352–355] and CO_2 reduction.[356–360] Advanced, lightweight batteries and their electrochemical properties will also play a crucial role in powering electric vehicles and other sectors in the future.[361–363]

Photochemistry (*via* photocatalysts and photosensitizers) allows for direct transformations through advanced chemical reactions into new building blocks for materials,[186,364] energy carriers,[365–373] and improved organic chemistry.[374–378] Thereby the number of synthetic steps are reduced, and atom economy is enhanced. Furthermore, photochemistry enables advanced medical applications such as imaging techniques and photopharmacology for smart drug design.[379]

Combining renewable energy sources with the aspects of sustainable solvents, solvent-free reactions through mechanochemistry not only drastically reduces chemical waste but also only requires green electricity to enable mixing and shearing of compounds to access new building blocks.[380,381]

3.7.3 Green Process Design and Continuous Processing in Flow

The application of the 12 *Principles of Green Chemistry* in industrial processes can already be exemplified through several examples, such as the

Haber-Bosch process for ammonia manufacture which turns atmospheric nitrogen into a sustainable feedstock for nitrate production, reducing the use of nitrate minerals mined in remote, environmentally sensitive locations.[15] Another example is the synthesis of sertraline, where application of Green Chemistry greatly reduced the waste in manufacturing at Pfizer for the antidepressive drug Zoloft.[15]

Some of the many historic precursors to Green Chemistry include Alice Hamilton's pioneering work in industrial toxicology,[382,383] Rachel Carson's research on synthetic pesticides,[384] and Theo Colborn's studies on the health impacts of endocrine-disrupting chemicals.[385–387] Another historical contribution appeared in an article titled '*What you don't have, can't leak*' by Trevor Kletz in 1978, who suggested that chemical plants should be designed for alternative processes using safer raw materials and intermediates, and only use hazardous ones at lower temperatures and pressures, to avoid rather than solve problems and accidents.[388,389]

Another approach (a precursor to the Circular Economy) widely implemented in the chemical industry and plants is the use of the waste from one process, as a feedstock or reagent in another one (Verbund sites).[15] In this way, not only is 'waste' repurposed and viewed as a renewable feedstock for several processes at a chemical site, but also transport costs from one site to the next are kept at a minimum. As such, the waste greenhouse gas N_2O from the manufacturing of adipic acid (a building block for nylon) is being reused as an oxidant in a greener route to phenol industrially.[15]

Large-scale catalytic processes, such as petroleum refining, have achieved notable atom efficiency through heterogeneous catalysis. The transformation and upgrading of a variety of products is preferably mediated by heterogeneous catalysts.[227] A few instances, though not comprehensive, include:

- Sulfuric acid production
- Cellulose, hemicellulose, and lignin biorefineries[390,391]
- Water–gas shift reaction (copper, zinc oxide, and alumina)
- Fischer-Tropsch synthesis (cobalt)[392]
- Waste air, water, and VOC treatments[393]
- Photocatalysis (TiO_2)[337,394]
- Solar photon conversion in solar panels[395]

Several metrics have already been covered to assess chemical processes in terms of waste, energy usage, and chemistry efficiency.[396–398] The metrics of Atom Economy, Step Economy, and E Factor stand prominently as pivotal parameters within the assessment of a chemical process. In tandem with conventional benchmarks such as yield, conversion, selectivity, and TON (Turnover Number) and TOF (Turnover Frequency), these metrics encapsulate the quintessence of quantitative evaluation. The integration of Life Cycle Assessments (LCAs) amplifies the arsenal available to green chemists and chemical engineers, furnishing them with a comprehensive toolkit for the improvement of the chemical sector towards sustainability.[399]

Efforts to improve the yield or purity of a step result in significant reduction in the overall environmental impact of the process as harmful purification steps are reduced or eliminated.[36]

For the facilitation and implementation of Green Chemistry into chemical engineering and process design, P. T. Anastas and J. B. Zimmerman developed the **12 Principles of Green Engineering** in 2003: [400,401]

Principle #1: Inherent Rather Than Circumstantial
Designers need to strive to ensure that all material and energy inputs and outputs are as inherently nonhazardous as possible.

Principle #2: Prevention Instead of Treatment
It is better to prevent waste than to treat or clean up waste after it is formed.

Principle #3: Design for Separation
Separation and purification operations should be designed to minimize energy consumption and materials use.

Principle #4: Maximize Efficiency
Products, processes, and systems should be designed to maximize mass, energy, space, and time efficiency.

Principle #5: Output-Pulled *Versus* Input-Pushed
Products, processes, and systems should be "output pulled" rather than "input pushed" through the use of energy and materials.

Principle #6: Conserve Complexity
Embedded entropy and complexity must be viewed as an investment when making design choices on recycle, reuse, or beneficial disposition.

Principle #7: Durability Rather Than Immortality
Targeted durability, not immortality, should be a design goal.

Principle #8: Meet Need, Minimize Excess
Design for unnecessary capacity or capability (e.g., "one size fits all") solutions should be considered a design flaw.

Principle #9: Minimize Material Diversity
Material diversity in multicomponent products should be minimized to promote disassembly and value retention.

Principle #10: Integrate Material and Energy Flows
Design of products, processes, and systems must include integration and interconnectivity with available energy and materials flows.

Principle #11: Design for Commercial 'Afterlife'
Products, processes, and systems should be designed for performance in a commercial "afterlife".

Principle #12: Renewable Rather than Depleting
Material and energy inputs should be renewable rather than depleting.

Recently, photochemical[402] and electrochemical transformations of building blocks into commodity chemicals and energy carriers are advanced through **continuous processing in flow**, allowing for enhanced light penetration, product separation from the active site of the catalyst, and higher production efficiencies.[403–411] Examples include continuous flow electrosynthesis of ammonia by nitrogen reduction with hydrogen.[342]

The concept of continuous **fixed/packed bed reactors** has been widely utilized in industry as standard practice over **batch processing**,[412] albeit not for photo- or electrochemical conversions yet.

Process and reaction design optimization is facilitated through **computer-assisted synthetic planning, artificial intelligence (AI)** and **robotics.**[413–416] Using high-throughput screening at minimal scale, the best reaction conditions can be established through automated self-optimization, and then transferred to intensification, and scale-up.[417,418] These 'laboratories of the future' drastically reduce waste production, enhance reproducibility, and minimize human error.[419] Allowing for sustainable and autonomous resource management, chemical processes will benefit in the future from these technologies.[420]

Finally, the utilization of renewable energy in combination with the development of energy carriers such as hydrogen, as well as sustainable transformations and platform chemicals will lead to the coupling of the chemical sector with, for example, agriculture,[421] steel,[422] or cement industries.[21] Hence, to find solutions for the environmental issues ahead, Green Chemistry and Green Engineering will be necessary and need to become core aspects of education (discussed in the following section).[423]

3.8 Green Chemistry Education

The inclusion of Green Chemistry throughout the education continuum can promote the achievement of a sustainable future at the design stage of a product lifecycle, and it has several advantages for students, educators, staff, and institutions willing to promote this systemic change. In addition to traditional safety improvements for faculty, staff, and students, using less hazardous materials in teaching and research labs allows more efficient processes leading to savings in energy consumption, chemical purchasing, and waste disposal. This approach underscores promotion of chemistry as a tool to provide the betterment of society; and, perhaps most importantly, the preparation of students with the required skills to address our current global societal and environmental challenges.[424,425]

By applying and implementing more sustainable practices at the molecular level through Green Chemistry education, products, and processes less hazardous to human health and the environment can be assessed and invented through the development of a knowledgeable citizenry and scientific workforce.[426] Green Chemistry education works as a catalyst for the design of

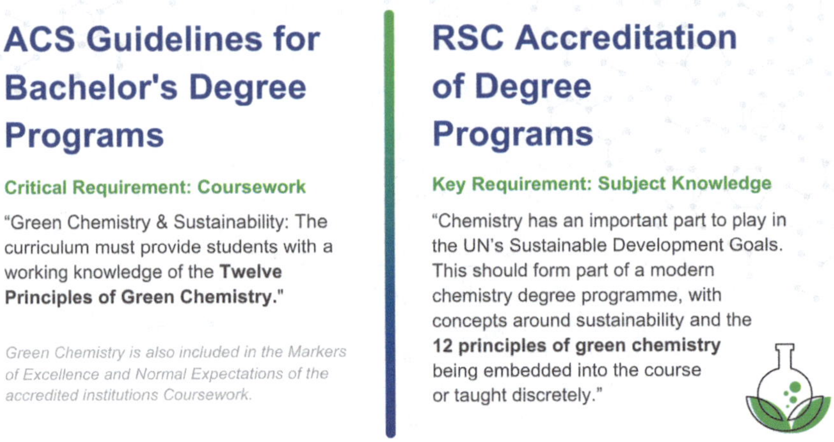

Figure 3.17 Current requirements of national organizations for the obtainment of their respective accreditations. Criteria defined by the ACS in the United States and RSC in the United Kingdom include the 12 Principles of Green Chemistry as a Critical or Key Requirement in their chemistry degree programs.

greener technological innovations with performance and economics comparable to the ones traditionally applied by the chemical industry. Since the creation of the 12 *Principles of Green Chemistry*, educators and scientific organizations around the world have recognized the long-lasting benefits and opportunities provided by Green Chemistry education. In 2023, the American Chemical Society (ACS) included Green Chemistry as a critical requirement in the ACS Guidelines for bachelor's degree programs, which provides specifications for approval and certification of undergraduate programs in the United States.[427] In the United Kingdom, chemistry degree programs are now required to embed the 12 Principles of Green Chemistry into courses to obtain accreditation from the Royal Society of Chemistry (RSC) (Figure 3.17).[428]

The guidelines released by the ACS and RSC are hopefully the beginning of a global movement led by national organizations to support the promotion of a more sustainable curriculum. The defiance of the status quo at the institutional level for preparing future scientists to address our modern challenges is a transformative step that can be successfully achieved by collaboration and empowerment of individuals, organizations, and sectors within our society.

3.8.1 The Importance of Green Chemistry Education

Surveys conducted by the Students Organizing for Sustainability (SOS) International, and the RSC demonstrate that the role of education in promoting a sustainable future is also common sense among the general public and students.[429,430] According to the RSC's *Green Shoots: A Sustainable Curriculum for a Sustainable Planet* report, students expect to learn about climate change and sustainability challenges in their chemistry courses, and they also believe that schools, colleges, and universities are the ones responsible for teaching them how to address these issues, followed by governments and scientific societies.[429]

In the *Students, Sustainability and Education* report published by the SOS, about 7000 students participated in a global survey to help understand their experiences regarding sustainability and education. Most of the interviewed students believe that sustainable development is something higher education institutions should actively incorporate and promote, whereas 40% of them report "low or no coverage of sustainable development concepts in their coursework".[430]

Results of these surveys highlight that educators and students across the world believe education is paramount to achieving a more sustainable future. They also show that even though education can provide students with the context and tools to address our global challenges, our future workforce is not being taught adequately on how to do so. There is a need to train students how to apply systems thinking (*i.e.* 'Systems Chemistry' and 'Systems Sustainability') to problems; that is, to think of the wider implications of their chemical activities, not just of the chemistry itself by incorporating the 12 *Principles of Green Chemistry*. This is particularly concerning when one considers the supply-demand for green skills around the world. Between 2022 and 2023, the growth in demand for green skills has surpassed the share of green talent in the workforce, a disconnection that is likely to continue increasing in the next years and further lead to an uneven and unjust transition to a sustainable future.[431] Investments in the workforce through the incorporation of Green Chemistry in higher education can help to address this issue, reducing the supply-demand disconnect and ensuring a smooth, comprehensive, inclusive, and equitable transition towards sustainability.

3.8.2 Inclusion of Green Chemistry in Higher Education

Although changes in chemistry education are urgently needed, barriers are associated with teaching students about Green Chemistry and preparing them for a sustainable future. In higher education, these challenges are typical when introducing any new subject into an already packed chemistry curriculum. The lack of training and expertise of faculty members in the field, the lack of incentive and departmental resources to teach Green Chemistry concepts, as well as the lack of curriculum resources to provide this incorporation is problematic. In order to overcome these barriers, researchers and educators worldwide have created, promoted, and shared several approaches and resources to facilitate the integration of Green Chemistry in higher education.[425,432–438] Although some approaches suggest a complete renovation of the chemistry curriculum (*e.g.*, development of brand-new Green Chemistry, toxicology, or environmental health science courses), other strategies are based on outreach, institutional external resources, and teaching traditional chemistry concepts through the 'lens of Green Chemistry' (*e.g.*, inclusion of the subject throughout existing courses, labs, research projects or programs).[439]

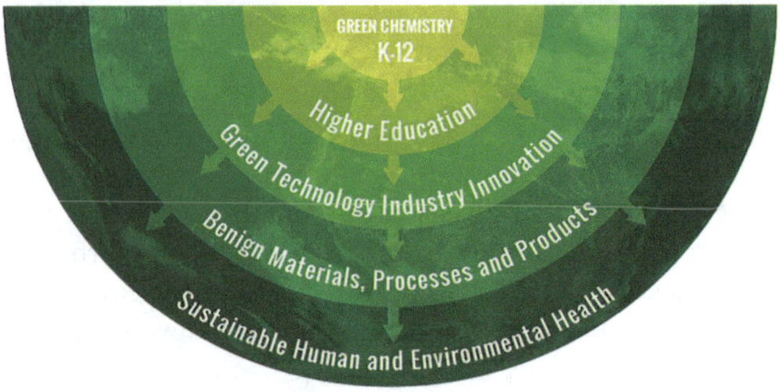

Figure 3.18 Beyond Benign's Theory of Change. Green Chemistry education works as a catalyst for the design of greener, successful, and cost-effective technological innovations that result in less harmful and hazardous materials, products, and processes, thus supporting a more sustainable future and healthier society.

The incorporation of Green Chemistry into undergraduate and graduate programs is supported by numerous local and global networks, including dedicated resources, organizations, and programs. Examples include the *Specialized Manual on Green and Sustainable Chemistry Education and Learning* developed by the United Nations Environment Programme (UNEP),[425] the *Green and Sustainable Chemistry Education Road Map* created by the American Chemical Society Green Chemistry Institute (ACS GCI),[432] and Beyond Benign and its *Green Chemistry Commitment (GCC).*[440]

Beyond Benign is a non-profit organization co-founded in 2007 by Dr. Amy Cannon and Dr. John Warner to host and foster a community dedicated to creating a systemic change in education at the molecular level through Green Chemistry. Beyond Benign works to empower individuals and provide resources for the inclusion of Green Chemistry across the education continuum. The ultimate goal is to support a stable environment, healthy population, and sustainable society by creating a citizenry and scientists able to choose and further design processes and products with our planet in mind (Figure 3.18).[119]

To support this transformation at undergraduate, graduate, and postgraduate levels, Beyond Benign created the Green Chemistry Commitment (GCC) program. The GCC was launched in 2013 with 13 participating higher education institutions based in the United States.[442] As of 2024, more than 185 institutions in more than 20 countries are part of this program, which works to empower a community confident to teach, research, and practice Green Chemistry by investing in educators, providing peer support networks, professional development opportunities, as well as developing and sharing open-access tools and resources.[441,443]

3.9 Green Chemistry in Research, Society and Beyond

When effectively communicated, the chemical sector, government, and society can adapt rapidly and collaborate to address environmental issues.[444]

Prime examples such as the Montreal Protocol,[445] leading to the replacement of chlorofluorocarbons with compounds that do not affect the ozone layer, indicate society's ability to adapt when life is at stake due to environmental problems.[15,27]

Through improved education and a change of mindset, Green Chemistry will be perceived as standard, not only in research and industry but especially applied unconsciously in life through consumer choices, offering a generational change for a sustainable society.[446]

3.9.1 Industry and Green Chemistry

An increasing number of consumers, retailers, policy makers, and legislators are demanding less hazardous processes and products in response to the escalating urgency of environmental challenges. Investment in Green Chemistry by manufacturers, retailers, brands, and R&D teams follows the same trend. Due to Green Chemistry's ability to be applied throughout the life cycle of our products and processes across multiple sectors, some industries even report Green Chemistry as "the future of all chemistry".[447]

A report published by the former *Green Chemistry & Commerce Council (GC3),* now *Change Chemistry,* shows that Green Chemistry-marketed products significantly outperform their conventional counterparts in consumer markets.[448] In fact, Green Chemistry-marketed products represented almost 61.5% of overall market growth between 2015 and 2019, growing 12.6 times faster than their conventional counterparts. This expansion is driven by consumers, institutional buyers, regulations, shareholders, and investors, as well as advocacy campaigns.[448]

Large brands and multinational corporations also play an important role in leading the way to establish audacious goals and further reduce their hazards, waste, and carbon footprints. One important example is the Zero Discharge of Hazardous Chemicals (ZDHC) organization, which is composed of about 100 global brands, suppliers, and others dedicated to enabling safer choices of chemical products for the textile, apparel, and footwear industries.[449]

The most visible Green Chemistry Principles applied by chemical manufacturing are *Waste Prevention* (Principle #1), *Less Hazardous Synthesis* (Principle #3), *Designing Safer Chemicals* (Principle #4), *Safer Solvents and Auxiliaries* (Principle #5), *Real-time analysis for pollution prevention* (Principle #11, is particularly important because one cannot control what you cannot measure), and *Inherently Safer Chemistry for Accident Prevention* (Principle #12).[450] Although a holistic integration of the 12 Principles of Green Chemistry in chemical manufacturing is ultimately required, industries and companies worldwide are actively engaged in producing more sustainable materials and processes by using some combination of individual principles. This integration is likely to enhance and improve as emerging policies and investor expectations, jobs and continued professional development, consumer knowledge, and safer products legislation continue to increase and fuel the growth of the Green Chemistry sector over the following years.[448]

3.9.2 Networks and Resources for a Sustainable Future

Young scientists play a fundamental role in driving a just and equitable sustainable transition by promoting scientific and behavioral changes at local and global levels. To amplify and leverage their impact, national and international networks, and organizations were created over the years to connect like-minded individuals, promote collaboration, and inspire them to act. Examples include the Network for Early-Career Sustainable Scientists and Engineers (NESSE),[451] the American Chemical Society Green Chemistry Institute (ACS GCI),[452] Beyond Benign[119] and its Green Chemistry Teaching and Learning Community (GCTLC),[453] the Sustainable Chemistry Network (GSCN) in Japan,[454] the Green Sciences for Sustainable Development Foundation (GSSD),[455] the Indian Green Chemistry Network Centre (GCNC),[456] the Chemicals & Waste Youth Platform of the United Nations Major Group for Children and Youth (MGCY)[457] as well as others. International chemistry networks such as the Global Young Academy (GYA),[458] the European Young Chemists' Network (EYCN)[459] and the International Younger Chemists Network[460] are also promoting initiatives to drive Green Chemistry and broader sustainability adoption across the world. The Global Conversation on Sustainability (GCS) is a project launched in 2021 by the IYCN and the International Union of Pure and Applied Chemistry (IUPAC) to drive awareness of our society's societal and environmental challenges through discussion and further implementation of more sustainable practices.[461]

A diversity of Green Chemistry resources to facilitate the learning and incorporation of Green Chemistry throughout research and teaching can be found in several online platforms, including the Green Chemistry Teaching and Learning Community (GCTLC),[453] the Green Chemistry and Engineering Learning Platform (GChELP), and the Green Chemistry for Sustainability platform.[462] The general public can also advocate for Green Chemistry daily by searching and purchasing products that meet the *Safer Choice Standard* of the United States Environmental Protection Agency (EPA).[463]

3.9.3 Sustainable Laboratory Practices

The adoption of **sustainable laboratory practices** is vital to complement the education of future scientists and promote the development of truly sustainable technologies powered by Green Chemistry.[464–467]

Scientists are crucial advocates for climate awareness, yet traditional lab practices contribute to environmental issues through excessive energy consumption, hazardous- and single-use waste generation, and resource depletion (*e.g.* water). Through their own operations, science, research, and laboratories have a significant carbon footprint contributing to the climate crisis. Grassroot movements across the sciences have recognized the overlooked impact of scientific research and so-called **Green Lab** initiatives emerged to address the environmental footprint of research.[468–473]

While Green Chemistry focuses on designing safer, more efficient chemical products and processes that minimize environmental impact at the molecular level, Green Labs or sustainable laboratory practices emphasize reducing the environmental footprint of laboratory operations—like energy use, water conservation, and waste management—regardless of the science being conducted.

Sustainability guidelines for laboratory research of all disciplines exist, as well as laboratory accreditation frameworks to reduce energy consumption, waste generation, or resource dependency.[474,475] Examples include increasing ultra-low temperature freezer temperatures, turning off equipment after use, avoiding unnecessary repetition of experiments through enhanced reproducibility, right-sizing experiments, sharing chemicals and equipment, or reducing the amount of single-use consumables in research.[122]

Adhering and following sustainable laboratory practices in turn generate cost benefits and savings in addition to drastically reducing the environmental impact and carbon emissions.[476,477]

3.9.4 Addressing Economic and Technological Barriers

As business strives to maximize profits the adoption of green chemistry processes is only viable when these processes outperform current methods at appropriate cost.[15,333] Ammonia and methanol play crucial roles in modern societies, yet their production relies on natural gas, contributing to greenhouse gas emissions and environmental issues. Despite extensive research into low-carbon and green production methods, their adoption has been impeded by elevated costs, rendering them economically unsustainable.[478] Recent developments in increasing energy pricing, however, have led to a unique opportunity for the transition to green production technologies.[260] Green ammonia, produced through using wind-powered water electrolysis, surpassed its fossil counterpart for a six-months period as of December 2021.[479,480] Green methanol, produced through CO_2 capture and wind-based water electrolysis also emerged as an economically viable alternative.[478,481,482]

The chemical industry only adopts fundamental changes in technology when they present genuine advantages in performance and/or cost. Green chemists need to develop and provide such technologies in a short time scale. Academia, governments, and industry need to increase education and collaboration to accelerate the implementation of Green and Sustainable Chemistry along the supply chain from chemical researchers, through industrial scale-up to commercial end-users.

A simple economic assessment allows for the calculation of avoidance costs to establish if a process is green and profitable:

Determination of avoidance costs for environmental production processes:[478]

$$AC = \frac{PC_{Green} - PC_{Fossil}}{CF_{Fossil} - PC_{Green}} \tag{3.8}$$

where AC is the avoidance cost in USD per kg CO_2e (CO_2-equivalents), $PC_{Fossil/Green}$ depicts the production cost in USD per kg for fossil or green routes, and $CF_{Fossil/Green}$ represents the carbon footprint in kg CO_2e per kg of product.[478]

It is vital to capitalize on the current momentum and energy pricing landscape for collaborative efforts from academia, industry, and policy-makers, where chemists have the potential to spearhead a significant transition towards more sustainable practices in the chemical industry.[478]

Universities need to teach and educate Green Chemistry, Life Cycle Assessments, and economic metrics to students to enhance the application of Green and Sustainable Chemistry in research and even more so when entering industry.

3.9.5 Global Collaboration for a Sustainable Society

While chemistry innovation will play an important role for all 17 of the United Nations SDGs, 12 of the 17 goals are directly relying on the chemical industry and chemical sector to enhance the energy-, feedstock-, and materials transition (Figure 3.4). Only when the chemical industry utilizes renewable feedstocks, enhances circularity, develops energy carriers, and designs new chemical building blocks, will our society be able to achieve and overcome the climate crisis by adhering to all 17 SDGs.[483]

The transformation towards a sustainable future can be depicted as a tree (Figure 3.19), where several concepts and definitions complement each other to provide a comprehensive framework for the sustainable chemistry of the future. With these tools at hand, a path toward achieving a sustainable future at the molecular level through chemistry is traced.

Through pioneers like Alice Hamilton,[383] Rachel Carson,[384] Trevor Kletz,[388] and Theo Colborn,[385] further roots and foundations for a sustainable future through chemistry were set with the establishment of the U.S. Environmental Protection Agency[484] and the publication of the **12 Principles of Green Chemistry** by Paul T. Anastas and John C. Warner in 1998 (Figure 3.6).[485,486] Green building blocks and chemicals, designed in research laboratories, require large-scale production by the chemical industry to meet the necessary quantities and demands essential for establishing a sustainable supply chain and procurement system. Consequently, the correlation between the 12 Principles of Green Chemistry and the **12 Principles of Green Engineering** facilitates seamless scale-up and production processes.[400,487] Building on these tools of designing chemicals and chemical processes with a holistic approach, the Green Chemistry Principles have been paraphrased several times, all following the core values of reduction of environmental impact, minimizing harm, (bio)degradability, circularity, renewable feedstocks, and energy efficiency.[333] Green Chemistry is the basis of future industry and chemical manufacturing and a foundational tool to provide benign products and goods, allowing for a circular economy and achieving net-zero targets.

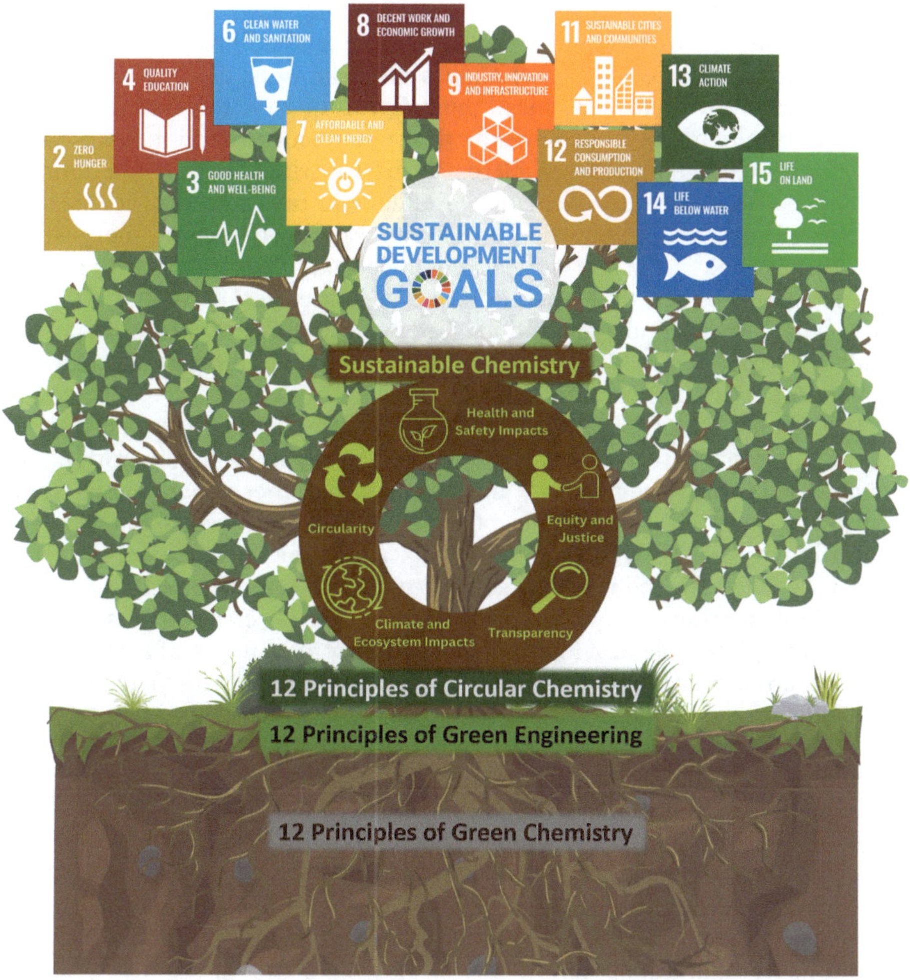

Figure 3.19 The interplay of different chemistry frameworks for the achievement of a sustainable future depicted as 'chemis-tree'. The 12 Principles of Green Chemistry can be seen as the foundation (roots) for the accomplishment of other approaches designed to promote a sustainable future, including Circular Chemistry and Sustainable Chemistry (branches). These strategies, together, are crucial for achieving the 17 United Nations Sustainable Development Goals (leaves), as the chemical sector is omnipresent in our daily life. Sustainable Development goal icons adapted from https://www.un.org/sustainabledevelopment/sustainable-development-goals/ with permission from United Nations (UN). The content of this publication has not been approved by the United Nations and does not reflect the views of the United Nations or its officials or Member States.

One concept of particular relevance for such a circular economy is the proposed Principle #10 of the **12 Principles of Circular Chemistry**: '*Sell service not product*',[59] which is essentially the underlying approach of the

UNIDO Chemical Leasing initiative.[333,488] The aim is a more efficient use of chemicals in the production process, where chemical producers should be selling a service or effect rather than a given amount of chemical.[333] This creates motivation for chemical manufacturers to reduce the necessary amount of chemicals to achieve the desired effect or function, as doing so will maximize their profit (**F factor**, F for function).[61] As the amount of chemicals needed to produce a given effect will be reduced over several iterations, ultimately, the burden on feedstocks and the environment will be reduced as well.[398] Furthermore, as chemical manufacturers provide service rather than product, they operate rather as a chemical transformation service, where waste and old product feedstock is converted into new value-added compounds, allowing for recycling schemes, Verbund sites through collaboration, and thus circularity.

In the end, all principles and definitions are part of the same tree, depicting its roots (**Green Chemistry**),[27] small grass (**Green Engineering**),[400,401] bushes (**Circular Chemistry**),[59] and finally, branches (**Sustainable Chemistry**)[489] as an overarching concept for environmentally conscious research and industry. All of which are needed for achieving and growing the leaves (*UN's 17 SDGs*) for the tree and society to survive (Figure 3.19).

By applying those frameworks, the chemical sciences need to target three pillars of sustainable development, being net zero in terms of *environmental* impact, *economic* viability, and social *equity* (*triple E* priorities), which target the following objectives:[260]

- Maximize material & energy circularity.
- Maximize material & energy renewability.
- Maximize material benignity.
- Maximize cost competitiveness.
- Maximize innovation capacity.
- Maximize social well-being (including health).
- Ensure data transparency.

Professionals of diverse disciplines, including researchers, scientists, technicians, and engineers, must collectively recognize the pivotal significance of Green and Sustainable Chemistry. A collaborative nexus between academia, industry, and policymakers needs to be established, facilitating a concerted effort towards advancing the required transitions in energy, feedstock, and materials within the chemical sector and therefore, society.[489] This coordinated endeavor is essential to fulfilling the imperative goals outlined in the global agenda for sustainable and circular development.

3.10 Conclusion

Most people on the planet want stronger climate action. Remarkably, 69% of the world's population is willing to allocate 1% of their personal income to the cause, while 86% support pro-climate social norms, and 89% advocate for increased political action.[490] Countries experiencing elevated vulnerability to climate change exhibit a notably high readiness to contribute.

Society relies on the chemical sector in every aspect of life, which in view of environmental issues requires a system-wide and systemic (re-)design with Green Chemistry at its core.[491] The concept of 'waste' needs to disappear from our design frameworks. Instead, our perspective should shift towards considering material and energy flows.[21] Biomass such as lignocellulose needs to be transformed into building blocks instead of practicing unsustainable burning of wood and wood waste for energy production.[492] Hazards of molecules, materials and processes should be viewed as critical design flaw and reassessed for (bio)degradability to minimize environmental impacts, allowing for circularity. Adopting a holistic systems thinking approach is key to unlocking the intrinsic functionalities of renewable feedstocks for enhanced properties of products as well as a circular and sustainable future.

Chemistry is key for the energy, feedstock, and material transition ahead, where industry needs to replace outdated processes with environmentally friendly ones. Through collaborative education and sustainable laboratory practices, future scientists will perceive the current paradigm shift on Green and Sustainable Chemistry as just good chemistry.[493]

References

1. J. Fahrenkamp-Uppenbrink, Chemistry Goes Green, *Science*, 2002, **297**, 798.
2. M. Poliakoff and P. Anastas, A principled stance, *Nature*, 2001, **413**(6853), 257.
3. V. G. Zuin Zeidler, Sustainable chemistry and food systems lessons—the same procedure as every year?, *Science*, 2024, **383**, eado2352.
4. J. Rockström, *et al.*, A safe operating space for humanity, *Nature*, 2009, **461**(7263), 472–475.
5. W. Steffen, *et al.*, Planetary boundaries: guiding human development on a changing planet, *Science*, 2015, **347**, 736–747.
6. T. M. Lenton, *et al.*, Climate tipping points — too risky to bet against, *Nature*, 2019, **575**, 592–595.
7. R. M. van Westen, M. Kliphuis and H. A. Dijkstra, New Physics-Based Early Warning Signal shows AMOC is on Tipping Course, *Sci. Adv.*, 2023, **10**, 1189.
8. S. Rahmstorf, A semi-empirical approach to projecting future sea-level rise, *Science*, 2007, **315**, 368–370.
9. T. M. Lenton, *et al.*, Tipping elements in the Earth's climate system, *Proc. Natl. Acad. Sci. U. S. A.*, 2008, **105**, 1786–1793.
10. D. Coumou and S. Rahmstorf, A decade of weather extremes, *Nat. Clim. Change*, 2012, **2**(7), 491–496.

11. K. Richardson, *et al.*, Earth beyond six of nine planetary boundaries, *Sci. Adv.*, 2023, **9**, eadh2458.
12. H. Lee and J. Romero, IPCC, 2003: Climate Change 2023: Synthesis Report. Contribution of Working Groups I, II and III to the Sixth Assessment Report of the Intergovernmental Panel on Climate Change, *IPCC*, 2023, 1–184.
13. A. Robinson, J. Lehmann, D. Barriopedro, S. Rahmstorf and D. Coumou, Increasing heat and rainfall extremes now far outside the historical climate, *npj Clim. Atmos. Sci.*, 2021, **4**(1), 1–4.
14. Planetary boundaries - Stockholm Resilience Centre. https://www.stockholmresilience.org/research/planetary-boundaries.html.
15. M. Poliakoff, J. M. Fitzpatrick, T. R. Farren and P. T. Anastas, Green chemistry: Science and politics of change, *Science*, 2002, **297**, 807–810.
16. L. Barreto, A. Makihira and K. Riahi, The hydrogen economy in the 21st century: a sustainable development scenario, *Int. J. Hydrogen Energy*, 2003, **28**, 267–284.
17. G. W. Crabtree, M. S. Dresselhaus and M. V. Buchanan, The Hydrogen Economy, *Phys. Today*, 2004, **57**, 39–44.
18. G. Marbán and T. Valdés-Solís, Towards the hydrogen economy?, *Int. J. Hydrogen Energy*, 2007, **32**, 1625–1637.
19. G. A. Olah, Beyond oil and gas: The methanol economy, *Angew. Chem., Int. Ed.*, 2005, **44**, 2636–2639.
20. G. A. Olah, A. Goeppert and G. K. Surya Prakash, *Beyond Oil and Gas: The Methanol Economy*, Wiley-VCH, 2018.
21. J. B. Zimmerman, P. T. Anastas, H. C. Erythropel and W. Leitner, Designing for a green chemistry future, *Science*, 2020, **367**, 397–400.
22. C. Niether, *et al.*, Improved water electrolysis using magnetic heating of FeC–Ni core–shell nanoparticles, *Nat. Energy*, 2018, **3**(6), 476–483.
23. J. K. Nørskov, *et al.*, Origin of the overpotential for oxygen reduction at a fuel-cell cathode, *J. Phys. Chem. B*, 2004, **108**, 17886–17892.
24. B. M. Trost, The Atom Economy—A Search for Synthetic Efficiency, *Science*, 1991, **254**, 1471–1477.
25. T. Freese, Integrating sustainable chemistry into research and industry for a circular economy, PhD Thesis – University of Groningen, Groningen, 2024, DOI: 10.33612/DISS.1119350631.
26. Home – United Nations Sustainable Development. https://www.un.org/sustainabledevelopment/.
27. P. T. Anastas and J. C. Warner, *Green Chemistry: Theory and Practice*, Oxford University Press, New York, 1998.
28. M. Lancaster, *Green Chemistry: An Introductory Text*, Royal Society of Chemistry, Cambridge, 2016.
29. H. C. Erythropel, *et al.*, The Green ChemisTREE: 20 years after taking root with the 12 principles, *Green Chem.*, 2018, **20**, 1929–1961.
30. P. Anastas and N. Eghbali, Green Chemistry: Principles and Practice, *Chem. Soc. Rev.*, 2010, **39**, 301–312.

31. P. Qi and R. A. Sheldon, Aqueous cross-coupling. The E Factor: fifteen years on, *Green Chem.*, 2007, **9**, 1273–1283.
32. R. A. Sheldon, The E factor 25 years on: the rise of green chemistry and sustainability, *Green Chem.*, 2017, **19**, 18–43.
33. R. A. Sheldon, The E factor at 30: a passion for pollution prevention, *Green Chem.*, 2023, **25**, 1704–1728.
34. R. A. Sheldon, Metrics of Green Chemistry and Sustainability: Past, Present, and Future, *ACS Sustainable Chem. Eng.*, 2018, **6**, 32–48.
35. C. Jimenez-Gonzalez, C. S. Ponder, Q. B. Broxterman and J. B. Manley, Using the right green yardstick: Why process mass intensity is used in the pharmaceutical industry to drive more sustainable processes, *Org. Process Res. Dev.*, 2011, **15**, 912–917.
36. P. G. Jessop and A. R. MacDonald, The need for hotspot-driven research, *Green Chem.*, 2023, **25**, 9457–9462.
37. G. Finnveden, *et al.*, Recent developments in Life Cycle Assessment, *J. Environ. Manage.*, 2009, **91**, 1–21.
38. M. A. Curran, Life Cycle Assessment: A review of the methodology and its application to sustainability, *Curr. Opin. Chem. Eng.*, 2013, **2**, 273–277.
39. R. G. Hunt and W. E. Franklin, LCA – How it Came about – Personal Reflections on the Origin and the Development of LCA in the USA, *Int. J. Life Cycle Assess.*, 1996, **1**, 4–7.
40. X. Domènech, J. A. Ayllón, J. Peral and J. Rieradevall, How green is a chemical reaction? Application of LCA to green chemistry, *Environ. Sci. Technol.*, 2002, **36**, 5517–5520.
41. S. Hellweg, E. Benetto, M. A. J. Huijbregts, F. Verones and R. Wood, Life-cycle assessment to guide solutions for the triple planetary crisis, *Nat. Rev. Earth Environ.*, 2023, **4**(7), 471–486.
42. D. Zhang, Z. Wang, C. Oberschelp, E. Bradford and S. Hellweg, Enhanced Deep-Learning Model for Carbon Footprints of Chemicals, *ACS Sustainable Chem. Eng.*, 2024, **12**, 2700–2708.
43. D. W. Pennington, *et al.*, Life cycle assessment Part 2: Current impact assessment practice, *Environ. Int.*, 2004, **30**, 721–739.
44. L. C. Dreyer, M. Z. Hauschild and J. Schierbeck, A framework for social life cycle impact assessment, *Int. J. Life Cycle Assess.*, 2006, **11**, 88–97.
45. M. Z. Hauschild, *et al.*, Identifying best existing practice for characterization modeling in life cycle impact assessment, *Int. J. Life Cycle Assess.*, 2013, **18**, 683–697.
46. M. A. J. Huijbregts, *et al.*, ReCiPe2016: a harmonised life cycle impact assessment method at midpoint and endpoint level, *Int. J. Life Cycle Assess.*, 2017, **22**, 138–147.
47. B. M. Trost, Atom Economy—A Challenge for Organic Synthesis: Homogeneous Catalysis Leads the Way, *Angew. Chem., Int. Ed.*, 1995, **34**, 259–281.
48. B. M. Trost, On inventing reactions for atom economy, *Acc. Chem. Res.*, 2002, **35**, 695–705.

49. W. Wang, J. Lü, L. Zhang and Z. Li, Real atom economy and its application for evaluation the green degree of a process, *Front. Chem. Eng. China*, 2011, **5**, 349–354.
50. P. A. Wender, V. A. Verma, T. J. Paxton and T. H. Pillow, Function-oriented synthesis, step economy, and drug design, *Acc. Chem. Res.*, 2008, **41**, 40–49.
51. M. C. Bryan, *et al.*, Sustainable practices in medicinal chemistry: Current state and future directions, *J. Med. Chem.*, 2013, **56**, 6007–6021.
52. N. V. Fedoroff, *et al.*, Radically rethinking agriculture for the 21st century, *Science*, 2010, **327**, 833–834.
53. K. Kümmerer, Antibiotics in the aquatic environment – A review – Part I, *Chemosphere*, 2009, **75**, 417–434.
54. P. W. Andersen, More is different, *Science*, 1972, **177**, 393–396.
55. K. C. Jones and P. De Voogt, Persistent organic pollutants (POPs): state of the science, *Environ. Pollut.*, 1999, **100**, 209–221.
56. P. T. Anastas, Beyond Reductionist Thinking in Chemistry for Sustainability, *Trends Chem.*, 2019, **1**, 145–148.
57. J. Liu, *et al.*, Systems integration for global sustainability, *Science*, 2015, **347**, 1258832.
58. S. A. Matlin, G. Mehta, H. Hopf and A. Krief, One-world chemistry and systems thinking, *Nat. Chem.*, 2016, **8**(5), 393–398.
59. T. Keijer, V. Bakker and J. C. Slootweg, Circular chemistry to enable a circular economy, *Nat. Chem.*, 2019, **11**, 190–195.
60. C. R. McElroy, A. Constantinou, L. C. Jones, L. Summerton and J. H. Clark, Towards a holistic approach to metrics for the 21st century pharmaceutical industry, *Green Chem.*, 2015, **17**, 3111–3121.
61. J. Clark, R. Sheldon, C. Raston, M. Poliakoff and W. Leitner, 15 years of Green Chemistry, *Green Chem.*, 2013, **16**, 18–23.
62. M. Beller, G. Centi and L. Sun, Chemistry Future: Priorities and Opportunities from the Sustainability Perspective, *ChemSusChem*, 2017, **10**, 6–13.
63. P. V. Petrovic and P. T. Anastas, *First Do No Harm: A Chemist's Guide to Molecular Design for Reduced Hazard*, Jenny Stanford Publishing Pte. Ltd., Singapore, 2023.
64. A. DeVierno Kreuder, *et al.*, A Method for Assessing Greener Alternatives between Chemical Products Following the 12 Principles of Green Chemistry, *ACS Sustainable Chem. Eng.*, 2017, **5**, 2927–2935.
65. A. S. Cannon and J. C. Warner, Noncovalent Derivatization: Green Chemistry Applications of Crystal Engineering, *Cryst. Growth Des.*, 2002, **2**, 255–257.
66. J. H. Clark, Green chemistry: Challenges and opportunities, *Green Chem.*, 1999, **1**, 1–8.
67. S. L. Y. Tang, R. L. Smith and M. Poliakoff, Principles of green chemistry: PRODUCTIVELY, *Green Chem.*, 2005, 7, 761–762.
68. J. Glüge, *et al.*, An overview of the uses of per- and polyfluoroalkyl substances (PFAS), *Environ. Sci.:Processes Impacts*, 2020, **22**, 2345–2373.

69. S. E. Fenton, *et al.*, Per- and Polyfluoroalkyl Substance Toxicity and Human Health Review: Current State of Knowledge and Strategies for Informing Future Research, *Environ. Toxicol. Chem.*, 2021, **40**, 606–630.
70. M. G. Evich, *et al.*, Per- and polyfluoroalkyl substances in the environment, *Science*, 2022, **375**, eabg9065.
71. P. T. Anastas, T. C. Williamson, D. Hjeresen and J. J. Breen, Promoting Green Chemistry Initiatives, *Environ. Sci. Technol.*, 2011, **33**, 116A.
72. J. J. Breen and M. J. Dellarco, Pollution Prevention, *ACS Symp. Ser.*, 1992, **508**, 2–12.
73. C. J. Li and B. M. Trost, Green chemistry for chemical synthesis, *Proc. Natl. Acad. Sci. U. S. A.*, 2008, **105**, 13197–13202.
74. M. J. Bu, C. Cai, F. Gallou and B. H. Lipshutz, PQS-enabled visible-light iridium photoredox catalysis in water at room temperature, *Green Chem.*, 2018, **20**, 1233–1237.
75. Y. Fu, *et al.*, Efficient, Near-Infrared Light-Induced Photoclick Reaction Enabled by Upconversion Nanoparticles, *Adv. Funct. Mater.*, 2023, 2306531, DOI: 10.1002/ADFM.202306531.
76. S. Sharma, J. B. Jasinski, W. M. Braje and S. Handa, Ultrasmall CuI Nanoparticles Stabilized on Surface of HPMC: An Efficient Catalyst for Fast and Organic Solvent-Free Tandem Click Chemistry in Water, *ChemSusChem*, 2023, **16**, e202201826.
77. J. M. Baskin, *et al.*, Copper-free click chemistry for dynamic *in vivo* imaging, *Proc. Natl. Acad. Sci. U. S. A.*, 2007, **104**, 16793–16797.
78. E. Saxon and C. R. Bertozzi, Cell Surface Engineering by a Modified Staudinger Reaction, *Science*, 2000, **287**, 2007–2010.
79. E. M. Sletten and C. R. Bertozzi, Bioorthogonal Chemistry: Fishing for Selectivity in a Sea of Functionality, *Angew. Chem., Int. Ed.*, 2009, **48**, 6974–6998.
80. N. J. Agard, J. A. Prescher and C. R. Bertozzi, A strain-promoted [3 + 2] azide-alkyne cycloaddition for covalent modification of biomolecules in living systems, *J. Am. Chem. Soc.*, 2004, **126**, 15046–15047.
81. P. T. Anastas and J. J. Breen, Design for the environment and Green Chemistry: The heart and soul of industrial ecology, *J. Cleaner Prod.*, 1997, **5**, 97–102.
82. H. Flerlage and J. C. Slootweg, Modern chemistry is rubbish, *Nat. Rev. Chem.*, 2023, 7, 593–594.
83. T. Rastogi, C. Leder and K. Kümmerer, A sustainable chemistry solution to the presence of pharmaceuticals and chemicals in the aquatic environment – the example of re-designing β-blocker Atenolol, *RSC Adv.*, 2014, **5**, 27–32.
84. H. C. Erythropel, *et al.*, Designing green plasticizers: Influence of molecule geometry and alkyl chain length on the plasticizing effectiveness of diester plasticizers in PVC blends, *Polymer*, 2016, **89**, 18–27.

85. G. Chyr and J. M. DeSimone, Review of high-performance sustainable polymers in additive manufacturing, *Green Chem.*, 2023, **25**, 453–466.
86. R. A. Gross and B. Kalra, Biodegradable polymers for the environment, *Science*, 2002, **297**, 803–807.
87. N. S. Sarai, *et al.*, Directed evolution of enzymatic silicon-carbon bond cleavage in siloxanes, *Science*, 2024, **383**, 438–443.
88. E. M. N. Polman, G. J. M. Gruter, J. R. Parsons and A. Tietema, Comparison of the aerobic biodegradation of biopolymers and the corresponding bioplastics: A review, *Sci. Total Environ.*, 2021, **753**, 141953.
89. Y. Wang, *et al.*, Relationship between Composition and Environmental Degradation of Poly(isosorbide-*co*-diol oxalate) (PISOX) Copolyesters, *Environ. Sci. Technol.*, 2024, **58**, 2293–2302.
90. G.-J. M. Gruter, L. Sipos and M. Adrianus Dam, Accelerating Research into Bio-Based FDCA-Polyesters by Using Small Scale Parallel Film Reactors, *Comb. Chem. High Throughput Screening*, 2011, **15**, 180–188.
91. A. F. Sousa, *et al.*, Biobased polyesters and other polymers from 2,5-furandicarboxylic acid: A tribute to furan excellency, *Polym. Chem.*, 2015, **6**, 5961–5983.
92. A. Caretto, *et al.*, Fully Biobased Polyesters Based on an Isosorbide Monomer for Coil Coating Applications, *ACS Sustainable Chem. Eng.*, 2018, **6**, 14125–14134.
93. S. D. Mürtz, *et al.*, Electrochemical depolymerisation of polylactic acid, *Green Chem.*, 2024, **26**, 6423–6428.
94. Y. Cui, C. Deng, L. Fan, Y. Qiu and L. Zhao, Progress in the biosynthesis of bio-based PET and PEF polyester monomers, *Green Chem.*, 2023, **25**, 5836–5857.
95. X. Wu, *et al.*, Closed-loop recyclability of a biomass-derived epoxy-amine thermoset by methanolysis, *Science*, 2024, **384**, eadj9989.
96. Q. Zhang, *et al.*, Dual closed-loop chemical recycling of synthetic polymers by intrinsically reconfigurable poly(disulfides), *Matter*, 2021, **4**, 1352–1364.
97. Q. Zhang, D. H. Qu, B. L. Feringa and H. Tian, Disulfide-Mediated Reversible Polymerization toward Intrinsically Dynamic Smart Materials, *J. Am. Chem. Soc.*, 2022, **144**, 2022–2033.
98. Q. Zhang, *et al.*, Exploring a naturally tailored small molecule for stretchable, self-healing, and adhesive supramolecular polymers, *Sci. Adv.*, 2018, **4**, 8192–8219.
99. Y. Deng, Q. Zhang and B. L. Feringa, Dynamic Chemistry Toolbox for Advanced Sustainable Materials, *Adv. Sci.*, 2024, 2308666, DOI: 10.1002/ADVS.202308666.
100. K. S. Toohey, N. R. Sottos, J. A. Lewis, J. S. Moore and S. R. White, Self-healing materials with microvascular networks, *Nat. Mater.*, 2007, **6**(8), 581–585.
101. S. R. White, *et al.*, Autonomic healing of polymer composites, *Nature,*, 2001, **409**(6822), 794–797.

102. Y. Deng, *et al.*, A Chemically Recyclable Crosslinked Polymer Network Enabled by Orthogonal Dynamic Covalent Chemistry, *Angew. Chem., Int. Ed.*, 2022, **61**, e202209100.
103. R. A. Sheldon, Green solvents for sustainable organic synthesis: state of the art, *Green Chem.*, 2005, **7**, 267–278.
104. J. M. DeSimone, Practical Approaches to Green Solvents, *Science*, 2002, **297**, 799–803.
105. C. M. Alder, *et al.*, Updating and further expanding GSK's solvent sustainability guide, *Green Chem.*, 2016, **18**, 3879–3890.
106. G. Szekely, M. F. Jimenez-Solomon, P. Marchetti, J. F. Kim and A. G. Livingston, Sustainability assessment of organic solvent nanofiltration: from fabrication to application, *Green Chem.*, 2014, **16**, 4440–4473.
107. S. Narayan, *et al.*, "On Water": Unique Reactivity of Organic Compounds in Aqueous Suspension, *Angew. Chem., Int. Ed.*, 2005, **44**, 3275–3279.
108. Y. M. Tian, W. Silva, R. M. Gschwind and B. König, Accelerated photochemical reactions at oil–water interface exploiting melting point depression, *Science*, 2024, **383**, 750–756.
109. B. H. Lipshutz and A. R. Abela, Micellar catalysis of Suzuki-Miyaura cross-couplings with heteroaromatics in water, *Org. Lett.*, 2008, **10**, 5329–5332.
110. B. H. Lipshutz, G. T. Aguinaldo, S. Ghorai and K. Voigtritter, Olefin cross-metathesis reactions at room temperature using the nonionic amphiphile 'PTS': Just add water, *Org. Lett.*, 2008, **10**, 1325–1328.
111. D. Prat, *et al.*, Sanofi's Solvent Selection Guide: A Step Toward More Sustainable Processes, *Org. Process Res. Dev.*, 2013, **17**, 1517–1525.
112. D. Prat, J. Hayler and A. Wells, A survey of solvent selection guides, *Green Chem.*, 2014, **16**, 4546–4551.
113. D. Prat, *et al.*, CHEM21 selection guide of classical- and less classical-solvents, *Green Chem.*, 2015, **18**, 288–296.
114. L. J. Diorazio, D. R. J. Hose and N. K. Adlington, Toward a More Holistic Framework for Solvent Selection, *Org. Process Res. Dev.*, 2016, **20**, 760–773.
115. Solvent Selection Tool - American Chemical Society. https://www.acs.org/greenchemistry/research-innovation/tools-for-green-chemistry/solvent-tool.html.
116. R. K. Henderson, *et al.*, Expanding GSK's solvent selection guide – embedding sustainability into solvent selection starting at medicinal chemistry, *Green Chem.*, 2011, **13**, 854–862.
117. F. P. Byrne, *et al.*, Tools and techniques for solvent selection: green solvent selection guides, *Sustainable Chem. Processes*, 2016, (1)**4**, 1–24.
118. K. Alfonsi, *et al.*, Green chemistry tools to influence a medicinal chemistry and research chemistry based organisation, *Green Chem.*, 2008, **10**, 31–36.
119. Beyond Benign. https://www.beyondbenign.org/.

120. L. Cseri, S. Kumar, P. Palchuber and G. Szekely, NMR Chemical Shifts of Emerging Green Solvents, Acids, and Bases for Facile Trace Impurity Analysis, *ACS Sustainable Chem. Eng.*, 2023, **11**, 5696–5725.
121. J. L. Vidal, *et al.*, Green Solvents for the Liquid-Phase Exfoliation of Biochars, *ACS Sustainable Chem. Eng.*, 2021, **9**, 9114–9125.
122. T. Freese *et al.* A guidebook for sustainability in laboratories, *ChemRxiv*, 2023, pp. 1–76.
123. M. Poliakoff and P. King, Phenomenal fluids, *Nature*, 2001, **412**(6843), 125.
124. P. G. Jessop, T. Ikariya and R. Noyori, Homogeneous Catalysis in Supercritical Fluids, *Chem. Rev.*, 1999, **99**, 475–493.
125. P. Stephenson, P. Licence, S. K. Ross and M. Poliakoff, Continuous catalytic asymmetric hydrogenation in supercritical CO_2, *Green Chem.*, 2004, **6**, 521–523.
126. T. Welton, Room-Temperature Ionic Liquids. Solvents for Synthesis and Catalysis, *Chem. Rev.*, 1999, **99**, 2071–2083.
127. C. Reichardt and T. Welton, *Solvents and Solvent Effects in Organic Chemistry*, Wiley-VCH, Weinheim, 2011.
128. A. Brandt, J. Gräsvik, J. P. Hallett and T. Welton, Deconstruction of lignocellulosic biomass with ionic liquids, *Green Chem.*, 2013, **15**, 550–583.
129. P. G. Jessop, D. J. Heldebrant, X. Li, C. A. Eckert and C. L. Liotta, Reversible nonpolar-to-polar solvent, *Nature*, 2005, **436**(7054), 1102.
130. E. F. Smith, F. J. M. Rutten, I. J. Villar-Garcia, D. Briggs and P. Licence, Ionic liquids in vacuo: Analysis of liquid surfaces using ultra-high-vacuum techniques, *Langmuir*, 2006, **22**, 9386–9392.
131. J. P. Armstrong, *et al.*, Vapourisation of ionic liquids, *Phys. Chem. Chem. Phys.*, 2007, **9**, 982–990.
132. W. Leitner, Supercritical carbon dioxide as a green reaction medium for catalysis, *Acc. Chem. Res.*, 2002, **35**, 746–756.
133. A. I. Cooper, Polymer synthesis and processing using supercritical carbon dioxide, *J. Mater. Chem.*, 2000, **10**, 207–234.
134. R. Sheldon, Catalytic reactions in ionic liquids, *Chem. Commun.*, 2001, **1**, 2399–2407.
135. Y. Liu, *et al.*, Ionic liquids as a new cornerstone to support hydrogen energy, *Green Chem.*, 2023, **25**, 4981–4994.
136. M. Marchini, *et al.*, What makes every work perfect is cooking and grinding": the ancient roots of mechanochemistry, *RSC Mechanochem.*, 2024, **1**, 123–129.
137. R. Abbasi, A. Mitchell, P. G. Jessop and M. F. Cunningham, Cross-linking CO_2-switchable polymers for paints and coatings applications, *RSC Appl. Polym.*, 2024, **2**, 214–223.
138. J. Ho, *et al.*, Water-borne coatings that share the mechanism of action of oil-based coatings, *Green Chem.*, 2018, **20**, 1899–1905.

139. K. Pieters and T. H. Mekonnen, Progress in waterborne polymer dispersions for coating applications: commercialized systems and new trends, *RSC Sustainability*, 2024, **2**, 3704–3729.
140. J. Van Haveren, *et al.*, Resins and additives for powder coatings and alkyd paints, based on renewable resources, *J. Coat. Technol. Res.*, 2007, **4**, 177–186.
141. A. Jensma, N. Elders, K. J. van den Berg and B. L. Feringa, Waterborne polymers and coatings from bio-based butenolides, *Green Chem.*, 2024, **26**, 9676–9681.
142. J. L. Do and T. Friščić, Mechanochemistry: A Force of Synthesis, *ACS Cent. Sci.*, 2017, **3**, 13–19.
143. T. Friščič and W. Jones, Recent advances in understanding the mechanism of cocrystal formation *via* grinding, *Cryst. Growth Des.*, 2009, **9**, 1621–1637.
144. T. Friščić, C. Mottillo and H. M. Titi, Mechanochemistry for Synthesis, *Angew. Chem., Int. Ed.*, 2020, **132**, 1030–1041.
145. L. Takacs, The historical development of mechanochemistry, *Chem. Soc. Rev.*, 2013, **42**, 7649.
146. F. Hajiali, *et al.*, Extraction of Chitin from Green Crab Shells by Mechanochemistry and Aging, *ACS Sustainable Chem. Eng.*, 2022, **10**, 11348–11357.
147. S. L. James, *et al.*, Mechanochemistry: opportunities for new and cleaner synthesis, *Chem. Soc. Rev.*, 2011, **41**, 413–447.
148. K. Kubota, T. Endo, M. Uesugi, Y. Hayashi and H. Ito, Solid-State C–N Cross-Coupling Reactions with Carbazoles as Nitrogen Nucleophiles Using Mechanochemistry, *ChemSusChem*, 2022, **15**, e202102132.
149. L. Konnert, *et al.*, Mechanochemical preparation of hydantoins from amino esters: Application to the synthesis of the antiepileptic drug phenytoin, *J. Org. Chem.*, 2014, **79**, 10132–10142.
150. R. Takahashi, *et al.*, Mechanochemical synthesis of magnesium-based carbon nucleophiles in air and their use in organic synthesis, *Nat. Commun.*, 2021, **12**(1), 1–10.
151. R. Takahashi, T. Seo, K. Kubota and H. Ito, Palladium-Catalyzed Solid-State Polyfluoroarylation of Aryl Halides Using Mechanochemistry, *ACS Catal.*, 2021, **11**, 14803–14810.
152. Y. Gao, C. Feng, T. Seo, K. Kubota and H. Ito, Efficient access to materials-oriented aromatic alkynes *via* the mechanochemical Sonogashira coupling of solid aryl halides with large polycyclic conjugated systems, *Chem. Sci.*, 2022, **13**, 430–438.
153. T. Seo, N. Toyoshima, K. Kubota and H. Ito, Tackling Solubility Issues in Organic Synthesis: Solid-State Cross-Coupling of Insoluble Aryl Halides, *J. Am. Chem. Soc.*, 2021, **143**, 6165–6175.
154. M. J. Eckelman, *et al.*, Applying green chemistry to raw material selection and product formulation at The Estée Lauder Companies, *Green Chem.*, 2022, **24**, 2397–2408.

155. C. O. Tuck, E. Pérez, I. T. Horváth, R. A. Sheldon and M. Poliakoff, Valorization of biomass: Deriving more value from waste, *Science*, 2012, **337**, 695–699.
156. C. Yan, *et al.*, Waste to wealth: direct utilization of spent materials for electrocatalysis and energy storage, *Green Chem.*, 2023, **25**, 3816–3846.
157. A. J. Ragauskas, *et al.*, The path forward for biofuels and biomaterials, *Science*, 2006, **311**, 484–489.
158. J. Langanke, *et al.*, Carbon dioxide (CO_2) as sustainable feedstock for polyurethane production, *Green Chem.*, 2014, **16**, 1865–1870.
159. M. Wang, *et al.*, CO_2 electrochemical catalytic reduction with a highly active cobalt phthalocyanine, *Nat. Commun.*, 2019, **10**(1), 1–8.
160. Q. Liu, L. Wu, R. Jackstell and M. Beller, Using carbon dioxide as a building block in organic synthesis, *Nat. Commun.*, 2015, **6**(1), 1–15.
161. M. M. Abu-Omar, *et al.*, Guidelines for performing lignin-first biorefining, *Energy Environ. Sci.*, 2021, **14**, 262–292.
162. A. Afanasenko and K. Barta, Pharmaceutically relevant (hetero)cyclic compounds and natural products from lignin-derived monomers: Present and perspectives, *iScience*, 2021, **24**, 102211.
163. S. Van Den Bosch, *et al.*, Reductive lignocellulose fractionation into soluble lignin-derived phenolic monomers and dimers and processable carbohydrate pulps, *Energy Environ. Sci.*, 2015, **8**, 1748–1763.
164. W. Schutyser, *et al.*, Chemicals from lignin: An interplay of lignocellulose fractionation, depolymerisation, and upgrading, *Chem. Soc. Rev.*, 2018, **47**, 852–908.
165. A. J. Ragauskas, *et al.*, Lignin valorization: Improving lignin processing in the biorefinery, *Science*, 2014, **344**, 1246843.
166. Z. Sun, B. Fridrich, A. De Santi, S. Elangovan and K. Barta, Bright Side of Lignin Depolymerization: Toward New Platform Chemicals, *Chem. Rev.*, 2018, **118**, 614–678.
167. C. Somerville, H. Youngs, C. Taylor, S. C. Davis and S. P. Long, Feedstocks for lignocellulosic biofuels, *Science*, 2010, **329**, 790–792.
168. R. Geyer, J. R. Jambeck and K. L. Law, Production, use, and fate of all plastics ever made, *Sci. Adv.*, 2017, **3**, e1700782.
169. A. Corma Canos, S. Iborra and A. Velty, Chemical routes for the transformation of biomass into chemicals, *Chem. Rev.*, 2007, **107**, 2411–2502.
170. C. O. Tuck, E. Pérez, I. T. Horváth, R. A. Sheldon and M. Poliakoff, Valorization of Biomass: Deriving More Value from Waste, *Science*, 2012, **337**, 695–699.
171. B. Program, T. Werpy and G. Petersen, *Top Value Added Chemicals from Biomass Volume I - Results of Screening for Potential Candidates from Sugars and Synthesis Gas Produced by the Staff at Pacific Northwest National Laboratory (PNNL), National* Renewable Energy Laboratory (NREL), Office of Biomass Program, Office of the Energy

Efficiency and Renewable Energy *(EERE)*, 2004. http://www.osti.gov/bridge.
172. J. J. Bozell and G. R. Petersen, Technology development for the production of biobased products from biorefinery carbohydrates—the US Department of Energy's "Top 10" revisited, *Green Chem.*, 2010, **12**, 539–554.
173. R. J. Van Putten, *et al.*, Hydroxymethylfurfural, a versatile platform chemical made from renewable resources, *Chem. Rev.*, 2013, **113**, 1499–1597.
174. J. U. Hernández-Beltrán, *et al.*, Insight into Pretreatment Methods of Lignocellulosic Biomass to Increase Biogas Yield: Current State, Challenges, and Opportunities, *Appl. Sci.*, 2019, **9**, 3721.
175. A. Akhtar, V. Krepl and T. Ivanova, A Combined Overview of Combustion, Pyrolysis, and Gasification of Biomass, *Energy Fuels*, 2018, **32**, 7294–7318.
176. P. McNeice, *et al.*, Cellulose modification for sustainable polymers: overcoming problems of solubility and processing, *RSC Sustainability*, 2024, **2**, 369–376.
177. B. Thomas, *et al.*, Nanocellulose, a Versatile Green Platform: From Biosources to Materials and Their Applications, *Chem. Rev.*, 2018, **118**, 11575–11625.
178. S. C. Maßmann, *et al.*, Regioselective palladium-catalysed aerobic oxidation of dextran and its use as bio-based binder in paperboard coatings, *Green Chem.*, 2024, **26**, 4005–4012.
179. H. S. Siebe, *et al.*, Formation of substituted dioxanes in the oxidation of gum arabic with periodate, *Green Chem.*, 2023, **25**, 4058–4066.
180. B. Girisuta, L. P. B. M. Janssen and H. J. Heeres, Green Chemicals: A Kinetic Study on the Conversion of Glucose to Levulinic Acid, *Chem. Eng. Res. Des.*, 2006, **84**, 339–349.
181. P. Mäki-Arvela, I. L. Simakova, T. Salmi and D. Y. Murzin, Production of lactic acid/lactates from biomass and their catalytic transformations to commodities, *Chem. Rev.*, 2014, **114**, 1909–1971.
182. W. Leitner, J. Klankermayer, S. Pischinger, H. Pitsch and K. Kohse-Höinghaus, Advanced Biofuels and Beyond: Chemistry Solutions for Propulsion and Production, *Angew. Chem., Int. Ed.*, 2017, **56**, 5412–5452.
183. R. Hu, *et al.*, Bio-based platform chemicals synthesized from lignin biorefinery, *Green Chem.*, 2023, **25**, 8970–9000.
184. C. Chio, M. Sain and W. Qin, Lignin utilization: A review of lignin depolymerization from various aspects, *Renewable Sustainable Energy Rev.*, 2019, **107**, 232–249.
185. I. Delidovich, *et al.*, Alternative Monomers Based on Lignocellulose and Their Use for Polymer Production, *Chem. Rev.*, 2016, **116**, 1540–1599.
186. J. G. H. Hermens, T. Freese, K. J. Van Den Berg, R. Van Gemert and B. L. Feringa, A coating from nature, *Sci. Adv.*, 2020, **6**, 26–42.
187. J. G. H. Hermens, *et al.*, A sustainable polymer and coating system based on renewable raw materials, *Green Chem.*, 2022, **24**, 9772–9780.

188. J. G. H. Hermens, A. Jensma and B. L. Feringa, Highly Efficient Bio-based Synthesis of Acrylic Acid, *Angew. Chem., Int. Ed.*, 2022, **61**, e202112618.
189. M. L. Lepage, *et al.*, Electron-Poor Butenolides: The Missing Link between Acrylates and Maleic Anhydride in Radical Polymerization, *J. Am. Chem. Soc.*, 2023, **145**, 17211–17219.
190. J. Zakzeski, P. C. A. Bruijnincx, A. L. Jongerius and B. M. Weckhuysen, The catalytic valorization of lignin for the production of renewable chemicals, *Chem. Rev.*, 2010, **110**, 3552–3599.
191. S. Gillet, *et al.*, Lignin transformations for high value applications: towards targeted modifications using green chemistry, *Green Chem.*, 2017, **19**, 4200–4233.
192. L. Shuai, *et al.*, Formaldehyde stabilization facilitates lignin monomer production during biomass depolymerization, *Science*, 2016, **354**, 329–333.
193. J. S. Luterbacher, *et al.*, Nonenzymatic sugar production from biomass using biomass-derived γ-valerolactone, *Science*, 2014, **343**, 277–280.
194. J. Wildschut, F. H. Mahfud, R. H. Venderbosch and H. J. Heeres, Hydrotreatment of Fast Pyrolysis Oil Using Heterogeneous Noble-Metal Catalysts, *Ind. Eng. Chem. Res.*, 2009, **48**, 10324–10334.
195. P. J. De Wild, W. J. J. Huijgen and H. J. Heeres, Pyrolysis of wheat straw-derived organosolv lignin, *J. Anal. Appl. Pyrolysis*, 2012, **93**, 95–103.
196. P. J. Deuss, *et al.*, Aromatic Monomers by in Situ Conversion of Reactive Intermediates in the Acid-Catalyzed Depolymerization of Lignin, *J. Am. Chem. Soc.*, 2015, **137**, 7456–7467.
197. K. Barta, *et al.*, Catalytic disassembly of an organosolv lignin *via* hydrogen transfer from supercritical methanol, *Green Chem.*, 2010, **12**, 1640–1647.
198. J. S. Luterbacher, D. Martin Alonso and J. A. Dumesic, Targeted chemical upgrading of lignocellulosic biomass to platform molecules, *Green Chem.*, 2014, **16**, 4816–4838.
199. C. W. Lahive, *et al.*, Advanced Model Compounds for Understanding Acid-Catalyzed Lignin Depolymerization: Identification of Renewable Aromatics and a Lignin-Derived Solvent, *J. Am. Chem. Soc.*, 2016, **138**, 8900–8911.
200. Z. Sun, *et al.*, Complete lignocellulose conversion with integrated catalyst recycling yielding valuable aromatics and fuels, *Nat. Catal.*, 2018, **1**, 82–92.
201. C. W. Lahive, P. C. J. Kamer, C. S. Lancefield and P. J. Deuss, An Introduction to Model Compounds of Lignin Linking Motifs; Synthesis and Selection Considerations for Reactivity Studies, *ChemSusChem*, 2020, **13**, 4238–4265.
202. P. J. Deuss and K. Barta, From models to lignin: Transition metal catalysis for selective bond cleavage reactions, *Coord. Chem. Rev.*, 2016, **306**, 510–532.

203. Z. Luo, *et al.*, Carbon–carbon bond cleavage for a lignin refinery, *Nat. Chem. Eng.*, 2024, **1**(1), 61–72.
204. R. Rinaldi, *et al.*, Paving the Way for Lignin Valorisation: Recent Advances in Bioengineering, Biorefining and Catalysis, *Angew. Chem., Int. Ed.*, 2016, **55**, 8164–8215.
205. T. Freese, *et al.*, A molecular motor from lignocellulose, *Green Chem.*, 2022, **24**, 3689–3696.
206. F. Brienza, D. Cannella, D. Montesdeoca, I. Cybulska and D. P. Debecker, A guide to lignin valorization in biorefineries: traditional, recent, and forthcoming approaches to convert raw lignocellulose into valuable materials and chemicals, *RSC Sustainability*, 2024, **2**, 37–90.
207. G. Szekely, New era of sustainable design for molecular motors, *Chem Catal.*, 2022, **2**, 1532–1533.
208. J. Huo, Z. Wang, C. Oberschelp, G. Guillén-Gosálbez and S. Hellweg, Net-zero transition of the global chemical industry with CO_2-feedstock by 2050: feasible yet challenging, *Green Chem.*, 2023, **25**, 415–430.
209. J. Artz, *et al.*, Sustainable Conversion of Carbon Dioxide: An Integrated Review of Catalysis and Life Cycle Assessment, *Chem. Rev.*, 2018, **118**, 434–504.
210. C. Hepburn, *et al.*, The technological and economic prospects for CO_2 utilization and removal, *Nature*, 2019, **575**(7781), 87–97.
211. M. Peters, *et al.*, Chemical Technologies for Exploiting and Recycling Carbon Dioxide into the Value Chain, *ChemSusChem*, 2011, **4**, 1216–1240.
212. H. O. LeClerc, *et al.*, The CO_2 Tree: The Potential for Carbon Dioxide Utilization Pathways, *ACS Sustainable Chem. Eng.*, 2025, **13**(1), 5–29.
213. P. G. Jessop, T. Ikariya and R. Noyori, Homogeneous Hydrogenation of Carbon Dioxide, *Chem. Rev.*, 1995, **95**, 259–272.
214. E. T. C. Vogt and B. M. Weckhuysen, The refinery of the future, *Nature*, 2024, **629**(8011), 295–306.
215. W. Leitner, Carbon Dioxide as a Raw Material: The Synthesis of Formic Acid and Its Derivatives from CO_2, *Angew. Chem., Int. Ed.*, 1995, **34**, 2207–2221.
216. Q. Liu, L. Wu, R. Jackstell and M. Beller, Using carbon dioxide as a building block in organic synthesis, *Nat. Commun.*, 2015, **6**(1), 1–15.
217. J. Klankermayer and W. Leitner, Love at second sight for CO_2 and H_2 in organic synthesis, *Science*, 2015, **350**, 629–630.
218. J. L. Vidal, V. P. Andrea, S. L. MacQuarrie and F. M. Kerton, Oxidized Biochar as a Simple, Renewable Catalyst for the Production of Cyclic Carbonates from Carbon Dioxide and Epoxides, *ChemCatChem*, 2019, **11**, 4089–4095.
219. J. R. Vanhoof, S. Spittaels and D. E. De Vos, A comparative overview of the electrochemical valorization and incorporation of CO_2 in industrially relevant compounds, *EES Catal.*, 2024, **2**, 753–779.

220. A. Correa, T. León and R. Martin, Ni-catalyzed carboxylation of C(sp^2)- and C(sp^3)–O bonds with CO_2, *J. Am. Chem. Soc.*, 2014, **136**, 1062–1069.
221. A. Tortajada, F. Juliá-Hernández, M. Börjesson, T. Moragas and R. Martin, Transition-Metal-Catalyzed Carboxylation Reactions with Carbon Dioxide, *Angew. Chem., Int. Ed.*, 2018, **57**, 15948–15982.
222. A. Correa and R. Martín, Palladium-catalyzed direct carboxylation of aryl bromides with carbon dioxide, *J. Am. Chem. Soc.*, 2009, **131**, 15974–15975.
223. P. Markewitz, *et al.*, Worldwide innovations in the development of carbon capture technologies and the utilization of CO_2, *Energy Environ. Sci.*, 2012, **5**, 7281–7305.
224. M. Schreier, *et al.*, Solar conversion of CO_2 to CO using Earth-abundant electrocatalysts prepared by atomic layer modification of CuO, *Nat. Energy*, 2017, **2**(7), 1–9.
225. I. Vollmer, *et al.*, Beyond Mechanical Recycling: Giving New Life to Plastic Waste, *Angew. Chem., Int. Ed.*, 2020, **59**, 15402–15423.
226. R. Geyer, J. R. Jambeck and K. L. Law, Production, use, and fate of all plastics ever made, *Sci. Adv.*, 2017, **3**, e1700782.
227. A. J. Martín, C. Mondelli, S. D. Jaydev and J. Pérez-Ramírez, Catalytic processing of plastic waste on the rise, *Chem*, 2021, **7**, 1487–1533.
228. S. D. Jaydev, *et al.*, Consumer Grade Polyethylene Recycling *via* Hydrogenolysis on Ultrafine Supported Ruthenium Nanoparticles, *Angew. Chem., Int. Ed.*, 2024, **63**(11), e202317526.
229. A. Ong, *et al.*, The global burden of plastics in oral health: prospects for circularity, sustainable materials development and practice, *RSC Sustainability*, 2024, **2**, 881–902.
230. R. Kol, *et al.*, Increasing the Dissolution Rate of Polystyrene Waste in Solvent-Based Recycling, *ACS Sustainable Chem. Eng.*, 2024, **12**, 4619–4630.
231. J. R. J. Strien, H. C. Genuino, M. C. P. van Eijk, P. J. Deuss and H. J. Heeres, Pyrolysis of Polyolefin-Enriched Mixed Plastic Waste Streams: Effects of Pretreatments and Presence of Hydrogen during Pyrolysis, *Energy Fuels*, 2025, **39**(1), 686–698.
232. M. van Akker, *et al.*, Full Utilization of Hard-to-Recycle Mixed Plastic Waste by Conversion toward Pyrolysis Oil and BTX Aromatics on a Pilot Scale, *Energy Fuels*, 2025, **39**(13), 6438–6451.
233. X. Jiang, B. Zhu and M. Zhu, An overview on the recycling of waste poly (vinyl chloride), *Green Chem.*, 2023, **25**, 6971–7025.
234. E. Colacino, V. Isoni, D. Crawford and F. García, Upscaling Mechanochemistry: Challenges and Opportunities for Sustainable Industry, *Trends Chem.*, 2021, **3**, 335–339.
235. J. F. Reynes, V. Isoni and F. García, Tinkering with Mechanochemical Tools for Scale Up, *Angew. Chem., Int. Ed.*, 2023, **135**, e202300819.
236. D. Ferrari, *et al.*, PVDC multilayer films: Chemical recycling by means of catalytic conversion into aromatics (BTX), *Sustainable Chem. Environ.*, 2023, **2**, 100010.

237. H. Li, *et al.*, Aliphatic amines from waste polyolefins by tandem pyrolysis, hydroformylation, and reductive amination, *Green Chem.*, 2024, **26**, 8718–8727.
238. S. Zhang, M. Li, Z. Zuo and Z. Niu, Recent advances in plastic recycling and upgrading under mild conditions, *Green Chem.*, 2023, **25**, 6949–6970.
239. J.-P. Lange, S. R. A. Kersten, S. De Meester, M. C. P. van Eijk and K. Ragaert, Plastic recycling stripped naked – from circular product to circular industry with recycling cascade, *ChemSusChem*, 2024, e202301320, DOI: 10.1002/CSSC.202301320.
240. A. Heeres, *et al.*, Synthesis of Bio-aromatics from Black Liquors Using Catalytic Pyrolysis, *ACS Sustainable Chem. Eng.*, 2018, **6**, 3472–3480.
241. S. He, *et al.*, Catalytic pyrolysis of crude glycerol over shaped ZSM-5/bentonite catalysts for bio-BTX synthesis, *Appl. Catal., B*, 2018, **235**, 45–55.
242. B. Trang, *et al.*, Low-temperature mineralization of perfluorocarboxylic acids, *Science*, 2022, **377**, 839–845.
243. S. Joudan and R. J. Lundgren, Taking the "F" out of forever chemicals, *Science*, 2022, **377**, 816–817.
244. P. T. Anastas, M. M. Kirchhoff and T. C. Williamson, Catalysis as a foundational pillar of green chemistry, *Appl. Catal., A*, 2001, **221**, 3–13.
245. A. Abbas, *et al.*, Catalysis at the intersection of sustainable chemistry and a circular economy, *One Earth*, 2024, **7**, 738–741.
246. D. J. C. Constable, *et al.*, Green chemistry measures for process research and development, *Green Chem.*, 2001, **3**, 7–9.
247. M. A. Murphy, Early Industrial Roots of Green Chemistry and the history of the BHC Ibuprofen process invention and its Quality connection, *Found. Chem.*, 2018, **20**, 121–165.
248. T. Bligaard, *et al.*, Toward Benchmarking in Catalysis Science: Best Practices, Challenges, and Opportunities, *ACS Catal.*, 2016, **6**, 2590–2602.
249. G. Zante, *et al.*, A toolbox for improved recycling of critical metals and materials in low-carbon technologies, *RSC Sustainability*, 2024, **2**, 320–347.
250. M. He, *et al.*, Combined pyro-hydrometallurgical technology for recovering valuable metal elements from spent lithium-ion batteries: a review of recent developments, *Green Chem.*, 2023, **25**, 6561–6580.
251. F. Luo, T. Lyu, D. Wang and Z. Zheng, A review on green and sustainable carbon anodes for lithium ion batteries: utilization of green carbon resources and recycling waste graphite, *Green Chem.*, 2023, **25**, 8950–8969.
252. R. E. Ciez and J. F. Whitacre, Examining different recycling processes for lithium-ion batteries, *Nat. Sustainability*, 2019, **2**, 148–156.
253. M. Chen, *et al.*, Recycling End-of-Life Electric Vehicle Lithium-Ion Batteries, *Joule*, 2019, **3**, 2622–2646.

254. G. Harper, *et al.*, Recycling lithium-ion batteries from electric vehicles, *Nature*, 2019, **575**, 75–86.
255. Element Scarcity - EuChemS Periodic Table - EuChemS. https://www.euchems.eu/euchems-periodic-table/.
256. T. F. Ramspoth, J. Flapper, K. J. van den Berg, B. L. Feringa and S. R. Harutyunyan, A highly efficient and sustainable catalyst system for terminal epoxy-carboxylic acid ring opening reactions, *Green Chem.*, 2024, **26**, 3346–3355.
257. M. Bystrzanowska, P. Petkov and M. Tobiszewski, Ranking of Heterogeneous Catalysts Metals by Their Greenness, *ACS Sustainable Chem. Eng.*, 2019, **7**, 18434–18443.
258. T. Yan, B. L. Feringa and K. Barta, Iron catalysed direct alkylation of amines with alcohols, *Nat. Commun.*, 2014, **5**(1), 1–7.
259. R. A. Sheldon, Homogeneous catalysts to solid catalysts, *Curr. Opin. Solid State Mater. Sci.*, 1996, **1**, 101–106.
260. S. Mitchell, A. J. Martín, G. Guillén-Gosálbez and J. Pérez-Ramírez, The Future of Chemical Sciences is Sustainable, *Angew. Chem., Int. Ed.*, 2024, **136**(26), e202318676.
261. J. H. Clark, Solid acids for green chemistry, *Acc. Chem. Res.*, 2002, **35**, 791–797.
262. I. Fechete, Y. Wang and J. C. Védrine, The past, present and future of heterogeneous catalysis, *Catal. Today*, 2012, **189**, 2–27.
263. C. M. Friend and B. Xu, Heterogeneous catalysis: A central science for a sustainable future, *Acc. Chem. Res.*, 2017, **50**, 517–521.
264. C. Descorme, P. Gallezot, C. Geantet and C. George, Heterogeneous Catalysis: A Key Tool toward Sustainability, *ChemCatChem*, 2012, **4**, 1897–1906.
265. D. P. Debecker and P. H. Mutin, Non-hydrolytic sol–gel routes to heterogeneous catalysts, *Chem. Soc. Rev.*, 2012, **41**, 3624–3650.
266. D. P. Debecker, E. M. Gaigneaux and G. Busca, Exploring, Tuning, and Exploiting the Basicity of Hydrotalcites for Applications in Heterogeneous Catalysis, *Chem. – Eur. J.*, 2009, **15**, 3920–3935.
267. Z. Guo, *et al.*, Recent advances in heterogeneous selective oxidation catalysis for sustainable chemistry, *Chem. Soc. Rev.*, 2014, **43**, 3480–3524.
268. R. Sheldon and J. K. Kochi, *Metal-Catalyzed Oxidations of Organic Compounds: Mechanistic Principles and Synthetic Methodology Including Biochemical Processes*, Elsevier, 1981.
269. D. I. Enache, *et al.*, Solvent-free oxidation of primary alcohols to aldehydes using Au–Pd/TiO_2 catalyst, *Science*, 2006, **311**, 362–365.
270. D. S. Su, *et al.*, Metal-Free Heterogeneous Catalysis for Sustainable Chemistry, *ChemSusChem*, 2010, **3**, 169–180.
271. J. C. Védrine, Metal Oxides in Heterogeneous Oxidation Catalysis: State of the Art and Challenges for a More Sustainable World, *ChemSusChem*, 2019, **12**, 577–588.

272. M. T. Sabatini, L. T. Boulton, H. F. Sneddon and T. D. Sheppard, A green chemistry perspective on catalytic amide bond formation, *Nat. Catal.*, 2019, **2**(1), 10–17.
273. S. H. Y. S. Abdullah, *et al.*, A review of biomass-derived heterogeneous catalyst for a sustainable biodiesel production, *Renewable Sustainable Energy Rev.*, 2017, **70**, 1040–1051.
274. Y. Zhang, *et al.*, Nanostructured single-atom catalysts derived from natural building blocks, *EES Catal.*, 2024, **2**, 475–506.
275. X. Wang, S. Blechert and M. Antonietti, Polymeric graphitic carbon nitride for heterogeneous photocatalysis, *ACS Catal.*, 2012, **2**, 1596–1606.
276. Z. Teng, *et al.*, Atomically dispersed antimony on carbon nitride for the artificial photosynthesis of hydrogen peroxide, *Nat. Catal.*, 2021, **4**, 374–384.
277. X. Zhang, *et al.*, Unraveling the dual defect sites in graphite carbon nitride for ultra-high photocatalytic H_2O_2 evolution, *Energy Environ. Sci.*, 2022, **15**, 830–842.
278. M. Nikitin, *et al.*, C(sp^2)–S cross-coupling reactions with nickel, visible light, and mesoporous graphitic carbon nitride, *Green Chem.*, 2024, **26**, 5845–5851.
279. Y. Kofuji, *et al.*, Carbon Nitride-Aromatic Diimide-Graphene Nanohybrids: Metal-Free Photocatalysts for Solar-to-Hydrogen Peroxide Energy Conversion with 0.2% Efficiency, *J. Am. Chem. Soc.*, 2016, **138**, 10019–10025.
280. Y. Wang, X. Wang and M. Antonietti, Polymeric graphitic carbon nitride as a heterogeneous organocatalyst: From photochemistry to multipurpose catalysis to sustainable chemistry, *Angew. Chem., Int. Ed.*, 2012, **51**(1), 68–89.
281. Z. Wei, *et al.*, Efficient visible-light-driven selective oxygen reduction to hydrogen peroxide by oxygen-enriched graphitic carbon nitride polymers, *Energy Environ. Sci.*, 2018, **11**, 2581–2589.
282. D. Bhanderi, P. Lakhani and C. K. Modi, Graphitic carbon nitride (g-C_3N_4) as an emerging photocatalyst for sustainable environmental applications: a comprehensive review, *RSC Sustainability*, 2024, **2**, 265–287.
283. H. Li, B. Cheng, J. Xu, J. Yu and S. Cao, Crystalline carbon nitrides for photocatalysis, *EES Catal.*, 2024, **2**, 411–447.
284. S. Enthaler, K. Junge and M. Beller, Sustainable Metal Catalysis with Iron: From Rust to a Rising Star?, *Angew. Chem., Int. Ed.*, 2008, **47**, 3317–3321.
285. J. Woo, *et al.*, Fe–N/C catalysts with tunable mesoporous structures and carbon layer numbers reveal the role of interlayer O_2 activation, *EES Catal.*, 2023, **1**, 62–73.
286. F. Liu, *et al.*, Manipulating the spin state to activate the atomically dispersed Fe–N–C catalyst for oxygen reduction, *EES Catal.*, 2023, **1**, 562–570.
287. J. Li, *et al.*, OER highly active encapsulants to improve the electrochemical anticorrosion of Fe–N–C for ultralong-lifespan and high-rate rechargeable zinc–air batteries, *EES Catal.*, 2023, **1**, 987–997.

288. T. Zeng, *et al.*, Fe_3O_4 nanoparticles: a robust and magnetically recoverable catalyst for three-component coupling of aldehyde, alkyne and amine, *Green Chem.*, 2010, **12**, 570–573.
289. T. Freese, *et al.*, Iron oxide-promoted photochemical oxygen reduction to hydrogen peroxide (H_2O_2), *EES Catal.*, 2024, **2**, 262–275.
290. R. Hudson, Y. Feng, R. S. Varma and A. Moores, Bare magnetic nanoparticles: sustainable synthesis and applications in catalytic organic transformations, *Green Chem.*, 2014, **16**, 4493–4505.
291. S. Iravani, Surfactant-free synthesis of metal and metal oxide nanomaterials: a perspective, *RSC Sustainability*, 2023, **1**, 72–82.
292. G. Ginnakakis, S. Mitchell and J. Pérez-Ramírez, Single-atom heterogeneous catalysts for sustainable organic synthesis, *Trends Chem.*, 2022, **4**, 264–276.
293. S. K. Kaiser, Z. Chen, D. Faust Akl, S. Mitchell and J. Pérez-Ramírez, Single-Atom Catalysts across the Periodic Table, *Chem. Rev.*, 2020, **120**, 11703–11809.
294. X. Li, *et al.*, Advances in heterogeneous single-cluster catalysis, *Nat. Rev. Chem.*, 2023, 7(11), 754–767.
295. X. F. Yang, *et al.*, Single-atom catalysts: A new frontier in heterogeneous catalysis, *Acc. Chem. Res.*, 2013, **46**, 1740–1748.
296. S. Mitchell, A. J. Martín and J. Pérez-Ramírez, Transcending scales in catalysis for sustainable development, *Nat. Chem. Eng.*, 2024, **1**(1), 13–15.
297. P. P. Pescarmona and M. Taherimehr, Challenges in the catalytic synthesis of cyclic and polymeric carbonates from epoxides and CO_2, *Catal. Sci. Technol.*, 2012, **2**, 2169–2187.
298. A. J. Kamphuis, F. Picchioni and P. P. Pescarmona, CO_2-fixation into cyclic and polymeric carbonates: principles and applications, *Green Chem.*, 2019, **21**, 406–448.
299. E. Schuler, *et al.*, Oxalic acid hydrogenation to glycolic acid: heterogeneous catalysts screening, *Green Chem.*, 2023, **25**, 2409–2426.
300. E. Schuler, L. Grooten, P. Oulego, N. R. Shiju and G. J. M. Gruter, Selective reduction of oxalic acid to glycolic acid at low temperature in a continuous flow process, *RSC Sustainability*, 2023, **1**, 2072–2080.
301. F. De Clippel, *et al.*, Fast and selective sugar conversion to alkyl lactate and lactic acid with bifunctional carbon–silica catalysts, *J. Am. Chem. Soc.*, 2012, **134**, 10089–10101.
302. J. Bueno Moron, G. van Klink and G. J. M. Gruter, Production and Downstream Integration of 5-(Chloromethyl)furfural from Lignocellulose, *ACS Sustainable Chem. Eng.*, 2023, **11**, 17492–17509.
303. I. Van Zandvoort, *et al.*, Formation, Molecular Structure, and Morphology of Humins in Biomass Conversion: Influence of Feedstock and Processing Conditions, *ChemSusChem*, 2013, **6**, 1745–1758.
304. W. Gao, *et al.*, Revisiting group 4–7 transition metals for heterogeneous ammonia synthesis, *EES Catal.*, 2024, **2**, 780–788.

305. W. Qiu, *et al.*, Structural engineering of catalysts for ammonia electrosynthesis from nitrate: recent advances and challenges, *EES Catal.*, 2024, **2**, 202–219.
306. H. Wang, *et al.*, Rational element-doping of FeOOH-based electrocatalysts for efficient ammonia electrosynthesis, *EES Catal.*, 2024, **2**, 324–334.
307. E. B. Pinxterhuis, M. Giannerini, V. Hornillos and B. L. Feringa, Fast, greener and scalable direct coupling of organolithium compounds with no additional solvents, *Nat. Commun.*, 2016, **7**(1), 1–7.
308. J. L. Osorio-Tejada, F. Ferlin, L. Vaccaro and V. Hessel, The sustainability impact of Nobel Prize Chemistry: life cycle assessment of C–C cross-coupling reactions, *Green Chem.*, 2023, **25**, 9760–9778.
309. D. Faust Akl, *et al.*, Assessing the environmental benefit of palladium-based single-atom heterogeneous catalysts for Sonogashira coupling, *Green Chem.*, 2022, **24**, 6879–6888.
310. S. Zaman, *et al.*, Oxygen Reduction Electrocatalysts toward Practical Fuel Cells: Progress and Perspectives, *Angew. Chem., Int. Ed.*, 2021, **60**, 17832–17852.
311. L. Han, S. Guo, P. Wang and S. Dong, Light-driven, membraneless, hydrogen peroxide based fuel cells, *Adv. Energy Mater.*, 2015, **5**, 1400424.
312. I. Staffell, *et al.*, The role of hydrogen and fuel cells in the global energy system, *Energy Environ. Sci.*, 2019, **12**, 463–491.
313. Y. Yamada, M. Yoneda and S. Fukuzumi, High and robust performance of H_2O_2 fuel cells in the presence of scandium ion, *Energy Environ. Sci.*, 2015, **8**, 1698–1701.
314. T. S. Andrade, B. A. C. Sá, F. G. E. Nogueira, L. C. A. Oliveira and M. C. Pereira, Unassisted photocatalytic hydrogen peroxide fuel cell based on dual photoelectrodes with high performance and stability, *J. Appl. Electrochem.*, 2023, **53**, 435–444.
315. A. F. Lee, J. A. Bennett, J. C. Manayil and K. Wilson, Heterogeneous catalysis for sustainable biodiesel production *via* esterification and transesterification, *Chem. Soc. Rev.*, 2014, **43**, 7887–7916.
316. D. Kovačič, *et al.*, A comparative study of palladium-gold and palladium-tin catalysts in the direct synthesis of H_2O_2, *Green Chem.*, 2023, **25**, 10436–10446.
317. J. K. Edwards, *et al.*, Switching Off Hydrogen Peroxide Hydrogenation in the Direct Synthesis Process, *Science*, 2009, **323**, 1037–1041.
318. S. J. Freakley, *et al.*, Palladium-tin catalysts for the direct synthesis of H_2O_2 with high selectivity, *Science*, 2016, **351**, 965–968.
319. M. Sankar, *et al.*, The benzaldehyde oxidation paradox explained by the interception of peroxy radical by benzyl alcohol, *Nat. Commun.*, 2014, **5**, 3332.
320. J. K. Edwards, S. J. Freakley, R. J. Lewis, J. C. Pritchard and G. J. Hutchings, Advances in the direct synthesis of hydrogen peroxide from hydrogen and oxygen, *Catal. Today*, 2015, **248**, 3–9.

321. X. Wang, *et al.*, A metal-free polymeric photocatalyst for hydrogen production from water under visible light, *Nat. Mater.*, 2009, **8**, 76–80.
322. G. M. Lari, B. Puértolas, M. Shahrokhi, N. López and J. Pérez-Ramírez, Hybrid Palladium Nanoparticles for Direct Hydrogen Peroxide Synthesis: The Key Role of the Ligand, *Angew. Chem., Int. Ed.*, 2017, **129**, 1801–1805.
323. L. F. De Freitas, *et al.*, Tunable Catalytic Performance of Palladium Nanoparticles for H_2O_2 Direct Synthesis *via* Surface-Bound Ligands, *ACS Catal.*, 2020, **10**, 5202–5207.
324. L. Li, T. I. Korányi, B. F. Sels and P. P. Pescarmona, Highly-efficient conversion of glycerol to solketal over heterogeneous Lewis acid catalysts, *Green Chem.*, 2012, **14**, 1611–1619.
325. E. M. Van Schrojenstein Lantman, T. Deckert-Gaudig, A. J. G. Mank, V. Deckert and B. M. Weckhuysen, Catalytic processes monitored at the nanoscale with tip-enhanced Raman spectroscopy, *Nat. Nanotechnol.*, 2012, 7(9), 583–586.
326. Y. Kim, *et al.*, A strong regioregularity effect in self-organizing conjugated polymer films and high-efficiency polythiophene:fullerene solar cells, *Nat. Mater.*, 2006, **5**(3), 197–203.
327. M. Campoy-Quiles, *et al.*, Morphology evolution *via* self-organization and lateral and vertical diffusion in polymer:fullerene solar cell blends, *Nat. Mater.*, 2008, 7(2), 158–164.
328. J. Burschka, *et al.*, Sequential deposition as a route to high-performance perovskite-sensitized solar cells, *Nature*, 2013, **499**(7458), 316–319.
329. G. Min, *et al.*, Multisource Energy Harvester on Textile and Plants for Clean Energy Generation from Wind and Rainwater Droplets, *ACS Sustainable Chem. Eng.*, 2024, **12**(2), 695–705.
330. J. Bisquert and G. Garcia-Belmonte, On voltage, photovoltage, and photocurrent in bulk heterojunction organic solar cells, *J. Phys. Chem. Lett.*, 2011, **2**, 1950–1964.
331. L. Marzo, S. K. Pagire, O. Reiser and B. König, Visible-Light Photocatalysis: Does It Make a Difference in Organic Synthesis?, *Angew. Chem., Int. Ed.*, 2018, **57**, 10034–10072.
332. G. Ciamician, The Photochemistry of the Future, *Science*, 1912, **36**, 385–394.
333. M. Poliakoff and M. W. George, Manufacturing chemicals with light: any role in the circular economy?, *Philos. Trans. R. Soc., A*, 2020, **378**, 20190260.
334. M. Grätzel, Photoelectrochemical cells, *Nature*, 2001, **414**(6861), 338–344.
335. T. R. Harris-Lee, F. Marken, C. L. Bentley, J. Zhang and A. L. Johnson, A chemist's guide to photoelectrode development for water splitting – the importance of molecular precursor design, *EES Catal.*, 2023, **1**, 832–873.
336. A. Tsyganok, P. Monroy-Castillero, Y. Piekner, A. Yochelis and A. Rothschild, Parallel water photo-oxidation reaction pathways in

hematite photoanodes: implications for solar fuel production, *Energy Environ. Sci.*, 2022, **15**, 2445–2459.
337. B. O'Regan and M. Grätzel, A low-cost, high-efficiency solar cell based on dye-sensitized colloidal TiO_2 films, *Nature*, 1991, **353**(6346), 737–740.
338. O. Schmidt, *et al.*, Future cost and performance of water electrolysis: An expert elicitation study, *Int. J. Hydrogen Energy*, 2017, **42**, 30470–30492.
339. Z. W. Seh, *et al.*, Combining theory and experiment in electrocatalysis: Insights into materials design, *Science*, 2017, **355**, eaad4998.
340. J. K. Nørskov, *et al.*, Trends in the Exchange Current for Hydrogen Evolution, *J. Electrochem. Soc.*, 2005, **152**, J23.
341. H. Liu, *et al.*, CeO_2 supported high-valence Fe oxide for highly active and stable water oxidation, *EES Catal.*, 2023, **1**, 720–729.
342. X. Fu, *et al.*, Continuous-flow electrosynthesis of ammonia by nitrogen reduction and hydrogen oxidation, *Science*, 2023, **379**, 707–712.
343. Y. Ni, Z. Han, Y. Chai, G. Wu and L. Li, Catalytic hydrogen storage in liquid hydrogen carriers, *EES Catal.*, 2023, **1**, 459–494.
344. M. Miola, D. Chillé, G. Papanikolaou, P. Lanzafame and P. P. Pescarmona, Optimisation of the electrochemical conversion of CO_2 into formate in a flow cell configuration using a bismuth-based electrocatalyst, *Green Chem.*, 2023, **25**, 1875–1883.
345. C. Ampelli, *et al.*, An artificial leaf device built with earth-abundant materials for combined H_2 production and storage as formate with efficiency >10%, *Energy Environ. Sci.*, 2023, **16**, 1644–1661.
346. I. Slobodkin, E. Davydova, M. Sananis, A. Breytus and A. Rothschild, Electrochemical and chemical cycle for high-efficiency decoupled water splitting in a near-neutral electrolyte, *Nat. Mater.*, 2024, **23**, 398–405.
347. B. Rausch, M. D. Symes, G. Chisholm and L. Cronin, Decoupled catalytic hydrogen evolution from a molecular metal oxide redox mediator in water splitting, *Science*, 2014, **345**, 1326–1330.
348. N. C. Kani, S. Olusegun, R. Chauhan, J. A. Gauthier and M. R. Singh, High-pressure electrochemistry: a new frontier in decarbonization, *EES Catal.*, 2024, **2**, 507–521.
349. S. A. Lee, *et al.*, Single atom catalysts for water electrolysis: from catalyst-coated substrate to catalyst-coated membrane, *EES Catal.*, 2024, **2**, 49–70.
350. N. Daems, X. Sheng, I. F. J. Vankelecom and P. P. Pescarmona, Metal-free doped carbon materials as electrocatalysts for the oxygen reduction reaction, *J. Mater. Chem. A*, 2014, **2**, 4085–4110.
351. J. Creus, M. Miola and P. P. Pescarmona, Unravelling and overcoming the challenges in the electrocatalytic reduction of fructose to sorbitol, *Green Chem.*, 2023, **25**, 1658–1671.
352. Y. Chen, *et al.*, Sustainable waste-nitrogen upcycling enabled by low-concentration nitrate electrodialysis and high-performance ammonia electrosynthesis, *EES Catal.*, 2023, **1**, 504–515.

353. H. He, *et al.*, Review and perspectives on carbon-based electrocatalysts for the production of H_2O_2 *via* two-electron oxygen reduction, *Green Chem.*, 2023, **25**, 9501–9542.
354. S. Aslam, *et al.*, Electrochemical hydrogen production: sustainable hydrogen economy, *Green Chem.*, 2023, **25**, 9543–9573.
355. Y. Sun, L. Han and P. Strasser, A comparative perspective of electrochemical and photochemical approaches for catalytic H_2O_2 production, *Chem. Soc. Rev.*, 2020, **49**, 6605–6631.
356. E. V. Kondratenko, G. Mul, J. Baltrusaitis, G. O. Larrazábal and J. Pérez-Ramírez, Status and perspectives of CO_2 conversion into fuels and chemicals by catalytic, photocatalytic and electrocatalytic processes, *Energy Environ. Sci.*, 2013, **6**, 3112–3135.
357. P. Lamagni, *et al.*, Restructuring Metal–Organic Frameworks to Nanoscale Bismuth Electrocatalysts for Highly Active and Selective CO_2 Reduction to Formate, *Adv. Funct. Mater.*, 2020, **30**, 1910408.
358. M. Miola, *et al.*, Ligand-free gold nanoparticles supported on mesoporous carbon as electrocatalysts for CO_2 reduction, *J. CO_2 Util.*, 2018, **28**, 50–58.
359. M. F. Philips, G. J. M. Gruter, M. T. M. Koper and K. J. P. Schouten, Optimizing the Electrochemical Reduction of CO_2 to Formate: A State-of-the-Art Analysis, *ACS Sustainable Chem. Eng.*, 2020, **8**, 15430–15444.
360. E. Schuler, M. Demetriou, N. R. Shiju and G. J. M. Gruter, Towards Sustainable Oxalic Acid from CO_2 and Biomass, *ChemSusChem*, 2021, **14**, 3636–3664.
361. M. W. Haynes, R. C. González and M. C. Hatzell, Optimizing geographic locations for electric vehicle battery recycling preprocessing facilities in California, *RSC Sustainability*, 2024, **2**, 377–389.
362. D. L. Thompson, *et al.*, The importance of design in lithium ion battery recycling-a critical review, *Green Chem.*, 2020, **22**, 7585–7603.
363. V. Muralidharan, S. Jayasubramaniyan and H.-W. Lee, Surpassing water-splitting potential in aqueous redox flow batteries: insights from kinetics and thermodynamics, *EES Catal.*, 2024, **2**, 522–544.
364. G. S. Yedase, S. Kumar, J. Stahl, B. König and V. R. Yatham, Cerium-photocatalyzed aerobic oxidation of benzylic alcohols to aldehydes and ketones, *Beilstein J. Org. Chem.*, 2021, **17**, 1727–1732.
365. B. C. Moon, B. Bayarkhuu, K. A. I. Zhang, D. K. Lee and J. Byun, Solar-driven H_2O_2 production *via* cooperative auto- and photocatalytic oxidation in fine-tuned reaction media, *Energy Environ. Sci.*, 2022, **15**, 5082–5092.
366. L. Liu, *et al.*, Linear Conjugated Polymers for Solar-Driven Hydrogen Peroxide Production: The Importance of Catalyst Stability, *J. Am. Chem. Soc.*, 2021, **143**, 19287–19293.
367. R. S. Sprick, *et al.*, Tunable organic photocatalysts for visible-light-driven hydrogen evolution, *J. Am. Chem. Soc.*, 2015, **137**, 3265–3270.

368. W. Zhao, *et al.*, Accelerated Synthesis and Discovery of Covalent Organic Framework Photocatalysts for Hydrogen Peroxide Production, *J. Am. Chem. Soc.*, 2022, **144**, 9902–9909.
369. W. Zhao, *et al.*, Accelerated Synthesis and Discovery of Covalent Organic Framework Photocatalysts for Hydrogen Peroxide Production, *J. Am. Chem. Soc.*, 2022, **144**, 9902–9909.
370. H. Yang, *et al.*, Packing-induced selectivity switching in molecular nanoparticle photocatalysts for hydrogen and hydrogen peroxide production, *Nat. Nanotechnol.*, 2023, **18**, 307–315.
371. T. Freese, J. T. Meijer, B. L. Feringa and S. B. Beil, An organic perspective on photocatalytic production of hydrogen peroxide, *Nat. Catal.*, 2023, **6**, 553–558.
372. R. S. Sprick, *et al.*, Visible-Light-Driven Hydrogen Evolution Using Planarized Conjugated Polymer Photocatalysts, *Angew. Chem., Int. Ed.*, 2016, **55**, 1792–1796.
373. A. Gopakumar, *et al.*, Lignin-Supported Heterogeneous Photocatalyst for the Direct Generation of H_2O_2 from Seawater, *J. Am. Chem. Soc.*, 2022, **144**, 2603–2613.
374. C. K. Prier, D. A. Rankic and D. W. C. MacMillan, Visible light photoredox catalysis with transition metal complexes: Applications in organic synthesis, *Chem. Rev.*, 2013, **113**, 5322–5363.
375. D. A. Nagib and D. W. C. Macmillan, Trifluoromethylation of arenes and heteroarenes by means of photoredox catalysis, *Nature*, 2011, **480**(7376), 224–228.
376. D. A. Nicewicz and D. W. C. MacMillan, Merging photoredox catalysis with organocatalysis: The direct asymmetric alkylation of aldehydes, *Science*, 2008, **322**, 77–80.
377. M. H. Shaw, J. Twilton and D. W. C. MacMillan, Photoredox Catalysis in Organic Chemistry, *J. Org. Chem.*, 2016, **81**, 6898–6926.
378. D. W. C. MacMillan, The advent and development of organocatalysis, *Nature*, 2008, **455**(7211), 304–308.
379. W. A. Velema, W. Szymanski and B. L. Feringa, Photopharmacology: Beyond proof of principle, *J. Am. Chem. Soc.*, 2014, **136**, 2178–2191.
380. K. Kubota, T. Seo, K. Koide, Y. Hasegawa and H. Ito, Olefin-accelerated solid-state C–N cross-coupling reactions using mechanochemistry, *Nat. Commun.*, 2019, **10**(1), 1–11.
381. R. Tedjini, R. Viveiros, T. Casimiro, V. D. B. Bonifácio and B. Bonifácio, Iron-free mechanochemical limonene inverse vulcanization, *RSC Mechanochem.*, 2024, **1**, 176–180.
382. A. Hamilton, Industrial Poisons in the United States, *Nature*, 1925, **116**, 604.
383. A. Hamilton, *Industrial Toxicology*, Harper, Michigan, 1934, vol. 4.
384. R. Carson, *Silent Spring*, Houghton Mifflin, Boston, 1962.
385. T. Colborn, F. S. Vom Saal and A. M. Soto, Developmental effects of endocrine-disrupting chemicals in wildlife and humans, *Environ. Health Perspect.*, 1993, **101**, 378–384.

386. T. Colborn, C. Kwiatkowski, K. Schultz and M. Bachran, Natural Gas Operations from a Public Health Perspective, *Hum. Ecol. Risk Assess.: Int. J.*, 2011, **17**, 1039–1056.
387. L. N. Vandenberg, *et al.*, Hormones and Endocrine-Disrupting Chemicals: Low-Dose Effects and Nonmonotonic Dose Responses, *Endocr. Rev.*, 2012, **33**, 378–455.
388. T. Kletz, What you don't have, can't leak, *Chem. Ind.*, 1978, **287**, 292.
389. T. Kletz, A lifetime spent saving lives, *Chem. Eng.*, 2012, **53**, 56.
390. F. M. Kerton, Y. Liu, K. W. Omari and K. Hawboldt, Green chemistry and the ocean-based biorefinery, *Green Chem.*, 2013, **15**, 860–871.
391. J. H. Clark, *et al.*, Green chemistry and the biorefinery: a partnership for a sustainable future, *Green Chem.*, 2006, **8**, 853–860.
392. E. De Smit and B. M. Weckhuysen, The renaissance of iron-based Fischer–Tropsch synthesis: on the multifaceted catalyst deactivation behaviour, *Chem. Soc. Rev.*, 2008, **37**, 2758–2781.
393. P. Liu, *et al.*, Anthraquinone-catalyzed H_2O_2 electrosynthesis coupled with an advanced oxidation process for water treatment, *RSC Sustainability*, 2024, **2**, 483–490.
394. S. Bagheri, N. Muhd Julkapli and S. Bee Abd Hamid, Titanium dioxide as a catalyst support in heterogeneous catalysis, *Sci. World J.*, 2014, **2014**, 1–21.
395. H. Mirletz, H. Hieslmair, S. Ovaitt, T. L. Curtis and T. M. Barnes, Unfounded concerns about photovoltaic module toxicity and waste are slowing decarbonization, *Nat. Phys.*, 2023, **19**(10), 1376–1378.
396. K. Van Aken, L. Strekowski, L. Patiny and L. Strekowski, EcoScale, a semi-quantitative tool to select an organic preparation based on economical and ecological parameters, *Beilstein J. Org. Chem.*, 2006, **2**, 3.
397. N. Fantozzi, *et al.*, Green metrics in mechanochemistry, *Chem. Soc. Rev.*, 2023, **52**, 6680–6714.
398. M. Poliakoff, P. Licence and M. W. George, A New Approach to Sustainability: A Moore's Law for Chemistry, *Angew. Chem., Int. Ed.*, 2018, **57**, 12590–12591.
399. R. A. Sheldon, Fundamentals of green chemistry: efficiency in reaction design, *Chem. Soc. Rev.*, 2012, **41**, 1437–1451.
400. P. T. Anastas and J. B. Zimmerman, Design through the 12 principles of green engineering, *Environ. Sci. Technol.*, 2003, **37**, 94A–101A.
401. M. A. Abraham and N. Nguyen, "Green engineering: Defining the principles"— resdts from the sandestin conference, *Environ. Prog.*, 2003, **22**, 233–236.
402. C. A. Clark, D. S. Lee, S. J. Pickering, M. Poliakoff and M. W. George, A Simple and Versatile Reactor for Photochemistry, *Org. Process Res. Dev.*, 2016, **20**, 1792–1798.
403. D. Cambié, C. Bottecchia, N. J. W. Straathof, V. Hessel and T. Noël, Applications of Continuous-Flow Photochemistry in Organic Synthesis,

Material Science, and Water Treatment, *Chem. Rev.*, 2016, **116**, 10276–10341.
404. C. Sambiagio and T. Noël, Flow Photochemistry: Shine Some Light on Those Tubes!, *Trends Chem.*, 2020, **2**, 92–106.
405. F. Lévesque and P. H. Seeberger, Highly efficient continuous flow reactions using singlet oxygen as a 'Green' reagent, *Org. Lett.*, 2011, **13**, 5008–5011.
406. D. S. Lee, *et al.*, Continuous Photo-Oxidation in a Vortex Reactor: Efficient Operations Using Air Drawn from the Laboratory, *Org. Process Res. Dev.*, 2017, **21**, 1042–1050.
407. T. Noël, Y. Cao and G. Laudadio, The Fundamentals behind the Use of Flow Reactors in Electrochemistry, *Acc. Chem. Res.*, 2019, **52**, 2858–2869.
408. D. S. Lee, *et al.*, Scalable Continuous Vortex Reactor for Gram to Kilo Scale for UV and Visible Photochemistry, *Org. Process Res. Dev.*, 2020, **24**, 201–206.
409. L. Buglioni, F. Raymenants, A. Slattery, S. D. A. Zondag and T. Noël, Technological Innovations in Photochemistry for Organic Synthesis: Flow Chemistry, High-Throughput Experimentation, Scale-up, and Photoelectrochemistry, *Chem. Rev.*, 2022, **122**(2), 2752–2906.
410. T. Freese, J. T. Meijer, M. Miola, P. P. Pescarmona and B. L. Feringa, Photochemical on-demand production of hydrogen peroxide in a modular flow reactor, *Sustainable Energy Fuels*, 2024, **9**, 141–151.
411. M. D. Edwards, *et al.*, Process Intensification of the Continuous Synthesis of Bio-Derived Monomers for Sustainable Coatings Using a Taylor Vortex Flow Reactor, *Org. Process Res. Dev.*, 2024, **28**(5), 1917–1928.
412. Z. Gang, J. Huan-Feng and C. Ming-Cai, Chemical reactions in supercritical carbon dioxide: from laboratory to commercial plant, *Green Chem.*, 2003, **5**, 99–104.
413. S. Szymkuć, *et al.*, Computer-Assisted Synthetic Planning: The End of the Beginning, *Angew. Chem., Int. Ed.*, 2016, **55**, 5904–5937.
414. S. Steiner, *et al.*, Organic synthesis in a modular robotic system driven by a chemical programming language, *Science*, 2019, **363**, eaav2211.
415. J. M. Granda, L. Donina, V. Dragone, D. L. Long and L. Cronin, Controlling an organic synthesis robot with machine learning to search for new reactivity, *Nature*, 2018, **559**(7714), 377–381.
416. B. Burger, *et al.*, A mobile robotic chemist, *Nature*, 2020, **583**(7815), 237–241.
417. A. Slattery, *et al.*, Automated self-optimization, intensification, and scale-up of photocatalysis in flow, *Science*, 2024, **383**, eadj1817.
418. S. M. Mennen, *et al.*, The Evolution of High-Throughput Experimentation in Pharmaceutical Development and Perspectives on the Future, *Org. Process Res. Dev.*, 2019, **23**, 1213–1242.
419. N. L. Bell, *et al.*, Autonomous execution of highly reactive chemical transformations in the Schlenkputer, *Nat. Chem. Eng.*, 2024, **1**(2), 180–189.

420. M. Kamkar, *et al.*, Artificial Intelligence (AI) for Sustainable Resource Management and Chemical Processes, *ACS Sustainable Resour. Manage.*, 2024, **1**, 178–180.
421. J. García-Martínez and R. M. Hartshorn, Feeding the Planet: The Main Contribution and Challenge of Chemistry, *ACS Agric. Sci. Technol.*, 2023, **3**, 457–459.
422. M. Jovičević-Klug, I. R. Souza Filho, H. Springer, C. Adam and D. Raabe, Green steel from red mud through climate-neutral hydrogen plasma reduction, *Nature*, 2024, **625**(7996), 703–709.
423. J. Gong, *et al.*, Rethinking chemical engineering education, *Nat. Chem. Eng.*, 2024, **1**(2), 127–133.
424. K. B. Aubrecht, M. Bourgeois, E. J. Brush, J. Mackellar and J. E. Wissinger, Integrating Green Chemistry in the Curriculum: Building Student Skills in Systems Thinking, Safety, and Sustainability, *J. Chem. Educ.*, 2019, **96**, 2872–2880.
425. Specialized Manual on Green and Sustainable Chemistry Education and Learning. https://www.unep.org/resources/report/specialized-manual-green-and-sustainable-chemistry-education-and-learning, 2023.
426. F. A. Etzkorn and J. L. Ferguson, Integrating Green Chemistry into Chemistry Education, *Angew. Chem., Int. Ed.*, 2023, **135**, e202209768.
427. ACS Committee on Professional Training, 2023 ACS Guidelines for Undergraduate Chemistry Programs, 2023. https://www.acs.org/education/policies/acs-approval-program/guidelines.html.
428. Royal Society of Chemistry, Accredited Degree Programmes. https://www.rsc.org/membership-and-community/degree-accreditation/.
429. Green Shoots: A Sustainable Chemistry Curriculum for a Sustainable Planet, 2021, www.rsc.org/new-perspectives.
430. *Students,* Sustainability and Education, 2021. https://sos.earth/.
431. Global Green Skills Report, 2023. https://economicgraph.linkedin.com/research/global-green-skills-report.
432. J. J. MacKellar, D. J. C. Constable, M. M. Kirchhoff, J. E. Hutchison and E. Beckman, Toward a Green and Sustainable Chemistry Education Road Map, *J. Chem. Educ.*, 2020, **97**, 2104–2113.
433. A. E. Marteel-Parrish, Teaching green and sustainable chemistry: A revised one-semester course based on inspirations and challenges, *J. Chem. Educ.*, 2014, **91**, 1084–1086.
434. K. Grieger and A. Leontyev, Promoting student awareness of green chemistry principles *via* student-generated presentation videos, *J. Chem. Educ.*, 2020, **97**, 2657–2663.
435. A. P. Dicks, *et al.*, A Systems Thinking Department: Fostering a Culture of Green Chemistry Practice among Students, *J. Chem. Educ.*, 2019, **96**, 2836–2844.
436. L. B. Armstrong, *et al.*, Developing a Green Chemistry Focused General Chemistry Laboratory Curriculum: What Do Students Understand and Value about Green Chemistry?, *J. Chem. Educ.*, 2019, **96**, 2410–2419.

437. S. M. Reed and J. E. Hutchison, Green Chemistry in the Organic Teaching Laboratory: An Environmentally Benign Synthesis of Adipic Acid, *J. Chem. Educ.*, 2000, 77, 1627–1629.
438. V. G. Zuin, *et al.*, Integrating Green and Sustainable Chemistry into Undergraduate Teaching Laboratories: Closing and Assessing the Loop on the Basis of a Citrus Biorefinery Approach for the Biocircular Economy in Brazil, *J. Chem. Educ.*, 2019, **96**, 2975–2983.
439. T. E. Goodwin, An Asymptotic Approach to the Development of a Green Organic Chemistry Laboratory, *J. Chem. Educ.*, 2004, **81**, 1187–1190.
440. The Green Chemistry Commitment|Beyond Benign. https://www.beyondbenign.org/he-green-chemistry-commitment/.
441. A. Cannon, *et al.*, A Promise to a Sustainable Future: 10 Years of the Green Chemistry Commitment at Beyond Benign, *Green Chem.*, 2024, **26**, 6983–6993.
442. A. S. Cannon and I. J. Levy, The Promise of Chemical Education: Addressing our Students' Needs, American Chemical Society, in *ACS Symposium Series*, American Chemical Society, 2015, vol. 1193, pp. 115–125.
443. The Green Chemistry Commitment – Who's Committed|Beyond Benign. https://www.beyondbenign.org/he-whos-committed/.
444. M. S. Prévot, *et al.*, An anthropocene-framed transdisciplinary dialog at the chemistry-energy nexus, *Chem. Sci.*, 2024, **15**, 9054–9086.
445. M. J. Molina and F. S. Rowland, Stratospheric sink for chlorofluoromethanes: chlorine atom-catalysed destruction of ozone, *Nature*, 1974, **249**(5460), 810–812.
446. P. T. Anastas and M. M. Kirchhoff, Origins, Current Status, and Future Challenges of Green Chemistry, *Acc. Chem. Res.*, 2002, **35**, 686–694.
447. K. Betts, *How industrial applications in green chemistry are changing our world*, American Chemical Society, Washington, DC, 2015, https://www.acs.org/content/dam/acsorg/membership/acs/benefits/discovery-reports/green-chemistry-applications.pdf (accessed Jan 02, 2026).
448. J. Golden, R. Handfield, J. Daystar, R. Kronthal-Sacco and J. Tickner, Green Chemistry - A Strong Driver of Innovation, Growth, and Business Opportunity, *Ind. Biotechnol.*, 2021, **17**(6), 311–315.
449. ZDHC, Roadmap to Zero. https://www.roadmaptozero.com/.
450. R. J. Giraud, *et al.*, Implementing green chemistry in chemical manufacturing: A survey report, *ACS Sustainable Chem. Eng.*, 2014, **2**, 2237–2242.
451. Network of Early-Career Sustainable Scientists and Engineers (NESSE). https://www.sustainablescientists.org/.
452. American Chemical Society - Green Chemistry Institute (ACS GCI). https://www.acs.org/greenchemistry.html.
453. Welcome to Your Green Chemistry Teaching & Learning Community | Green Chemistry Teaching and Learning Community (GCTLC). https://gctlc.org/.
454. Green & Sustainable Chemistry Network, Japan. https://www.jaci.or.jp/english/gscn/page_01.html.

455. Green Sciences for Sustainable Development Foundation. https://www.gssd-foundation.org/.
456. Green Chemistry Network Centre (GCNC). https://www.gcnc.in/.
457. Youth Forum on Chemicals Governance 2023|IISD Earth Negotiations Bulletin. https://enb.iisd.org/youth-forum-chemicals-governance-2023.
458. GYA Home - Global Young Academy. https://globalyoungacademy.net/.
459. European Young Chemists' Network - EuChemS. https://www.euchems.eu/divisions/european-young-chemists-network/.
460. International Younger Chemists Network. https://www.iycnglobal.com/.
461. J. L. Vidal and J. Borges, The Global Conversation on Sustainability: An IYCN/IUPAC Joint Effort to Creating a Sustainable Future Worldwide, *Chem. Int.*, 2023, **45**, 10–16.
462. Center for Green Chemistry & Green Engineering at Yale, American Chemical Society & Beyond Benign. Green Chemistry For Sustainability. https://chemistryforsustainability.org/.
463. Search Products that Meet the Safer Choice Standard|US EPA. https://www.epa.gov/saferchoice/products.
464. T. Freese, N. Elzinga, M. Heinemann, M. M. Lerch and B. L. Feringa, The relevance of sustainable laboratory practices, *RSC Sustainability*, 2024, **2**, 1300–1336.
465. S. M. Kernaghan, T. Coady, M. Kinsella and C. M. Lennon, A tutorial review for research laboratories to support the vital path toward inherently sustainable and green synthetic chemistry, *RSC Sustainability*, 2024, **2**, 578–607.
466. M. Poliakoff, P. Licence and M. W. George, UN sustainable development goals: How can sustainable/green chemistry contribute? By doing things differently, *Curr. Opin. Green Sustainable Chem.*, 2018, **13**, 146–149.
467. C. Woolston, The trials and triumphs of sustainable science, *Nature*, 2024, **633**, S69–S71.
468. Royal Society of Chemistry, *Sustainable Laboratories: A Community-Wide Movement toward Sustainable Laboratory Practices*, 2022. https://www.rsc.org/policy-evidence-campaigns/environmental-sustainability/sustainability-reports-surveys-and-campaigns/sustainable-laboratories/.
469. P. Smith *et al.*, Wellcome Report: Advancing Environmentally Sustainable Health Research, 2023. https://wellcome.org/reports/advancing-environmentally-sustainable-health-research.
470. N. Borgermann, A. Schmidt and J. Dobbelaere, Preaching water while drinking wine: Why universities must boost climate action now, *One Earth*, 2022, **5**, 18–21.
471. J. Dobbelaere, J. B. Heidelberger and N. Borgermann, Achieving sustainable transformation in science - green grassroots groups need nurturing from the top, *J. Cell Sci.*, 2022, **135**, jcs259645.
472. N. Winter, *et al.*, The paradox of the life sciences, *EMBO Rep.*, 2023, **24**, e56683.

473. J. Durgan, M. Rodríguez-Martínez and B. Rouse, Green Labs: a guide to developing sustainable science in your organization, *Immunol. Cell Biol.*, 2023, **101**, 289–301.
474. I. Ragazzi Id, M. Farley, K. J. Id and I. B. Id, Using life cycle assessments to guide reduction in the carbon footprint of single-use lab consumables, *PLOS Sustainability Transform.*, 2023, **2**, e0000080.
475. M. Farley and B. P. Nicolet, Re-use of laboratory utensils reduces CO_2 equivalent footprint and running costs, *PLoS One*, 2023, **18**, 1–11.
476. M. Farley, How green is your science? The race to make laboratories sustainable, *Nat. Rev. Mol. Cell Biol.*, 2022, **23**, 517.
477. M. Luck and M. Farley, Good science shouldn't cost the earth, *Res. Prof. News*, 2023, https://www.researchprofessionalnews.com/rr-news-uk-views-of-the-uk-2023-3-good-science-shouldn-t-cost-the-earth/ (accessed Jan 02, 2026).
478. A. Nabera, I. R. Istrate, A. J. Martín, J. Pérez-Ramírez and G. Guillén-Gosálbez, Energy crisis in Europe enhances the sustainability of green chemicals, *Green Chem.*, 2023, **25**, 6603–6611.
479. A. Klerke, C. H. Christensen, J. K. Nørskov and T. Vegge, Ammonia for hydrogen storage: Challenges and opportunities, *J. Mater. Chem.*, 2008, **18**, 2304–2310.
480. S. C. D'Angelo, *et al.*, Environmental and economic potential of decentralised electrocatalytic ammonia synthesis powered by solar energy, *Energy Environ. Sci.*, 2023, **16**, 3314–3330.
481. R. Dawson, E. Stöckel, J. R. Holst, D. J. Adams and A. I. Cooper, Microporous organic polymers for carbon dioxide capture, *Energy Environ. Sci.*, 2011, **4**, 4239–4245.
482. N. MacDowell, *et al.*, An overview of CO_2 capture technologies, *Energy Environ. Sci.*, 2010, **3**, 1645–1669.
483. P. Lozano and E. García-Verdugo, From green to circular chemistry paved by biocatalysis, *Green Chem.*, 2023, **25**, 7041–7057.
484. U.S. Environmental Protection Agency|US EPA. https://www.epa.gov/.
485. J. A. Linthorst, An overview: Origins and development of green chemistry, *Found. Chem.*, 2010, **12**, 55–68.
486. J. A. Linthorst, Research between science, society and politics: the history and scientific development of green chemistry, Maastricht University, 2023, DOI: 10.26481/DIS.20230209JL.
487. P. T. Anastas and J. B. Zimmerman, The periodic table of the elements of green and sustainable chemistry, *Green Chem.*, 2019, **21**, 6545–6566.
488. Chemical Leasing – A Performance-Based Business Model. https://chemicalleasing.com/.
489. A. Cannon, *et al.*, An actionable definition and criteria for "sustainable chemistry" based on literature review and a global multisectoral stakeholder working group, *RSC Sustainability*, 2023, **1**, 2092–2106.

490. P. Andre, T. Boneva, F. Chopra and A. Falk, Globally representative evidence on the actual and perceived support for climate action, *Nat. Clim. Change*, 2024, **14**, 253–259.
491. P. T. Anastas and J. B. Zimmerman, The molecular basis of sustainability, *Chem*, 2016, **1**, 10–12.
492. W. Arts, *et al.*, Feasibility of wood as a renewable carbon feedstock for the production of chemicals in Europe, *Biofuels, Bioprod. Biorefin.*, 2024, **18**(2), 365–377.
493. M. K. M. Lane, *et al.*, Green chemistry as just chemistry, *Nat. Sustainability*, 2023, **6**, 502–512.

CHAPTER 4

Estimating and Reducing the Carbon Footprint of Academic Travel[†]

TEUN BOUSEMA*[a] AND KATE WHITFIELD[b]

[a] Department of Medical Microbiology, Radboud University Medical Center, Nijmegen, The Netherlands; [b] Zeroverse, Okinawa, Japan
*Email: teun.bousema@radboudumc.nl

Academic institutions that estimate their carbon footprint commonly observe that air travel is a large contributor to emissions.[1] Work-related travel of employees and students is part of an academic organization's carbon footprint and categorized as Scope 3 emissions (Scope 1 being "direct emissions" from sources that are owned or controlled by the organization; Scope 2 being emissions associated with purchased electricity; Scope 3 being everything else, including emissions associated with purchases, work-related travel, *etc.*). As an example, air travel was responsible for 12–27% of the carbon footprint of seven universities in The Netherlands.[2] These estimates corroborate results of a multi-country carbon-footprinting effort where work-related travel accounted for up to 20% of the carbon footprint of universities in Germany, Australia and the United States.[1] The large variation in the importance of travel for the total carbon footprint of institutions will in part be due to missing data, differences in calculations and the precision with which different sources of emissions can be estimated.[1] In addition, the

The Sustainability of Science: How Science Impacts the Environment, and What Can Be Done
Edited by Martin Farley

Published by the Royal Society of Chemistry, www.rsc.org

nature of academic activities varies between institutions with some involving more travel than others. An institution of higher education that focuses on studying and advocating global health estimated that the majority of its total carbon footprint was due to air travel.[3] Despite considerable imprecision and heterogeneity, it is evident that air travel is an important contributor to carbon emissions of institutes of higher education[1–3] and of individual research activities.[4] Universities can cut emissions from energy use, for instance by insulating buildings, adopting energy efficient policies for data storage and lab practices, and switching to renewable and sustainable energy sources. Universities can also cut emissions from the products and services they purchase, for example by prioritizing plant-based food and beverages, and repairing and recycling equipment. These infrastructural and behavior change interventions are critical and in some cases are gaining momentum. However, it is also evident that air travel cannot be ignored as a source of greenhouse gas emissions in academic settings.

4.1 Measuring Greenhouse Gas Emissions Associated with Air Travel

4.1.1 Considerations on What Emissions to Include and Uncertainties in Estimates

When quantifying emissions associated with air travel, there are several things to consider. The most fundamental is probably what emissions are incorporated in travel-associated estimations. There is broad consensus that aviation contributes more to climate change than simply the CO_2 emissions from burning fuels.[5] The release of gases and participles into the upper troposphere (8–18 km altitude) and lower stratosphere (10–50 km altitude) impacts atmospheric composition through three processes, each with global warming potential. Firstly, there are direct emissions of CO_2 and water vapor, both radiatively active substances, whose impact on global warming is well characterized. Secondly, aviation results in the emission of chemical species that alter radiatively active substances such as NO_x which modifies ozone (O_3) and methane (CH_4) concentrations that in turn impact CO_2 and water vapor formation and heat retention. Thirdly, aviation emits substances that trigger the generation of aerosol particles that can have a warming effect through inducing changes in natural clouds that can trap heat.[5] Whilst cloud formation can have both a cooling and a warming effect, the latter dominates. Contrails of airplanes, for instance, can lead to the formation of cirrus clouds that have a significant net warming effect.[6] Whilst the global warming potential of the second and third processes are not as well quantified as direct CO_2 emission effects, they are highly relevant and best estimates of CO_2 equivalent emissions (CO_2e) of the three processes combined are 1.9–3 times that of CO_2 alone.[7] Estimators that ignore these impactful non-CO_2 effects, such as those presented on websites of some airlines as part of CO_2 compensation schemes,[8] are incomplete and 'optimistic' about the climate impact of air-travel. Other uncertainties that are relevant for estimating flight-associated emissions are the type (and thereby efficiency) of the airplane, occupancy rate, exact flight route, and weather conditions such as wind force and direction. Whilst relevant variables, these factors are evened out in estimators that are based on average conditions.

4.1.2 Approaches and Tools to Calculate the Carbon Footprint of Travel

There are numerous calculations in scientific articles on the carbon footprint associated with academic travel.[7,9–15] Many of these present the footprint of academic conferences. The most laborious estimates identified for each conference participant the reported or estimated airport of departure, their most likely travel route including connecting flights based on websites offering airline tickets, and the distance associated with each individual flight based on a geodesic (great circle) path.[10,13] A strength of this approach is that it allows incorporation of the differences in emissions between short haul flights, that are relatively inefficient due to excess fuel use during take-off, and more fuel efficient long-haul flights. Travel kilometers can be converted to emissions using established methods that incorporate or ignore non-CO_2 effects (Figure 4.1).

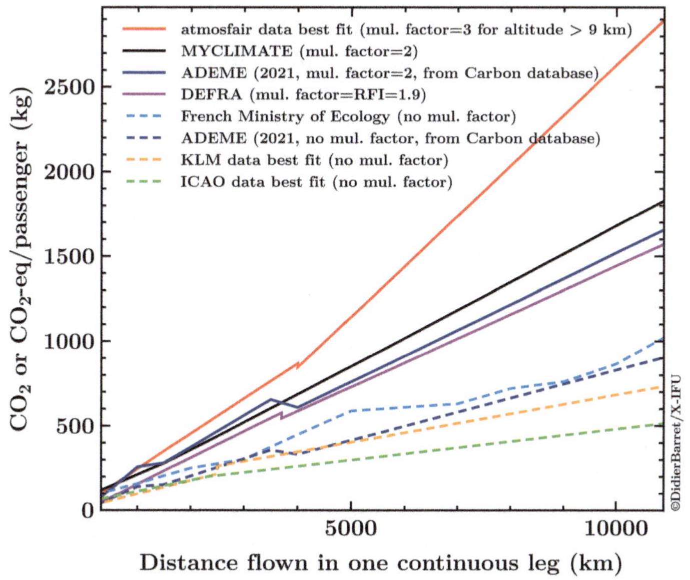

Figure 4.1 Emission factors as a function of the distance flown. Emission factors are given in kilograms as dashed lines (CO_2) or solid lines (CO_2e). Travel is given per kilometer; calculations assume economy class travel and one continuous leg. Seven calculation methods are presented. Reproduced from ref. 7 with permission from Springer Nature, Copyright 2020.

To mitigate the labor intensiveness of this approach and make these calculations more accessible, online calculators have been generated that allow users to simply upload an excel or csv file with the cities of departure and the destinations.[7,13] These online calculators use the same geodesic path and CO_2e calculations but ignore the fact that some journeys consist of multiple travel legs. Nevertheless, the differences between these approaches are generally small and not too relevant when a large number of flights is considered. The advantage of online tools for rapidly assessing travel-associated carbon emissions includes the option to easily explore different scenarios. One online planner for instance allows setting different thresholds for travel by train,[7] another allows high resolution maps with travel routes to be generated and downloaded.[13] This rapid generation of maps and associated summary measures can for instance be used when a choice has to be made between meeting locations and where the consequences of different options for travel-associated emissions can be used in decision-making.

4.2 The Carbon Footprint of Academic Activities

4.2.1 Examples of Conferences and a Rule of Thumb Estimate

Academic conferences are probably the largest meetings of researchers and have been the focus of several efforts to estimate travel-associated emissions. The vast majority of carbon emissions associated with conferences are due to air travel. Catering, ground transportation and accommodation together account for 4–10% of total emissions, dwarfed by emissions associated with international travel.[14,16] To provide one example of a relatively large conference with a wealth of associated data, we present the 2019 edition of the annual conference of the European Society of Clinical Microbiology and Infectious Diseases that welcomed 13 663 participants in Amsterdam (Figure 4.2).[13] Together, conference participants from six continents traveled a total of 85.9 million kilometers. When expressed as return trips to the moon, with an average distance of 384 400 km,[17] this total distance is equivalent to 113 return trips to the moon. Using a calculator that included non-CO_2 emissions, this travel was estimated to result in 16 558 tons CO_2e emitted.[13] For most audiences, carbon estimates are difficult to interpret. When expressed relative to the annual emissions of average European households, the conference travel emissions were equivalent to the annual emissions of 1160 European households. There have also been analyses on the amount of Arctic ice that is lost for each ton of CO_2 emitted, estimating that each ton equates to 3 m^2 of Arctic ice lost.[18] When using this estimate, the conference emissions would result in the loss of 51 139 m^2 of Arctic ice, a surface area equivalent to approximately 7 soccer pitches. These extrapolations have to be used with caution since they come with uncertainty but are more likely to capture the imagination than tons of CO_2.

In addition to the environmental arguments for thoughtful travel, it can be argued that travel is not the most efficient way of spending time that, for

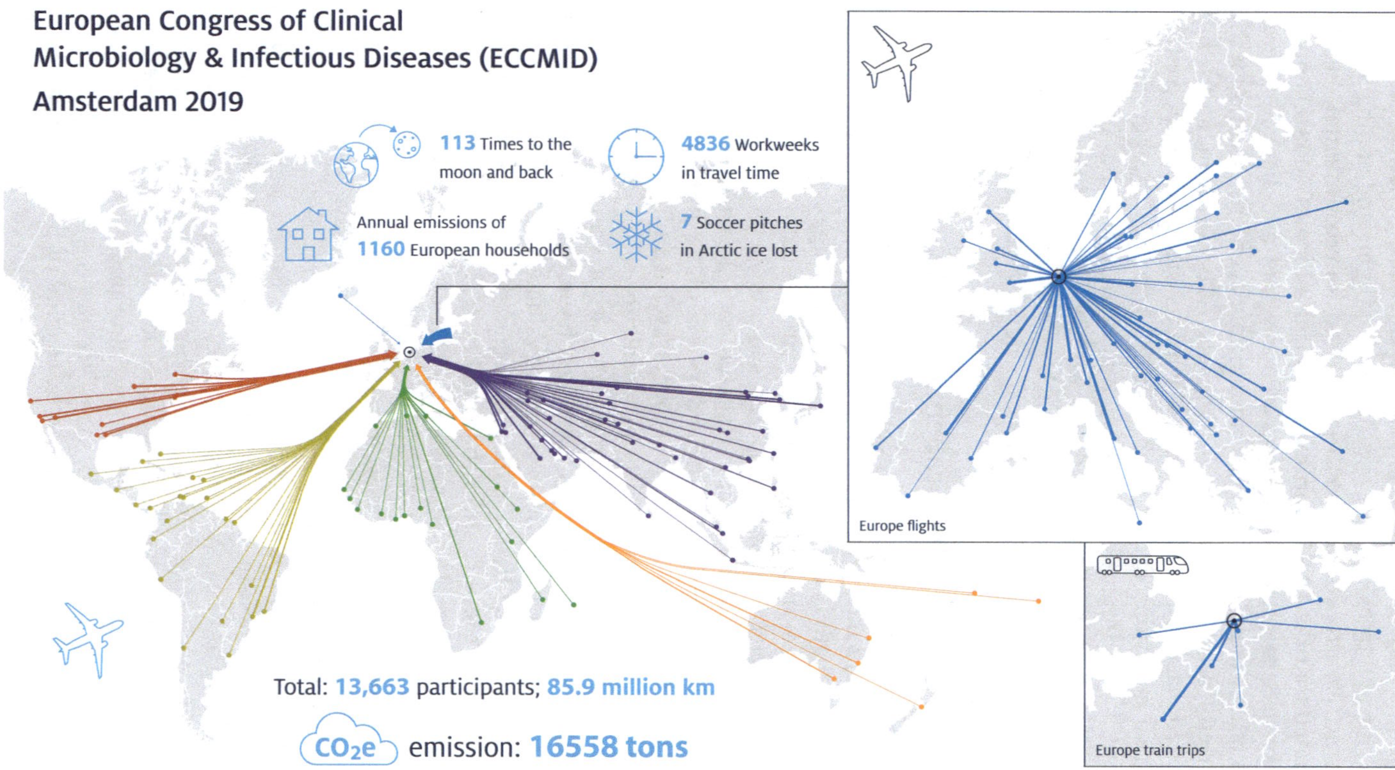

Figure 4.2 The travel carbon footprint of the 2019 European Congress of Clinical Microbiology & Infectious Diseases (ECCMID). The thickness of the line reflects the number of journeys, and colors indicate the region of origin. Details on the calculation are in the text of this chapter and in the original article.[13] Reproduced from ref. 13 with permission from Elsevier, Copyright 2024.

the participants of the above-mentioned conference, could have been used for research, patient care or teaching. When calculating the time investment related to air travel by calculating the duration of flights and adding a realistic waiting time at airports for continental and intercontinental flights, it was estimated that the 13 663 participants spent a total of ~193 000 hours or 4836 workweeks to travel to the conference venue and back.[13] There are numerous other examples of the carbon footprint of conferences and meetings, some of meetings as large as 52 000 participants.[15] The common finding here is that international conferences with participants from multiple continents typically have emissions in the range of 0.5–2 tons of CO_2e per conference participant.[10,13,15,19] In an opinion piece, this estimate was put in perspective indicating that if 7.8 million researchers worldwide each travel to one international conference per year, the emissions (~8 million tons of CO_2e per year) would be larger than the total carbon footprint of some nations.[19]

4.2.2 Other Reasons for Academic Travel

Whilst conferences are obvious examples of 'carbon-intensive' academic activities, there are more reasons for academic travel. One activity that has sparked some debate is the way by which funders organize grant reviewing processes. As an example, we present the carbon footprint associated with the process of selecting proposals for the European Research Council Starting Grants for early career researchers (ERC-StG; ~400 laureates each year). In 2019, applicants traveled a total of 1.8 million km for their 30-minute interviews while panel members traveled a total of 2.9 million km for two meetings, the first to select proposals for a full evaluation and the second for the actual selection of laureates. Carbon emissions combined to 1419 tons of CO_2e in a single year.[9] During the years of COVID-19 associated travel restrictions, ERC-StG proposals were evaluated by virtual means. Once travel restrictions were lifted, the old modus operandi of in-person meetings was resumed with the exception that pitches and interviews of applicants were now done by virtual means, saving ~33% of emissions as well as substantial travel funds and time. There are numerous similar funding meetings where in-person meetings were temporarily replaced by eco-friendly virtual meetings during the COVID-19 pandemic, but have become in-person again subsequently. An unbiased and well-considered decision is important in these meetings; fellowships can be decisive for the careers of (young) researchers. This importance is sometimes used as an argument for in-person panel discussions. At the same time, there is no evidence that the wrong decisions were made during the virtual panel meetings. In fact, several studies suggest that virtual meetings may allow for less biased[20] and more efficient interaction,[21] especially in culturally diverse teams.[22] Whilst in-person meetings are superior for creative processes (as described in some detail below),[23,24] decision-making can be as good or potentially superior by virtual means.[23] This argues that virtual meetings can be considered in decision making processes like grant reviewing processes.

4.3 Approaches to Reduce Travel-associated Carbon Emissions

As highlighted in the travel policy of the Wellcome Trust, a major funder of research in the UK, there are three pillars in reducing the environmental impact of travel: (i) minimizing the number of journeys taken by using (virtual) alternatives where possible; (ii) choosing travel that has a lower carbon impact, where practical; and (iii) offsetting carbon emissions of journeys that are still made.[25] We will discuss the first and last pillar in some detail. Regarding low carbon travel, there is an increasing number of institutions in Europe that impose a minimum distance required to fly and provide sustainable travel advice to promote low carbon travel.[2] Some tools to measure travel-related emissions and facilitate greener travel are listed in Box 4.1. Promoting travel by train or car for short distances is a feasible measure in Europe and other parts of the world where train connections are reasonably good. For destinations that can be reached within 7 or 8 hours by train, this mode of travel is mandatory at some European universities and recommended at others[2] although exceptions are regularly permitted. In addition to the

Box 4.1 Tools to measure travel-related emissions and encourage lower carbon modes of travel.

Travel emissions calculator

Estimates the total emissions between a point of origin and destination for one or multiple travelers. https://travel-footprint-calculator.irap.omp.eu/home.html

Travel emissions calculator and map

Estimates the total emissions between a point of origin and destination for one or multiple travelers and generates a travel map and data summary. https://bousema-lab.shinyapps.io/travel_calculator/

Greenhouse gas emissions converter

Estimates the total emissions between a point of origin and destination and compares different modes of transport. https://www.epa.gov/energy/greenhouse-gas-equivalencies-calculator

Global train travel guide – The Man in Seat 61

Provides information about train routes, train companies and how to purchase tickets. https://www.seat61.com/

Train journey planner for students – Go2Rail

Helps students to plan their European travel itineraries so that they can meet up on route and travel together. https://www.go2rail.eu/

inefficient train networks in large parts of the world, another relevant limitation of this measure is that long-haul flights often outweight short-haul flights. Over 60% of flights by employees of Wageningen University in The Netherlands were intercontinental and long-haul flights account for a much larger distance traveled (84% of all flight kilometers at the University of Groningen came from long-haul flights).[2] Whilst meaningful reductions can be achieved by low-carbon travel, this is unlikely to affect the majority of travel-associated carbon emissions. That brings us to the other two pillars.

4.4 Virtual Alternatives to In-person Meetings

Virtual conferencing has existed for many decades. Already in 1916, a virtual meeting of the American Institute of Electrical Engineers was organized. At that time, 5000 attendees across four time zones called in for a meeting that was considered a triumph of engineering and that was covered in major newspapers. Despite this early success, it took another century for most academic societies to be attracted to, or perhaps to be forced to, virtual conferencing. During the period when COVID-19 related travel restrictions were in place, there was a steep rise in the number of virtual conferences that were organized.[19,26,27] It was quickly acknowledged that virtual conferences can be an environmentally friendly alternative to in-person meetings. It should be noted that virtual meetings are also not carbon neutral. The major driver of the ecological impact of virtual meetings is the use of electricity associated with servers for streaming content and computers that are used by meeting participants.[11] There have been several studies that calculated the carbon footprint of large virtual meetings in a direct comparison with in-person meetings. These studies conclude that virtual meetings have a 1000–6000-fold lower carbon footprint per meeting participant.[9–11,13,28] The wide range is partially explained by variation in the electricity mix of different countries. CO_2 emissions per kW h of electricity are considerably lower in countries where renewable energy sources are prominent, compared to those that currently still rely on fossil fuels for electricity.[4] With the rapid increase in renewable energy across the globe, it is very likely that the gap between the footprints of in-person and virtual meetings will become even larger in the near future. In addition, virtual conferences typically attract a broader and more diverse population that may not have attended meetings for financial reasons, difficulties in obtaining visas and care duties.[14,27,29]

Whilst the environmental sustainability and accessibility benefits of virtual meetings are beyond dispute, in-person events are often considered more effective and enjoyable. Many academic societies that organize conferences explicitly advocate these conferences as prime opportunity to foster a sense of belonging to the society;[13,30] it is difficult to translate this feeling to a virtual format. On a social-level, in-person meetings with dinner and drinks and unexpected meetings are a great way to meet colleagues and allows building of friendships and relationships. Dr Anthony Fauci, former director of the National Institute of Allergy and Infectious Diseases (NIAID) at the National Institutes of Health (NIH) in the USA, commented that his favorite Keystone Symposia meeting was in Utah, USA in the early 1980s that he attended with his soon-to-be wife Dr Christine Grady.[31] At in-person meetings, participants can feel a sense of community, feel visible and acknowledged. Participants are able to participate in discussions and debates that are shaping their professional fields and contributions are recognized and maybe even celebrated. Where in-person conferences thus allow for both information sharing and social aspects, virtual conferences tend to primarily focus on information sharing and have largely failed to create a sense of community. Online formats that focus on sharing information where a few presenters broadcast their work to a passive audience fail to deliver on the social benefits. Whilst some elements of in-person meetings may be nearly impossible to cater for in online events, there are ample opportunities to innovate.

4.4.1 Building Community Online

To create interesting virtual meetings that participants enjoy, organizers need to build an engaged community. Some of the thoughts and recommendations below are already used in practice (but not published on) or are inspired by the book *Citizens* by Jon Alexander and Ariane Conrad.[32] Online conferences can improve to 'build community' and include multiple ways to engage participants. Here, it is important to consider that everyone is an active participant, not only invited speakers. Providing plenty of modes of participation for the community to get engaged during the virtual event is important. We list several examples. One interesting approach is to make presentations available several weeks prior to the event. Participants can then be prompted and encouraged to leave a comment or question for the presenters and incorporate this feedback into live discussion sessions in the main program; gatherings of Astronomers for Planet Earth (A4E) use YouTube for this. Organizers can also help people to connect and get to know one another. As very easy example, it is possible for registrants to introduce themselves using a digital sticky note or a video message that they post in the virtual conference space. At the beginning of a session, it is good practice to ask participants to say hello and share where they are connecting from in a chat box. During the conference, it is possible to explicitly value online contributions. One approach is to celebrate participants' contributions with gold, silver, and bronze medals voted on by fellow participants. For example, excellent presentations, online interaction, innovative work, and supportive mentorship can be awarded.

There are also many innovative virtual networking platforms that are specifically designed to humanize the online experience that was lacking in most of the hastily organized virtual conferences during the COVID-19 pandemic. Currently available innovations include 3D virtual conference environments with exhibiter boots, animated avatars for participants, games that are embedded in the social interaction space, spatial connect options and virtual tables where participants can join and benefit from private audio–video interaction. Developments are fast and opportunities for better online interaction and networking are improving rapidly. At the same time, younger generations are entering the academic playing field who grew up with a vibrant online social life; it is likely that these future academic leaders will be better adapted in using and enjoying virtual modes of social interaction.

4.4.2 Preparing a Virtual Event

When preparing a fully virtual or hybrid online and in-person event, it is important to make serious (time) investment. Whilst smaller events can sometimes be effectively organized using technologies that are already used in daily academic practices (*e.g.* Zoom, WebEx), professional conferencing platforms can improve the experience of (larger) meetings. It is beyond the scope of this chapter to review conferencing platforms. Heterogeneity in digital literacy of conference participants makes it important to help less adapted participants to navigate the online options. When preparing for an event, it is also important to consider the needs of the targeted community and provide fora that help address them. If a new technology or method is important in the field – work with experts to host online training. If professional development is a critical topic – create a booking platform for networking that runs throughout the program. For these networking events, that can take the form of online professional speed-dating, it can be valuable to ask leaders in the field to commit to a number of slots where they are available to meet with fellow participants. Learning about new opportunities is an important draw for people to attend conferences. Here, it is recommended to create an opportunities board in your virtual space. Before the event, one can ask funders, team leaders, editors, and research institutions to post information about forthcoming grants, job opportunities, calls for papers, visiting programs, *etc.*

It is also important to embrace some of the benefits or unique opportunities that virtual platforms bring. Using a voting option allows moderators to let the audience vote for questions to be discussed, improving the democratic element in academic discussions that can otherwise be dominated by the most vocal attendants. In addition, it is possible to let online sessions be followed by virtual breakout rooms where participants can continue the discussion with each other and a presenter. These small but valuable components can add great value to online gatherings. Given that virtual and in-person conferences will provide a different experience to users, it is important to explicitly celebrate the achievements and benefits of a successful virtual meeting (*e.g.* the number of participants and countries attending, the carbon emissions averted). This content works well on social media and can raise awareness about the event.

4.4.3 Practical Challenges with Virtual Events

Time zones are sometimes mentioned as a challenge in virtual conferences. Whilst virtual meetings are free of nighttime travel and jetlag, managing time zones for a virtual event is highly important. Meeting organizers are accommodating time zones in various effective ways.

- The American Society of Tropical Medicine and Hygiene (ASTMH) Green Task Force host a monthly meeting, which alternates between Asia-friendly and Africa-friendly times. Each meeting is recorded and a transcript is generated and shared with all members.
- Climate Action Network in International Education (CANIE) and Astronomers for Planet Earth (A4E) organize their annual virtual congresses by scheduling each session twice during the program. Ensuring that most participants can attend during (or close to) normal working hours.
- A4E allows participants to watch presentations at any time, 2 weeks prior to the scheduled program. This allows participants to manage their own time and allows the actual meeting to focus on discussion rather than broadcasting presentations.
- Tufts University and London School of Tropical Medicine and Hygiene hosted a hybrid meeting in 2 different locations simultaneously. The keynote talks were held at a time that was convenient for both locations and discussion involved participants in both locations. The social events and other sessions took place locally, somewhat like the example of Chicago on Schier (see Box 4.2 below).

Another challenge that is regularly mentioned in relation to virtual events is the difficulty to focus and truly commit to the event. Multi-tasking by attending a conference and at the same time continue other daily activities is a threat to a valuable conference experience. Workplaces and team leaders should adopt a policy whereby staff participating in virtual conferences (whether scheduled during normal working hours or during the night) take time away from their normal work duties. They are effectively 'out of the office' as if they had travelled to attend a conference. This allows participants to focus on the virtual event.

4.4.4 Hybrid Events

In addition to fully virtual meetings, there is of course also the option of hybrid events. Since the forced online conferences during the COVID-19 pandemic, many (large) conferences have retained the option to attend virtually. Whilst this is laudable and allows eco-minded participants, as well as those with care (or other) duties to attend virtually, it also brings challenges. There is a risk of feeling like a second-rate participant if discussions are dominated by those present in person; an elegant option that is adopted by meetings of the European Centers for Disease Control (ECDC) and of Coefficient Giving, is to only discuss questions that are posted online. The conference illustrated in Figure 4.1 also allows for online participation. Possibly related to the fact that conference fees are nearly identical for online

Box 4.2 Inspiring examples.

A low-carbon international conference

Whilst many top-quality medical conferences are international, much of the daily network of participants is national or regional. With that in mind, *Chicago on Schier* has been organized since 2012 as local hub on one of the Dutch islands. Each year, approximately 60 Dutch oncologists now choose to visit Schiermonnikoog instead of Chicago for the annual meeting of the American Society of Clinical Oncology (ASCO). The 4-day program on Schiermonnikoog lags one day behind the conference schedule in the US. In this manner, all content can be viewed on demand despite the time difference. The participants attend a plenary 'virtual ASCO', where the program committee makes a selection of sessions. In addition, there is room for interpretation, moderated in-depth discussion with (national) colleagues and a live connection with oncologists in Chicago. The scientific content is complemented with a social program in the evening to foster the conference feeling and allow for networking. The formula thus achieves both the knowledge dissemination and the networking objectives of a conference while avoiding jetlag and the large carbon footprint associated with international air travel.

A prize for the research group that embraces Thoughtful Academic Travel

The Department of Cognitive Neuroscience of the Donders Institute in Nijmegen, the Netherlands has taken up the challenge of reducing air travel while retaining a strong international focus. Since 2022, the department undertakes concrete efforts to reduce air travel for the many successful research groups it hosts. Its impact is monitored by annual reporting of work-related travels, including the destination, the staff member who traveled and the means of transportation. This registration serves as an entry for the Sustainability Team Award, with the winning group receiving €10 000 for their lab budget. The assessment values virtual conference attendance, green transportation choices, flight equity across career stages, and improvements in emission reductions over time. The focus of this competition is not on 'no travel' but on 'thoughtful travel'. In addition, annual seminars are organized to raise awareness and encourage sustainable travel behavior. In the future, the group aims to introduce a flight tax for an institutional Sustainability Fund that will be used to support local green initiatives.

Social and low-carbon travel

The Climate Action Network for International Educators (CANIE) collaborates with conference organizers and engages participants to reduce the carbon footprint of conferences.[52] In 2022, CANIE members initiated the 'Travel With CANIE' campaign to encourage colleagues to use low carbon modes of transport to travel to the Forum on Education Abroad in Milan, Italy. CANIE successfully made this behavior change initiative fun, attractive, and social for participants, which are key elements to engaging people in changing habits.[53] CANIE set an emissions reduction target and asked participants to pledge their commitment with an online registration form. Using the hashtag #TravelWithCANIE they leveraged social media to engage their community, track, and report on progress against their target. Low carbon travelers posted updates and their smiling faces as they boarded trains and meet with colleagues along the way. Each year, the initiative is becoming more popular. Various conference organizers promote the effort on their website and low carbon travelers are celebrated at participating conferences. At the 2022 European Association for International Education conference, CANIE was bestowed with the President's Award. In 2024, the year of the Olympics, CANIE used social media to create their own winners' podium and awarded medals to the countries with participants making the greatest emissions savings. For two consecutive years, the campaign has exceeded their emissions reductions targets and the CANIE team set up a poll on social media to ask their community what their new, more ambitious target should be. CANIE's approach is an inspiring example of how collaborating with key partners and celebrating low carbon choices can be very successful.

and in-person attendance, the online option is not very popular and accounts for only 6–8% of participants.[13] Whilst hybrid options can theoretically greatly reduce the carbon footprint of academic conferences,[10,33] there is room for improvement to make them truly attractive.

4.5 When Flying, is Carbon Offsetting an Easy Fix?

The third pillar of the Wellcome Trust travel policy is offsetting carbon emissions. One of the most common approaches to limit the climate impact of academic travel is indeed to compensate for CO_2 emissions through offsetting schemes. Whilst considered by some to be part of the solution and sometimes advocated or facilitated by meeting organizers to make meetings more environmentally sustainable,[7,13] the approach also meets considerable criticism. Three commonly mentioned points of criticism are that (i) prices that are used in offsetting schemes are unrealistically low; (ii) the achieved emission reductions are often much lower than those promised; and (iii) offsetting may create a situation where those who can afford to offset their emissions are no longer incentivized to reduce them.

So what is a realistic price for offsetting carbon emissions? Typical schemes that adhere to the Gold Standard, a Switzerland-based organization with relatively high standards in certified emission reductions, offer offsets in the range of €6–25 per ton of CO_2 emitted.[34,35] Assuming that these schemes achieve the promised emission reductions, some argue that this price is unrealistically low. The 2050 Foundation that investigates the true price of compensated emissions, argues that low-cost offsets are problematic. Low-cost offset programs are typically focused on avoiding additional emissions or actions that in the long run may remove CO_2 from the atmosphere such as planting trees or preventing forest loss. These offsets are much cheaper than the cost involved in directly neutralizing current emissions. The price of direct air capture and storage of CO_2, for instance, is in the range of €100–350 per ton CO_2[36] and requires considerable infrastructure to be used at scale. There is also another ethical challenge in depending on cheap offsets that assume future benefits. Using cheap offsets today will deplete their availability for future generations and will make carbon offsetting in the future increasingly expensive. One may still argue that offsets are better than nothing, provided they work and do not incentivize more air travel.

4.5.1 Do Carbon Offsets Deliver on their Promise?

As indicated above, a common carbon offsetting mechanism is based on reducing emissions from deforestation and forest degradation. The central concept is that improved monitoring, conservation and management of forested areas can remove CO_2 from the atmosphere or prevent its emission into the atmosphere. There is limited rigorous evidence for the performance of such schemes[37–39] and the robustness with which the impact of conservation efforts can be quantified is debatable. This quantification relies on scenarios and projections of deforestation and forest degradation in the absence of conservation efforts. In a recent study, the plausible impact of conservation efforts was estimated by comparing conservation areas with control areas that initially had similar characteristics and deforestation pressure.[40] From 26 analyzed project sites, the majority did not show significant improvements in conservation outcomes compared to control areas and for those where there was a beneficial impact, this impact was typically <10% of what was projected and on which offsets were based.[40] This sobering performance has repeatedly been demonstrated,[37–39] although there are also positive exceptions where impact was achieved.[40] There are several possible reasons for this poor performance of most carbon-funded forest conservation projects, including unrealistic (exaggerated) projections of deforestation in the absence of conservation and struggles to effectively implement conservation efforts. Directly planting trees would in some ways be better, or at least more direct, than preventing deforestation. Whilst there are high failure rates with tree-planting projects[41] and there are political and perhaps ethical challenges in planting trees in low-resource settings to offset carbon emissions from wealthy nations, the future will have to learn whether this type of offsetting can be improved to deliver on its promise.[42]

Another common approach is the promotion and distribution of energy efficient cooking stoves. With 2.4 billion people worldwide who rely on kerosene or smoky solid fuels like wood or charcoal, cooking may be responsible for 2% of global greenhouse gas emissions[43] and is thus a logical area for sustainable improvement. Efficient stoves can reduce emissions through better combustion and, by their efficiency, require less fuel for the same heat. Most of the carbon offset initiatives that involve energy efficient cooking stoves aim to achieve impact by replacing stone fires or other inefficient stoves with improved firewood stoves. Only a minority of projects distribute stoves that rely on solar energy, electricity, gas, ethanol or certain pellets that reduce smoke formation sufficiently to be labeled clean by the World Health Organization (WHO).[44,45] In an analysis of projects in 15 countries, Gill-Wiehl and colleagues quantified the usage of new stoves, the amount of cooking fuel that was used before and after obtaining the new stoves and the energy efficiency of the (new) cooking fuel. They conservatively estimated that the cookstove projects over-credited their impact approximately 9-fold. This was in part due to a net increase in overall cooking fuel consumption; the improved stoves made cooking easier and provided an extra (rather than replacement) of the original inefficient stove.[45] The authors indicate that it is important that offsetting projects that rely on energy efficient cooking stoves rely on stoves recommended by WHO.[45] At present, it is prudent to exert caution when interpreting the benefits and CO_2 emission reductions that are claimed by carbon offsetting schemes that promote and distribute energy efficient stoves.

4.5.2 Carbon Offsets as a 'Get Out of Jail for Free' Card

Carbon offsets may be part of thoughtful travel, especially if realistic prices are used for offsets. There is, however, a concern that the feeling that damage can be neutralized may stimulate damaging behavior.[46,47] Testing whether this fear is justified is complicated. One interesting attempt to quantify behavioral changes that may be induced by offsetting environmental damage involved a randomized trial in a German youth hostel. All guests received information on the electricity use and carbon cost of showering. Some of the guests received additional information that the hostel compensated all carbon emissions by financially contributing to an environmental project. The time that 7350 guests used the shower was subsequently measured and compared. Guests who received information that the carbon costs were compensated, used the shower 5–15% longer,[48] suggesting that damaging behavior may increase if there is a promise of offsetting this damage. Other studies similarly show that electricity consumption may increase when an offsetting scheme makes use 'guilt-free'.[49]

In conclusion, the price per ton of CO_2 is likely to be too low in most current carbon offsetting schemes, schemes typically fail to achieve the impact that is anticipated or promised and there are indications that carbon offsetting may stimulate damaging behavior or, putting it more mildly, dampen the incentive to reduce emissions. Whilst at an individual level, emission compensation schemes can be part of a sustainable lifestyle, offsetting is currently not a solution to global warming that should distract from the need to reduce carbon emissions.

4.6 Concluding Remarks: Towards Thoughtful Academic Travel

It is important to emphasize that science benefits from strong international collaborations and networks and it is equally important to acknowledge that for most people, there are evident advantages to in-person meetings and conferences. Virtual conferences and virtual meetings are – and in our opinion should be – advocated in a time when we have to drastically reduce our carbon emissions to prevent the worst of the climate crisis. Whilst a large consultation concluded that the vast majority (74%) of scientists also want virtual meetings to stay,[29] it is important to be honest about the (current) shortcomings of virtual meetings. In-person meetings allow for more enjoyable and bonding interactions. In addition, there is scientific evidence that creativity thrives better in direct personal interactions. A detailed study of human interaction, idea generation and decision making that directly compared in-person gatherings with videoconferencing, indicates that the latter may inhibit the production of creative ideas.[23] In contrast, decision-making and prioritization of ideas was at least as good, and possibly even more effective, online.[23] While the benefits of online and in-person are often contrasted, also in this chapter, this is a false dichotomy. There is good reason to combine virtual and in-person meetings and consider a future where both approaches are used in concert. Generating new ideas, brainstorming about new projects and building new collaborations clearly benefit from in-person interaction and this interaction should be catered for. At the same time, (smaller) meetings that are primarily organized to compare ideas or proposals and take decisions can often be organized efficiently online.

This chapter is also not a pledge against academic travel. Instead, it hopes to stimulate action and more thoughtful academic travel. If we acknowledge that in-person meetings and conferences have important roles in academic practices and that there is currently no good way of truly compensating carbon emissions, we can consider the frequency and scale at which these events should be organized. Some societies organize biennial conferences, others have established a system where in-person and virtual conferences alternate. Whilst location optimization models to select venues that minimize travel may often yield only small reductions in carbon emissions (<5%),[33] hub-and-spoke models and decentralized approaches where multiple regional venues are selected can have a much larger impact (reductions of 60–70% in some analyses[10,33]). The largest reductions are achieved by reducing the frequency and scale of in-person conferences. This would ideally be taken on by the conference organizers and there are initiatives where thousands of scientists call upon conference organizers to take this up as their responsibility.[50] Also at the level of institutes of higher education and (research) departments, it is possible to set carbon targets. This can translate in a maximum number of in-person participants to a given conference each year.

More generally speaking, it is important to define the objectives of travel when adopting a thoughtful travel approach. If visibility of the institute is one of the goals, a (small and potentially rotating) delegation of employees can be selected to attend a meeting. If a (junior) researcher hopes for international exposure and networking experiences at a conference abroad, it can be highly beneficial to combine this with a work-visit to a collaborating group in the same (or neighboring) country. If an established researcher is invited for a meeting, it may be possible to delegate a more junior colleague who generally travels less[51] and may benefit more from it. When networking and productive interaction with colleagues is the primary goal for traveling, smaller meetings are likely to be more productive than mega conferences.[13] There are numerous other examples of situations where a thoughtful approach to academic travel is possible. With the creative minds in academia, there is every reason to be optimistic that the international characteristic of science and environmental sustainability can go hand in hand.

References

1. E. Helmers, C. C. Chang and J. Dauwels, Carbon footprinting of universities worldwide: Part I—objective comparison by standardized metrics, *Environ. Sci. Eur.*, 2021, **33**(1), 30.
2. De Jonge Akademie, *Flying high but flying less: an overview study of Dutch university policies to reduce carbon emissions from research related air travel*, 2020.
3. K. Whitfield, A. Cretu, T. Bousema and J. Cohen, Environmentally sustainable practices in global health research and higher education institutions: Lessons from consultation with the TropEd Global Health institutions, *Trop. Med. Int. Health*, 2022, **27**(2), 122–128.
4. M. J. Smit, A. Mahamar and E. Kooistra, *et al.*, Life cycle assessment of a clinical malaria trial in Mali reveals large environmental impacts of electricity consumption and international trave, *PLOS Sustainability Transform.*, 2025, **4**(2), e0000131.
5. IPCC, Aviation and the Global Atmosphere. A Special Report of IPCC Working Groups I and II, 1999.
6. D. S. Lee, *The current state of scientific understanding of the non-CO_2 effects of aviation on climate*, Department of Transport UK government, 2018.
7. D. Barret, Estimating, monitoring and minimizing the travel footprint associated with the development of the Athena X-ray Integral Field Unit. An on-line travel footprint calculator released to the science community, *Exp. Astron.*, 2020, **49**, 183–216.
8. https://www.klm.nl/en/information/legal/extra-options/co2-impact-programme, 26-2-2025).
9. T. Bousema, L. Burtscher, R. P. van Rij, D. Barret and K. Whitfield, The critical role of funders in shrinking the carbon footprint of research, *Lancet Planet Health*, 2022, **6**(1), e4–e6.

10. T. Bousema, P. Selvaraj and A. A. Djimde, *et al.*, Reducing the Carbon Footprint of Academic Conferences: The Example of the American Society of Tropical Medicine and Hygiene, *Am. J. Trop. Med. Hyg.*, 2020, **103**(5), 1758–1761.
11. L. Burtscher, D. Barrett and A. P. Borkar, *et al.*, The carbon footprint of large astronomy meetings, *Nat. Astron.*, 2020, **4**, 823–825.
12. M. Klower, D. Hopkins, M. Allen and J. Higham, An analysis of ways to decarbonize conference travel after COVID-19, *Nature*, 2020, **583**(7816), 356–359.
13. T. Bousema, S. A. V. van Asten and J. Ramjith, *et al.*, Transforming ESCMID in a time of climate change: a call for sustainable conferencing, *Clin. Microbiol. Infect.*, 2024, **30**(11), 1347–1350.
14. K. E. Whitfield, A. Ares Pita, T. C. Jarvis, S. Weimen and T. Bousema, Geographic Equity and Environmental Sustainability of Conference Models: Results of a Comparative Analysis, *Am. J. Trop. Med. Hyg.*, 2024, **110**(5), 1039–1045.
15. D. Yakar and T. C. Kwee, Carbon footprint of the RSNA annual meeting, *Eur. J. Radiol.*, 2020, **125**, 108869.
16. E. B. Parker, A. Bluman and J. Pruneski, *et al.*, American Orthopaedic Foot and Ankle Society Annual Meeting All-in-person Attendance Results in Immense Carbon Expenditure, *Clin. Orthop. Relat. Res.*, 2023, **481**(12), 2469–2480.
17. https://spaceplace.nasa.gov/moon-distance/en/.
18. D. Notz and J. Stroeve, Observed Arctic sea-ice loss directly follows anthropogenic CO_2 emission, *Science*, 2016, **354**(6313), 747–750.
19. G. Viglione, How scientific conferences will survive the coronavirus shock, *Nature*, 2020, **582**(7811), 166–167.
20. S. Weisband and L. Atwater, Evaluating self and others in electronic and face-to-face groups, *J. Appl. Psychol.*, 1999, **84**(4), 632–639.
21. J. M. Denstadli, T. E. Julsrud and R. J. Hjorthol, Videoconferencing as a Mode of Communication:A Comparative Study of the Use of Videoconferencing and Face-to-Face Meetings, *J. Bus. Tech. Commun.*, 2012, **26**(1), 65–91.
22. D. S. Staples and L. Zhao, The Effects of Cultural Diversity in Virtual Teams Versus Face-to-Face Teams, *Group Decision Negotiation*, 2006, **15**(4), 389–406.
23. M. S. Brucks and J. Levav, Virtual communication curbs creative idea generation, *Nature*, 2022, **605**(7908), 108–112.
24. E. R. Zajdela, K. Huynh, A. L. Feig, R. J. Wiener and D. M. Abrams, Face-to-face or face-to-screen: A quantitative comparison of conferences modalities, *PNAS Nexus*, 2024, **4**(1), 522.
25. Wellcome Carbon offset policy for travel, https://wellcome.org/node/369, visited 28-2-2025.
26. https://spectrum.ieee.org/virtual-meeting.
27. G. Viglione, A year without conferences? How the coronavirus pandemic could change research, *Nature*, 2020, **579**(7799), 327–328.

28. C. Woolston, Learning to love virtual conferences in the coronavirus era, *Nature*, 2020, **582**, 135–136.
29. A. Remmel, Scientists want virtual meetings to stay after the pandemic, *Nature*, 2021, **591**, 185–186.
30. C. C. John, Something Big that Matters: The American Society of Tropical Medicine and Hygiene's Commitment to Combat Climate Change, *Am. J. Trop. Med. Hyg.*, 2020, **103**(5), 1751–1752.
31. https://keypoint.keystonesymposia.org/home/keypoint-newsletter-50th-anniversary-gala, visited 28-02-2025.
32. J. Alexander and A. Conrad, *Citizens. Why the Key to Fixing Everything is All of Us*, Canbury Press, Kingston upon Thames, UK, 2022.
33. K. E. Lichter, A. Sabbagh and S. Demeulenaere, *et al.*, Reducing the Environmental Impact of Health Care Conferences: A Study of Emissions and Practical Solutions, *JCO Glob. Oncol.*, 2024, **10**, e2300209.
34. https://news.klm.com/klm-doubles-co2-compensation/, accessed 12-1-2025.
35. https://co2.myclimate.org/, accessed 16-1-2025.
36. https://www.statista.com/statistics/1304575/global-carbon-capture-cost-by-technology/.
37. A. E. Duchelle, G. Simonet, W. D. Sunderlin and S. Wunder, What is REDD+ achieving on the ground?, *Curr. Opin. Environ. Sustainability*, 2018, **32**, 134–140.
38. E. O. Sills, C. de Sassi and P. Jagger, *et al.*, Building the evidence base for REDD +: Study design and methods for evaluating the impacts of conservation interventions on local well-being, *Global Environ. Change*, 2017, **43**, 148–160.
39. B. Haya, D. Cullenward and A. L. Strong, *et al.*, Managing uncertainty in carbon offsets: insights from California's standardized approach, *Clim. Policy*, 2020, **20**(9), 1112–1126.
40. T. A. P. West, S. Wunder and E. O. Sills, *et al.*, Action needed to make carbon offsets from forest conservation work for climate change mitigation, *Science*, 2023, **381**(6660), 873–877.
41. F. Fleischman, S. Basant and A. Chhatre, *et al.*, Pitfalls of Tree Planting Show Why We Need People-Centered Natural Climate Solutions, *BioScience*, 2020, **70**(11), 947–950.
42. https://www.ftm.nl/artikelen/carbify-de-carbo-cryptocoin.
43. R. Bailis, R. Drigo, A. Ghilardi and O. Masera, The carbon footprint of traditional woodfuels, *Nat. Clim. Change*, 2015, **5**(3), 266–272.
44. https://www.who.int/tools/clean-household-energy-solutions-toolkit/module-7-defining-clean, accessed 27-2-2025.
45. A. Gill-Wiehl, D. M. Kammen and B. K. Haya, Pervasive over-crediting from cookstove offset methodologies, *Nat. Sustainability*, 2024, 7(2), 191–202.
46. S. Becken, Tourists' Perception of International Air Travel's Impact on the Global Climate and Potential Climate Change Policies, *J. Sustainable Tourism*, 2007, **15**(4), 351–368.
47. R. E. Goodin, Selling Environmental Indulgences, *Kyklos*, 1994, **47**(4), 573–596.

48. S. A. Günther, T. Staake, S. Schöb and V. Tiefenbeck, The behavioral response to a corporate carbon offset program: A field experiment on adverse effects and mitigation strategies, *Global Environ. Change*, 2020, **64**, 102123.
49. M. Harding and D. Rapson, Does Absolution Promote Sin? A Conservationist's Dilemma, *Environ. Resour. Econ.*, 2019, **73**(3), 923–955.
50. https://sites.tufts.edu/flyingless/sample-page/, visited 2-3-2025.
51. S. Wynes, S. D. Donner, S. Tannason and N. Nabors, Academic air travel has a limited influence on professional success, *J. Cleaner Prod.*, 2019, **226**, 959–967.
52. https://canie.org/take-action/travel-with-canie.
53. https://www.unep.org/explore-topics/education-environment/what-we-do/little-book-green-nudges.

CHAPTER 5

Scientific Equipment Sustainability[†]

ANDY EVANS

Green Light Laboratories Ltd, UK
Email: office@greenlightlabs.co.uk

There is always the temptation to go into deep technical detail when talking about lab equipment. Not only does it demonstrate that I know what I am talking about, but it is also a great way of increasing the word count. However, I am not aiming to provide you with the largest chapter in the book. What I am hoping is that what follows will be one of the most useful chapters. I will go into technical detail only if there is a sustainable reason for doing so. I am also hoping that this chapter will be of interest to all three types of lab operators. The first being the lab users; the scientists and technical staff – without you there is no research or innovation. That said, operating a laboratory is a team effort requiring multiple areas of expertise. So, I hope that lab users will also be joined in reading this by team members from both estates/building management and procurement professionals. I am aiming through this chapter not only for the following information to be food for thought, but for it to be something you can all digest and incorporate into your specific roles to deliver more sustainable practices. If this does happen (and I sincerely hope so) then please do not credit me. Take all the glory. Save yourself money and our planet just by adjusting how you choose or use your laboratory equipment. This chapter will be broken down into four sections.

The Sustainability of Science: How Science Impacts the Environment, and What Can Be Done
Edited by Martin Farley

Published by the Royal Society of Chemistry, www.rsc.org

So, before we dive into these life-changing sections I want to impart one key piece of advice which is applicable to multiple equipment types. Before you choose an item of equipment, please do this: **specify your performance**. Now, performance can mean multiple things across a range of equipment types, but as an example we will use a −20 °C freezer. Performance for a freezer would include temperature uniformity (how cold from top to bottom inside the chamber) and how fast this unit can recover from a timed door opening (typically a 60 second door opening). So, decide what you are using the equipment for and how often the unit will be used for. Why is this so important? Well, a common first mistake many lab users make when pursuing sustainable procurement is that they overlook the performance bit and instead focus upon the lowest energy unit for the lowest price. The danger of this approach is a unit is bought that is not suitable for and cannot cope with the specific needs of a laboratory. The cheaper, low energy freezer may take 3–4 hours to recover from a single door opening, which is not suitable for a busy laboratory with multiple users accessing the freezer on a regular basis. The knock-on effects of this decision may also include samples or reagents being stored at temperatures detrimental to their quality or viability which then will have a knock-on effect upon experimental outcomes. Imagine all the wasted time, consumables and reagents that could result from making such a simple mistake in choosing just one item of equipment. So, decide upon the performance you need **first**. Once you have done that, you will be left with energy efficient models that perform as you require. It is then down to you and your procurement team to negotiate the best price. Now with no further ado, let us dive into the first section; lab equipment cost factors.

5.1 Equipment Running Cost Factors

There are in excess of forty lab equipment running cost factors. Don't worry, I am not going to list them all, otherwise I may need a couple of chapters. I will list some of the most prominent environmental and performance factors which are applicable to almost any item of lab equipment. These factors do not operate in isolation. Changing one factor will have knock-on effects upon other factors. Understanding these factors is of gargantuan benefit to all lab operators. These factors will determine the costs to operate your facility, the time taken to carry out your work and the carbon emissions associated with doing so. A select number of factors will now be explored. Any other factors they influence will be **bold**.

5.1.1 Origin

Definition: The place(s) where the item of equipment was manufactured. Furthermore, where are all components sourced from? Always ask for the address. From knowing this you can calculate the distance your equipment and any **spare parts** must travel to reach your point of **delivery**. This obviously has a direct influence on your **carbon emissions**. If you are willing to ask your supplier the mode(s) of transport used to deliver your equipment from the site of manufacture, then you can calculate the scope 3 **carbon emissions** of delivery. Furthermore, if manufactured overseas, what impact does this have on the lead **time** for the unit and, when needed, **spare parts.** The latter may also determine the length of time you may have to wait for this specific item of equipment to be **serviced** or **repaired**.

5.1.2 Capacity and Size

Definition: How much stuff can I fit in this thing? Seems like such a simple thing to ask right? But the impact of this factor is substantial. It impacts **energy efficiency**. A larger capacity oven, fridge, freezer, incubator will invariably draw more energy than a smaller unit. *However*, this can be misleading as the energy you are using to heat/cool that chamber per litre of capacity will decrease with larger units. Knowing the capacity of the units you are interested in allows for their energy consumption to be truly compared. **Dividing energy consumption by capacity** will give you a 'pound for pound' energy figure and is a step closer to identifying more energy efficient units. Now, some manufacturers will give you gross capacity and net capacity with the latter being the space you should have available to you for use. Sadly, this may not be the case. The reason why? Inadequate standards. And this is crucial if the scientific community desires tools which fairly identify sustainable products. Here is an example. Recent case studies I have carried out at the University of Bristol looked at ULT freezer performance. When Anna Lewis (University of Bristol, author of the chapter on Plastics and Carbon) and I put units of purportedly similar capacities next to one another we could see that they were considerable differences in size. So, as any curious scientist would, we measured the space actually available for storage. Each freezer compartment was individually measured to the nearest mm with the sum total of each compartment being referred to as the usable capacity (Figure 5.1).

As I hope you can appreciate from the example in Figure 5.2 the net capacity can greatly differ from the usable capacity, and it's the usable capacity we should be subscribing to. After all, this is the space you will use and is the true presentation of storage capacity. Amongst the ULT freezers we measured (and continue to do so) there is a range of differences between net capacity and usable capacity (Figure 5.2).

Let's be clear here. None of the manufacturers above are being intentionally misleading. They are just using a standard that does not represent what lab operators need to see. Often, capacity is measured using a standard such as AHAM HRF-1-2019. Instead, capacity should be measured as the total of the vertical projection of each individual shelf space as stated in BS EN ISO220412019 + A1 2019. In the absence of this data, you can measure this yourself. It's not hard. Nothing I have stated cannot be measured, and colleagues here in the UK have put my findings to the test and come to the same conclusion. The other consideration with capacity is footprint. This is the overall space the unit occupies in your facility. How much of the prime

Haier Biomedical

DW86L728J Cascade

External Dimensions

Height (mm)	Depth (mm)	Width (mm)	Footprint (L)
1980	980	1041	2019.9564

Anna Lewis

DW86L728J	Height (mm)	Depth (mm)	Width (mm)	Capacity (L)
Compartment 1	300	649	749	145.8303
Compartment 2	305	649	749	148.26081
Compartment 3	304	649	749	147.7747
Compartment 4	300	649	749	145.8303
			Total Litres	587.70

Andy Evans

DW86L728J	Height (mm)	Depth (mm)	Width (mm)	Capacity (L)
Compartment 1	302	649	748	146.6065
Compartment 2	305	649	748	148.06286
Compartment 3	305	649	748	148.06286
Compartment 4	302	649	748	146.6065
			Total Litres	589.34

Model	DW86L728J	
Mean U.C. (L)	588.52	Litres
Published Net (L)	728	Litres
Pub. Vs U.C (L)	139.48	Litres
Pub. Vs U.C (%) Diff.	19.16	% Smaller
Unit Footprint	2019.96	Litres
U.C. as % of Footprint	29.14	%

Figure 5.1 The published net capacity of this ULT was 728 L. When measured, the usable capacity was 588.52 L, 139.48 L (19.16%) smaller than the published net capacity. Data captured by independent testing carried out at the Learning and Research Centre, University of Bristol by Green Light Laboratories Ltd.

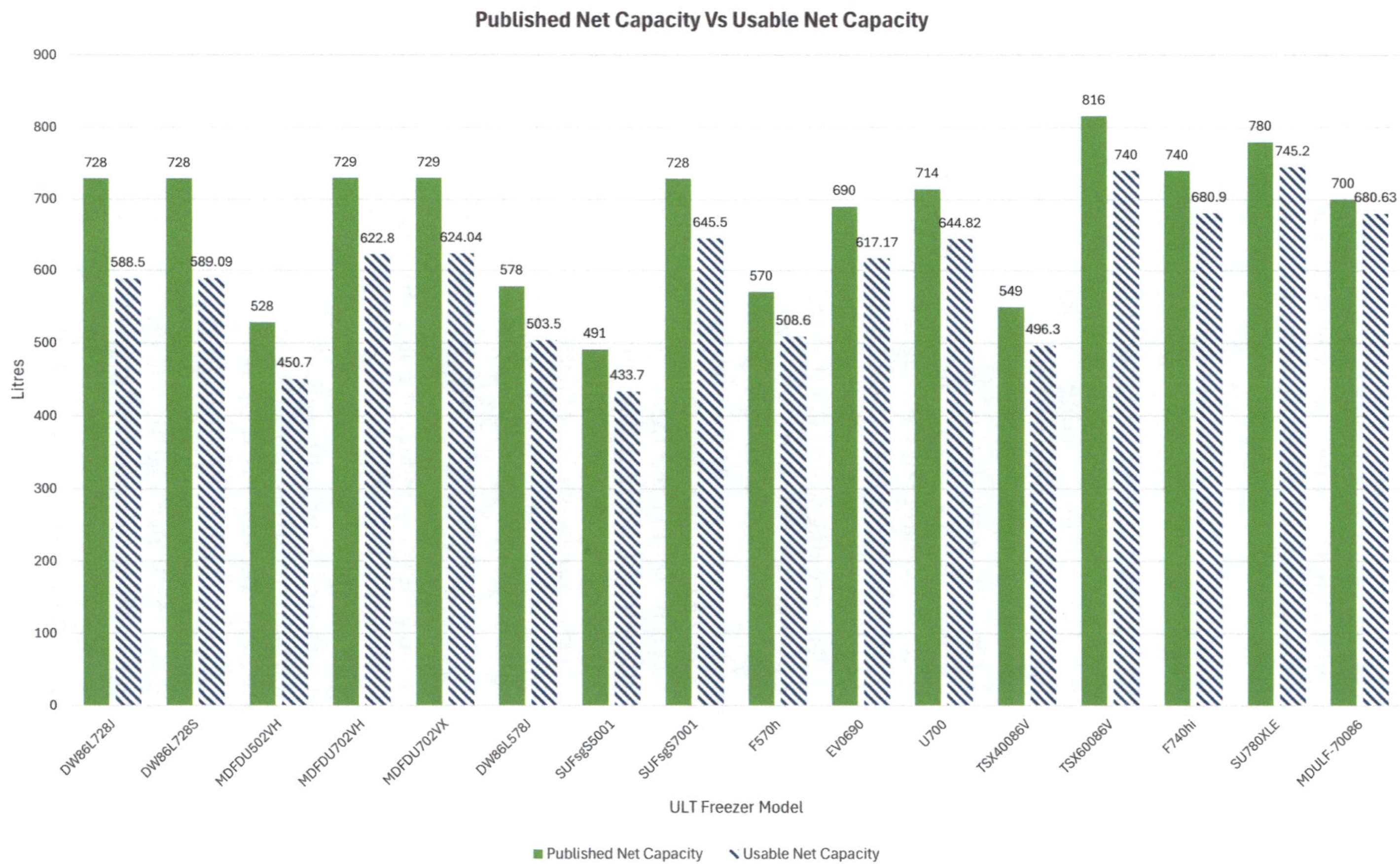

Figure 5.2 Net capacity *versus* measured usable capacity in ULT freezers. Data captured from independent testing carried out at the Learning and Research Centre, University of Bristol by Green Light Laboratories Ltd.

real estate that is lab space are you giving away for *X* amount of capacity? This data can help illuminate how efficiently or inefficiently an item of equipment is using your **space.** How much cold storage, incubation, heating capacity can you provide in the finite space you have available and is it enough for your current and future needs (**futureproofing**).

5.1.3 Performance

Definition:How well does this thing do its job(s), be it to measure, store, protect, or whatever specified function it is designed for. As mentioned earlier, performance is *key* when selecting the right item of equipment. Is this equipment suitable for your needs, your **usage?** Performance is closely tied to **energy consumption**; operating faster, colder, hotter requires **energy.** Performance will come in many forms. For an oven, incubator or cold storage it can, for example, be the temperature uniformity of the unit ($\pm$ *X* Celsius at the set temperature of *Y* throughout the piece of equipment). Performance can also be assessed in terms of temperature recovery, which is why one model of fridge may take 10 minutes to recover its temperature following a timed door opening whilst another may take well over an hour. This factor is so pivotal in identifying the right item of equipment that manufacturers sometimes have their *own* definitions of performance to show their item of equipment in the best light. One example would be the aforementioned door opening recovery times. Want a faster recovery time? Some suppliers do so by setting the threshold recovery temperature to be (in the case of cold storage), warmer than the pre-door-open temperature. You will find this detailed in the small print on datasheets and marketing materials. It is not a true representation of performance and leaves end users short-changed. Performance can also be altered through the use of **accessories or upgrades.** One example of an upgrade can be illustrated with a case study carried out at the University of Oxford which looked at autoclave performance (Table 5.1).

In this study two of the same model of autoclave was in use at the University with one difference. One unit had been upgraded with the vacuum drying feature. With identical loads the same cycle was run on both units. Initially, the cycle was completed more quickly, as it took 79.4% more time to finish in the non-vacuum drying unit. This resulted in a domino effect on the drying process, with vacuum dried tip boxes taking only a third of the **time** and energy to dry compared to their non-vacuum counterparts. When using the non-upgraded unit, the total processing **time** for the tip boxes more than doubled, and the overall **energy** consumption—and consequently, **carbon emissions**—were 39.5% greater. This case study serves as a reminder that if we had focused purely on the **energy** of running a cycle, the upgrade may have been adjudged ‘unsustainable’. However, when we considered other factors and the impact upon workflow there were savings to be made in **time, energy** and **carbon emissions.**

Table 5.1 Comparing the cycle times and costs to sterilise and dry the same load in the vacuum drying and standard versions of the Compact 40. Here the upgrade to vacuum drying feature resulted in both speeding up and reducing the energy cost of the process of sterilising the load and having it dry and ready for use. Data are from an independent case study carried out at the MRC Weatherall Institute of Molecular Medicine, University of Oxford. Case study commissioned by Priorclave Ltd, testing carried out by Green Light Laboratories Ltd.

	Maximum tip boxes load (24 boxes)	
	Compact 40 Vacuum	Compact 40
Cycle time	68 minutes	122 minutes
Cycle energy	1.701 kW h	1.329 kW h
Drying cabinet drying time (55 °C)	60 minutes	180 minutes
Drying cabinet energy (55 °C)	0.648 kW h	1.944 kW h
Cycle time + drying time	128 minutes	302 minutes
Cycle energy + drying energy	2.349 kW h	3.273 kW h
Energy: Kg/CO_2/Run	0.547	0.763

5.1.4 Consumables and Reagents

If I were to tell you I had an energy efficient computer printer you might be excited. Great. Another energy saving opportunity I hear you say with enthusiasm. However, many of you may also have a question "how much are the ink cartridges??". For those of you experienced in this field it would be fair to say that the cost of the consumable ink cartridges can dwarf the affordable and bargain price of buying the printer. Furthermore, the consumables are almost exclusively proprietary. This may be the case with numerous items of equipment. You may be offered a competitive, sometimes almost giveaway price, for an item of equipment. However, the potential hidden trade-off may be the price and availability of consumables and/or reagents. For some items of equipment their main running cost may be the consumables. One example is found in fluorescent microscopy. Traditionally, mercury bulbs have been used as the light source. When compared to LED based alternatives the bulb cost (Table 5.2) of the mercury vapor lamp was overwhelmingly higher during the warranty period of the LED system. Thus, despite the LED system having higher upfront costs for the purchase of the unit, overall, it would save the research group money through avoided purchases of bulb costs (while saving some energy). Notably, LED systems have other advantages, in that their light sources are more stable and do not wane, and they do not overheat samples through wasted energy.

5.1.5 Energy

Definition: It cannot be created or destroyed but it can be billed.

Table 5.2 Costs of operating a mercury vapor lamp compared to an LED based system. Both systems are compared for a period of operation of 25 000 hours as this was the warranty period of the LED system. Data are from an independent case study carried out at the School of Geographical Sciences, University of Bristol. Case study commissioned by CoolLED Ltd, testing carried out by Green Light Laboratories Ltd.

Data	Mercury-Vapor Lamp	L.E.D. CoolLED pE300ultra
A. Bulbs used in 25 000 hours	125 Bulbs	None
B. Cost of bulbs used in 25 000 hours	£250 each (bulb + installation):£31 250	£0.00
C. Time for microscope alignment per bulb	30 minutes per bulb, 125 bulbs: 62 hours 30 minutes	Not applicable
D. Energy consumed in 25 000 hours usage	0.114 $kW\,h\,h^{-1}$: 2 850 kW h	0.010 $kW\,h\,h^{-1}$: 250 kW h
E. Electricity cost of 25 000 hours (9.7 $p\,kW^{-1}\,h^{-1}$)	£276.45	£24.25
F. Warm up time	20 minutes	Not applicable (instant ON/OFF)
G. Warm up time energy consumption	0.117 $kW\,h\,h^{-1}$ × 20 minutes: 0.039 kW h	Not applicable (instant ON/OFF)
H. Warm up time for 8 hours of usage per bulb	24 hours (per 200 hours/ bulb)	Not applicable (instant ON/OFF)
I. Cost of unit being usable for 8 hours (adding warm up time of 20 minutes)	Warm up energy + 8 hours usage energy + bulb cost ((£250/200 hours per bulb) × 8.33): £10.50	8 hours energy usage: £0.01
J. Cost of 6 hours of viewing samples in the 8 hours the unit is on for.	Cost is the same due to time for warm up and cool down: £10.50	6 hours energy usage (off for 2 hours): £0.01
K. Cool down time	30 minutes	Not applicable (instant ON/OFF)
L. Total cost over 25 000 hours of usage	(25 000 hours/8.33 hours) × Cost of unit being usable for 8 hours: **£31 512.61 (£1.26 per hour)**	(25 000 hours × energy consumption in use per hour: **£24.25 (£0.00097 per h)**

Energy is the first step for most people on their journey towards sustainability. It is where I started 15 years ago. Fast forward to the present day and energy efficient products are widely available and popular. There is sometimes funding available to subsidise the procurement of energy efficient equipment to help support labs in reducing their running costs depending on the organisation. The scientific community will certainly benefit from less energy-hungry equipment. But, there are many pitfalls when it comes to energy. So, for the sake of fun and fun only, I am going to pretend to be an energy villain. I am hellbent on manipulating my energy consumption figures in my quest to be the number 1 energy efficient (insert equipment name of your choice here). Several tools are at my disposal:

5.1.5.1 Ambient Conditions

Before we even measure the energy consumption let's make the room as accommodating as possible. The colder the room the less energy your item of cold storage will need. Or *vice versa* with your incubators, ovens or drying cabinets in that a nice warm room will be of benefit. Whatever you do, never test your equipment in the lab environment under the same ambient conditions...

5.1.5.2 Modes

Modes are a great way of changing the performance of an item of equipment. Although a mode isn't naughty on its own, the data can be used to produce some fantastic marketing. Let's say you have one mode which saves energy and another mode that is a high-performance mode. Why not take the energy data from the energy saving mode and the performance data from the high-performance mode and present them as a single data set? Sure, these modes are mutually exclusive, but you can always point this out in some very well-placed small print.

5.1.5.3 Set Point

A really simple one for you, and although obvious it can still be overlooked. Why not test your item of equipment at a warmer or colder set point (equipment type dependent)? Just 5 °C can be the difference between you and that competitor's model.

5.1.5.4 Set Point Performance

Instead of changing the set point why not just tweak the performance at the commonly used set point? You can do this *via* a **mode**, or you can simply adjust how well your item of equipment reaches and maintains the set point. The more variance around the set points the better the energy savings.

5.1.5.5 Capacity

See previous section on "Capacity and Size", size matters so be as big as you can unrealistically be. Use an older standard that does not represent the space you get or normal usage. Normal usage? Give us an example? Well, in a drying cabinet, commonly used to dry glassware and plasticware, it is not advisable to place contents directly on the bottom of the chamber. Plasticware can melt or the heating element will cut out with nothing getting dried. But even though you should not use this space below the lowest shelf position why not still include it in your net capacity??

5.1.5.6 Usage

Although bordering on best practice we can employ usage to our advantage. Lots of equipment can be classed as being used upon demand, and turned off when not in use. Well, let's ignore that a certain item of equipment may be on 3 hours per day. Instead let's say it's on 24 hours per day because the energy you are then saving is 8 times higher!!

5.1.6 Price

Definition: You get what you pay for. Budgets must be adhered to, funds are finite. However, with few exceptions, you will get what you pay for. Hardly any items of equipment are created equal, and for that reason there's going to be a price difference. There are multiple explanations for differences in price and all of them will affect you. There tends to be a proportional relationship between price and the following lab cost factors (although not absolute), some of the most common factors that rise and fall according to price are:

- **Warranty** – Not only the length of warranty but what the warranty includes. A longer warranty is usually reflective of the **robustness** and **longevity** of the equipment you are procuring. Also be mindful of the money you would save on **service and maintenance** when your item of equipment has a longer, more comprehensive warranty.
- **Longevity and Robustness** – 'Buy cheap, buy twice' is a saying I have heard many times from both technical managers and some sales representatives. When both sides of the 'sales divide' say the same thing, they speak from experience. Having audited hundreds of laboratories capturing data on tens of thousands of items of equipment, this saying is certainly true in terms of longevity. On certain equipment types the lifespan of the higher priced model can be double that of the lowest priced options. If the item of equipment is robust and built to last, you will pay a little more.
- **Performance** – How well does this item of equipment do the job you want it to do? The less expensive the unit is, likely the poorer the performance will be. The ability to **reach**, **maintain** and **return** to the set point will influence the time taken to carry out your lab operations, the cost to do so and the viability, quality of your samples or product.
- **Functionality** – How many jobs can one item of equipment do? How broad is the set temperature range, can it generate humidity, detect all the gases or wavelengths of light you require both now and in the future? Can I easily **upgrade** this equipment? The more stuff this equipment can do or be upgraded to do, the higher the price will be. However, spending that bit of extra cash can help you **futureproof** your lab, allowing you to diversify your activities without having to buy a second item of equipment which will need its own bench space, service and maintenance contracts *etc.*

For procurement tenders or framework agreements we also need to be mindful of mark weighting. Understandably, procurement professionals want to ensure that their institution gets the best possible price for the equipment they want to buy. After all, it is an essential component of their job. Traditionally, it has been thought that the best way to do this is to assign 40%+ of marks to price alone. However, is this the sustainable way to procure equipment? My answer is no. Funnily enough, the cheapest model will almost always be the cheapest model regardless of the mark weighting. It seems disproportionate that price, being one out of >40 equipment running cost factors, gets to be the primary focus. Especially as the impact of the other factors can greatly outweigh the price difference gained by opting for the least expensive item of equipment. An example of why this is important can be seen when energy efficient equipment costs more upfront. So, while it would perform better in a tender where pricing is heavily weighted, long-term it will cost the owner more through poor energy efficiency. With success, we have applied a 15%-mark weighting to price for several procurement frameworks. What you do with that freed 25% is up to you, however, may I suggest you consider giving more weight to the following areas (Table 5.3) as they will benefit both your utility costs and individual lab running costs:

Table 5.3 Factors to consider incorporating into your procurement mark schemes alongside price.

Add marks for	Description	Benefits
Performance	How well does this item of equipment perform its function(s)	(1) Saves time. (2) Protects sample or product viability. (3) Reduction in wastage.
Longevity	If well maintained, how long will this item of equipment last for? Always ask for multiple reference sites.	(1) You will only buy it once *vs.* a cheaper model. (2) Reduced scope 3 carbon emissions.
Recyclability	By percentage weight how much of this is recyclable. If not 100% please detail the components that are not.	(1) Reduced scope 3 carbon emissions. (2) Identifying hidden end of life costs that may be costly to you.
Ergonomics demo	Want *X*% marks? Please submit your proposed model for examination by staff and end users.	(1) End users can identify features they like/dislike. (2) Robustness and ergonomics can be examined. (3) Technical questions can be asked of the manufacturers. (4) Performance and running costs can be independently measured.

5.2 Common Laboratory Equipment

We will now delve into several common equipment types. I am including kit which are widely used and numerous in research laboratories. We will be focusing on equipment types where there are numerous, simple steps you can take to make instant savings, both financially and in terms of energy efficiency.

So, if you are into movies I am going to venture into the spaghetti Western genre. I have combined aspects of features, technology, good and bad practice to give you the good, the bad and the ugly lab equipment. As I am not keen on repeating myself, if something is in the 'good' section of an item of equipment, then doing the opposite is likely to be in the 'bad' or even 'ugly' section. Let us start with fume cupboards.

5.2.1 Fume Cupboards

These chemistry essentials are certainly up there with those items of equipment with the highest financial running costs. Ducted fume cupboards extract vast volumes of air from your building using variations of fans which are likely in operation 24/7. That air has likely either been cooled or heated, so fume cupboards have a massive impact upon your energy costs. Yes, there is a variety of fume cupboards in terms of size (sash width and depth) and configuration (controller types and ducting extract systems), however, even when you have got a 'good', energy efficient fume cupboard they will still have relatively high running costs. Anything you can do to minimise fume cupboard running costs will have a significant impact long-term.

5.2.2 Fume Cupboards, The Good

In terms of airflow control when installing or refurbishing rooms with multiple fume cupboards, please go for a Variable Air Volume (VAV) controller. So much so, make this a non-negotiable feature for any procurement exercise, refurbishment or new build. The VAV controller will adjust volume extracted by the fume cupboard based upon sash position. Therefore, the simple act of shutting the sash will save you £100's per year per unit. To put this into perspective think about this. Imagine going for a jog. During this exercise you will breathe 16 $L\,min^{-1}$ so that is 960 $L\,h^{-1}$. Now let us take a VAV fume cupboard. It has a sash width of 1700 mm, worksurface width of 1900 mm and depth of 650 mm, and a design face velocity of 0.40 $m\,s^{-1}$. I will now open the sash to a height of 500 mm. Per minute this fume cupboard will extract 21 930 L of air from the lab. This is the same amount of air you would breathe if you went for a jog lasting 22.8 hours. Let us say I then nip off to get a quick coffee and I am back 15 minutes later, well, that's 328 950 L of air extracted (a 342.7-hour jog with no breaks). So, shutting the sash is good, leaving if needlessly open for a few minutes is bad, and going home without shutting is certainly an ugly thing to do. I will always champion best practice, it is by far the most cost-effective action a lab can take, and good habits also die hard travelling with the practitioner. However, you may also want to consider including auto sash closers with your VAV fume cupboard which will avoid the need for user good practice. That said, these will cost ≥£2000 per unit, and they will require their own level of servicing which can be a risk and added cost in time.

It is also fair to say that not all VAV controllers are created equal, here are some good features to look out for:

- Speed of response, how quickly does the unit get to a safe velocity once you have opened the sash? To keep things nice and safe, look for a response time of ≤5 seconds.
- Volumetric control features (volumetric sensing). If you have this capability, you can finely control the minimum volumes extracted by the fume cupboard during the nighttime and weekends. These times represent ~70% of fume cupboard operating times and so it is a great way to minimise your HVAC costs.
- Easy and accurate calibration. Firstly, it is great to have options when it comes to calibration so controllers that have an open calibration protocol give you a wider range of engineers who can service and calibrate your fume cupboards. Some controllers may require specific, proprietary software which can limit your choice of available engineers. Secondly, having a controller that facilitates a fast and finely tuned calibration saves you time and money. If an engineer can calibrate your VAV controller to adjust as close to design face velocity as possible in the shortest time possible, you will save multiple costs.

The other good thing to consider for any fume cupboard type is design face velocity. Why operate at 0.5 $\mathrm{m\,s^{-1}}$ when you can operate safely at 0.4 $\mathrm{m\,s^{-1}}$, saving 20% in running costs? In fact, you can legally operate a fume cupboard in the UK down to 0.3 $\mathrm{m\,s^{-1}}$ if you have had containment testing evidencing that they are safe (and typically using manufacturers that facilitate low-flow units). To be safe, some institutions target 0.35 $\mathrm{m\,s^{-1}}$, and will save approximately 30% in energy costs.

5.2.3 Fume Cupboards, The Bad

Let us start with airflow control. Try with every morsal of your being to avoid constant air volume (CAV) controlled fume cupboards in rooms with multiple units. As their name implies, regardless of sash position the volume of air removed from your lab, your building is constant. And so, their running costs compared to VAV are in between 2.3–3.5 times higher. I must stress that this range may vary further depending upon utilities costs and fume cupboard configuration. If we take the 1700 mm sash width VAV unit we used in the shut the sash example, its counterpart CAV unit would cost 3.2 times more to operate (Figure 5.3).

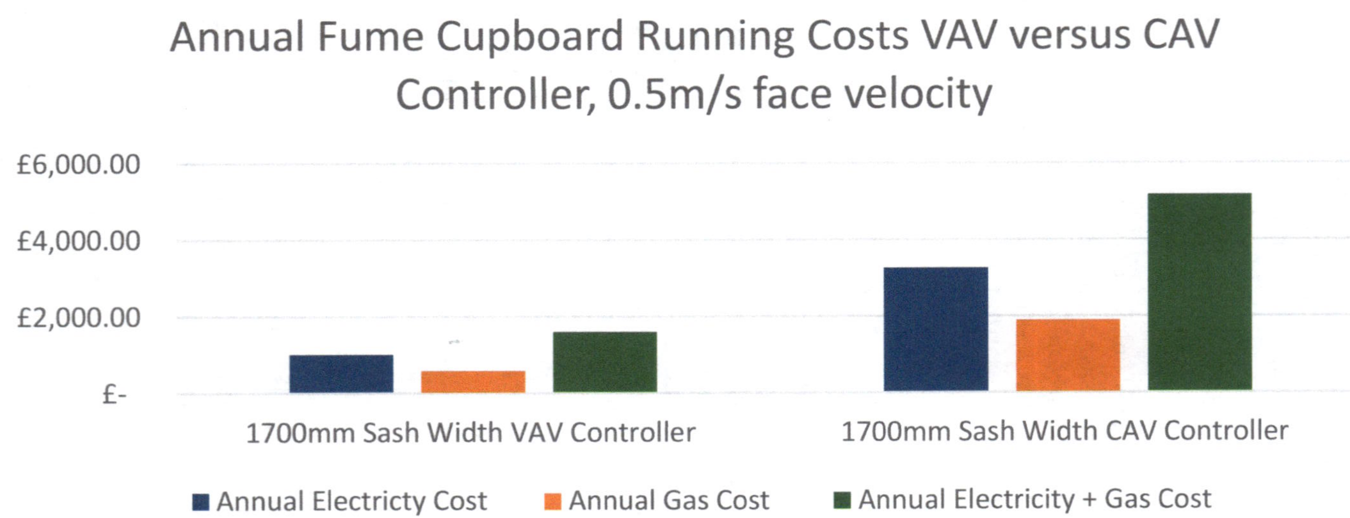

Figure 5.3 Fume cupboard running costs using VAV *vs.* CAV controllers. Utilities cost £0.23 per kW h electricity, £0.06 per kW h gas. Both units operate at design face velocity of 0.5 m s^{-1}.

So why do we have them? Well although they are becoming less commonplace, they are not rare. They are often found in older buildings or buildings that may only require a handful of fume cupboards. CAV controllers are less expensive to procure, install, maintain, and configure. It should be noted that in some rooms with single fume cupboards, a CAV unit may be beneficial to supplement the extract for a rooms ventilation system, though these situations in modern facilities are increasingly rare with more variable and energy efficient designs.

5.2.4 Fume Cupboards, The Ugly

Have you ever stored stuff inside your VAV fume cupboard? Bet you have. Especially if you are in a busy research lab. Why is this possibly an ugly thing? Where else should I be storing my bottles, rotary evaporator, hotplate stirrer, bath, oven *etc.*? Well, storing stuff inside a fume cupboard can increase its running costs. The number of items inside your fume cupboard and their placement will have a significant impact upon your running costs as well as risk reduced safety by impeding air flows. So, what ugly things can you avoid? How can I minimise my running costs by mastering the art of fume cupboard Feng Shui? Before we get into that, let's explore why loading can be a bad, and if unchecked, an ugly thing.

Up until this point, all the fume cupboard running cost data I have shared with you have been for empty fume cupboards. They are in the lab operating at a design face velocity, but so far, all I have done is opened and shut the sash. To highlight the importance of this point, I am going to reference a well-known UK research building with 300 VAV fume cupboards operating at a design face velocity of **0.4** $\mathbf{m\,s^{-1}}$ and the said width of 1700 mm. Now putting stuff inside the fume cupboard does not instantly increase its running costs. However, as you will know, fume cupboards are subject to annual safety tests that ensure that measured air speeds are within an acceptable, safe range. Well, when you put stuff inside the fume cupboard you essentially insert an obstacle course for your airflow, which can slow air down on its path from the front of the sash to the back of the units and off out into the big wide world. So, when an engineer finds that measured air speeds are two low in places and risks safety, there needs to be calibration. This calibration speeds up airflow so the slow areas or 'dead spots' are now fast enough to be deemed passable. The net result of this is the fume cupboard removes an increased volume of air from your lab (Table 5.4). And this air is valuable. You have spent money heating or cooling that air!

Now, you can negate some of the effects of loading by following good loading practice, do your utmost to avoid the loading infringements described in Table 5.5.

That said following these good practice steps is not always feasible, as you may need to store bulky equipment for research and safety purposes in the unit. As your fume cupboard becomes progressively more loaded, spacing

Table 5.4 Impact of loading a fume cupboard work surface upon running costs measured from a UK laboratory.

Work surface loading	Extracted volume increase	Annual running cost increase
35%	**12%**	**£154**
60%	**18%**	**£231**
80%	**40%**	**£514**

Table 5.5 Avoid these work surface loading infringements. They will help to facilitate airflow through your fume cupboard, keeping it safe and sustainable.

Loading infringement	Description
Front	There are items placed in the first 150 mm of the work surface.
Contents	There is an absence of 25 mm gaps between contents.
Back	Items are placed flat to the back surface of the chamber.
Side	Items are placed against the internal side walls of the chamber.

items appropriately may become more challenging. In these circumstances using a raised work surface provided by a plinth or tile loading system (Figure 5.4) will allow for air to flow freely, allowing you to keep your fume cupboard extracted volumes and running costs low.

One last thing regarding fume cupboards. It is an ugly thing that so many people miss, hiding in plain sight needlessly costing you money. It is a subject we have already mentioned and that is calibration. Calibration is not ugly but the absence of it is. So, what do I mean? Remember those annual safety tests, mandatory for each fume cupboard? Well, sometimes a fume cupboard will not just pass the test, it will crush it. Let us say the design face velocity of a CAV fume cupboard is 0.5 m s^{-1}. It has its annual test and passes it with an actual face velocity of 0.6 m s^{-1}. This means that this fume cupboard is operating 20% above its design face velocity, costing you more to run. Needlessly. Overtime, engineers may allow this rate to increase further by focusing on safety only and not energy efficiency, but appropriate calibration must balance the two. What should happen is the engineer should notice this and for CAV units where it is ≥15% above design face velocity they should be calibrated back down to design face velocity. With VAV units they can be more finely controlled and therefore when a VAV is ≥10% above design face velocity it is advised to calibrate back down to design face velocity. Quite often this step is missed and sometimes missed year after year. So have a look now and see if your fume cupboards need calibration, it could save you a significant amount of money (Table 5.6).

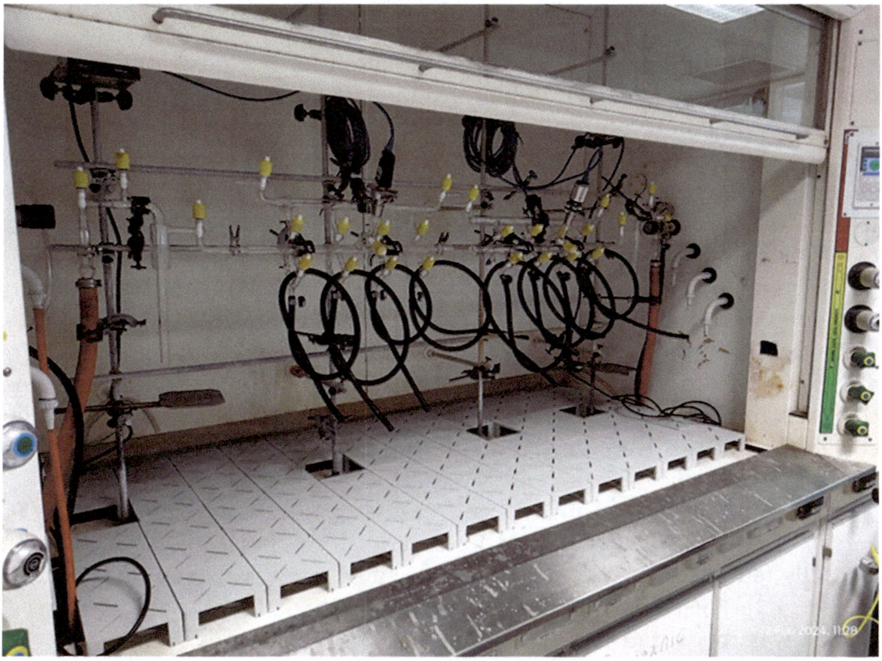

Figure 5.4 Here a 1700 mm sash with VAV fume cupboard has been fitted with a tile loading system. This raised worksurface permits airflow that is unimpeded by heavy loading which helps to minimise fume cupboard running costs.

Table 5.6 Differences in running costs when fume cupboard face velocity increases by 20% for a CAV controlled unit and 15% for a VAV controlled unit. Utilities cost £0.23 per kW h electricity, £0.06 per kW h gas.

	A. Design face velocity		B. Actual face velocity		Running cost
Fume cupboard type	Face velocity	Annual running cost	Face velocity	Annual running cost	Increase (B-A)
CAV 1700 mm Sash Width	0.50 m s^{-1}	£5182.23	0.60 m s^{-1}	£5637.90	**£455.67**
VAV 1700 mm Sash Width	0.40 m s^{-1}	£1185.30	0.46 m s^{-1}	£1363.09	**£177.79**

5.2.5 Cold Storage

'Keep cool and carry on' has tended to be the order of business when it comes to cold storage in the lab. However, in recent years there has been considerable efforts to quantify the impacts of behaviour and usage upon running costs and performance (we'll be looking at ULT freezers separately).

5.2.6 Cold Storage, The Good

As simple as this sounds, I'd be kicking myself for not mentioning it; please only buy models that have a digital temperature controller. A basic thing? Yes. However, there are still units in use and being procured that have a simple arbitrary dial (like most domestic fridges) that doesn't allow you to either set or get a visual indication of chamber temperature. Just being able to set your temperature accurately is a good thing, we will look at the impacts of set temperatures shortly.

The next good thing to do is to have an inventory or 'map' of your unit. This map needs to be stuck on the front door of the unit. Digital records are great, and I applaud those of you who keep detailed records. But where I am coming from is that when a researcher approaches the fridge or freezer they will know where they need to go to retrieve (and return) specific contents. Not until recently had the impact of having an inventory been measured.

To assess this, we examined two freezers operating at −20 °C at the University of Bristol's Learning and Research Centre, both filled with contents typically stored at that temperature such as Cryoboxes and samples (Figure 5.5). The total volume (measured in grams) of material stored in each freezer was measured and the results are summarised in Table 5.7.

We then had three users from the Learning and Research Centre retrieve specific tubes or Cryoboxes. On the first day, they had a detailed freezer map (Figure 5.6) showing the location of all contents allowing organised access. The users were asked to do the following:

- 9 am: Retrieve two specified items.
- 12 pm: Return the items taken at 9 am and retrieve others. After a 2-minute pause, return the items to their original locations.
- 3 pm: Retrieve specific items. After a 2-minute pause, return the items to their original locations.

This process was then repeated the next day without any freezer guide or indication of content locations leading to unorganised access.

The results are shown in Table 5.8. In this case, removing the freezer guide increased door opening times 97–191% and door opening recovery energy by 61–94%. Let us remember that as the energy consumption increases so does the heat output from the compressors of the freezers into the lab. If the lab is air conditioned, there's another running cost that just increased.

What about fridges I hear you ask? Well, a similar case study was also carried out on three fridges. Again, these units were loaded with labelled contents. This time there were three levels of inventory usage, shown in Table 5.9.

End users from the same Learning and Research Centre, University of Bristol, were tasked with retrieving named, labelled contents. During each procedure the same contents were retrieved and replaced but never by the same end user.

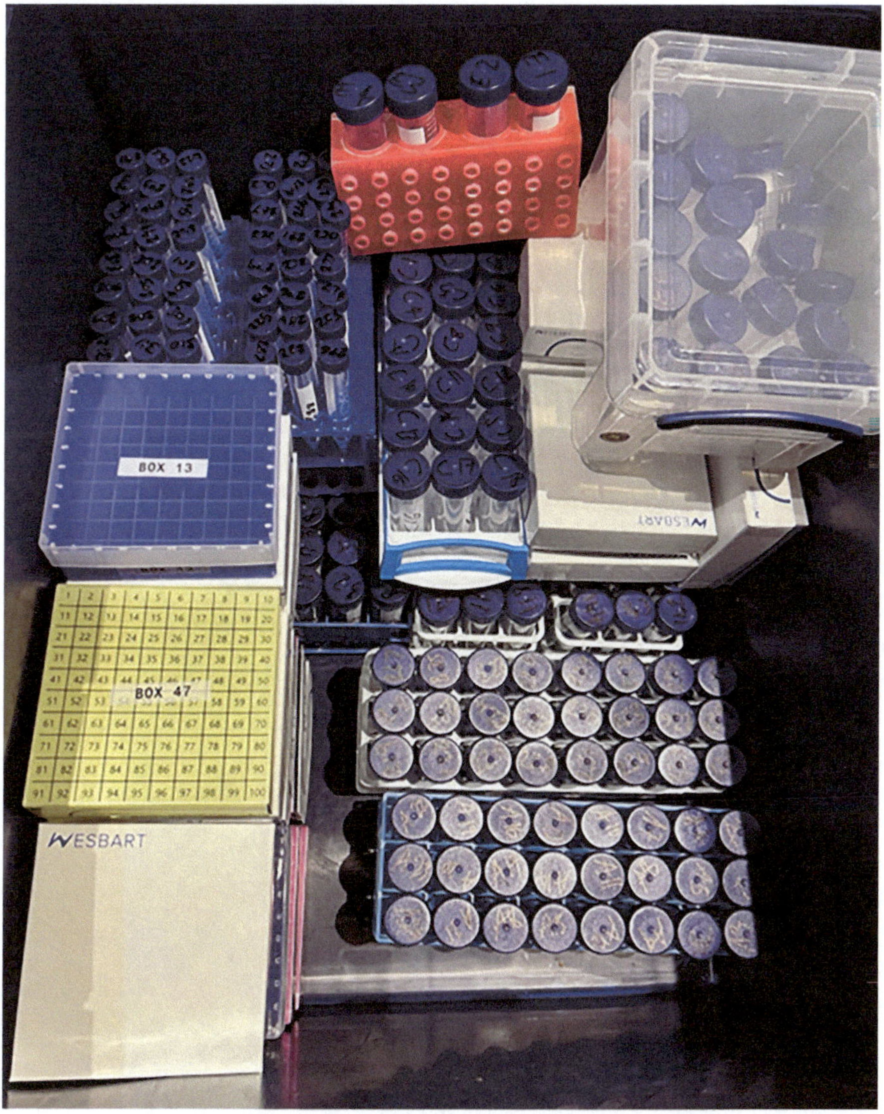

Figure 5.5 A variety of tubes and containers were used to provide representative loading for both freezers. Photo taken from an independent case study carried out at the Learning and Research Centre, University of Bristol. Case study commissioned by Calibre Scientific and Liebherr GB Ltd, testing carried out by Green Light Laboratories Ltd.

During each procedure the end users were given a note detailing the two items to be located and the name of the fictional owner of those items.

The first access of the day was carried out from 9 am onwards and involved the simple retrieval of two specific items, a 1-minute wait and then those items were then placed back in their original locations. The second

Table 5.7 Total amounts of material stored in each freezer. Data taken from an independent case study carried out at the Learning and Research Centre, University of Bristol. Case study commissioned by Calibre Scientific and Liebherr GB Ltd, testing carried out by Green Light Laboratories Ltd.

	Liebherr freezer model	
Contents data	SFFvh5501	SFFsg5501
Total space occupied (l)	88.48	50.78
A. Total plastics weight (g)	8773.44	4394.74
B. Total water (g)	14 720	9479
C. Total cardboard weight (g)	1305.22	1398.45
D. Total aluminium weight (g)	3910.86	1133.81
Total weight (A + B + C + D) (g)	28 709.52	16 406

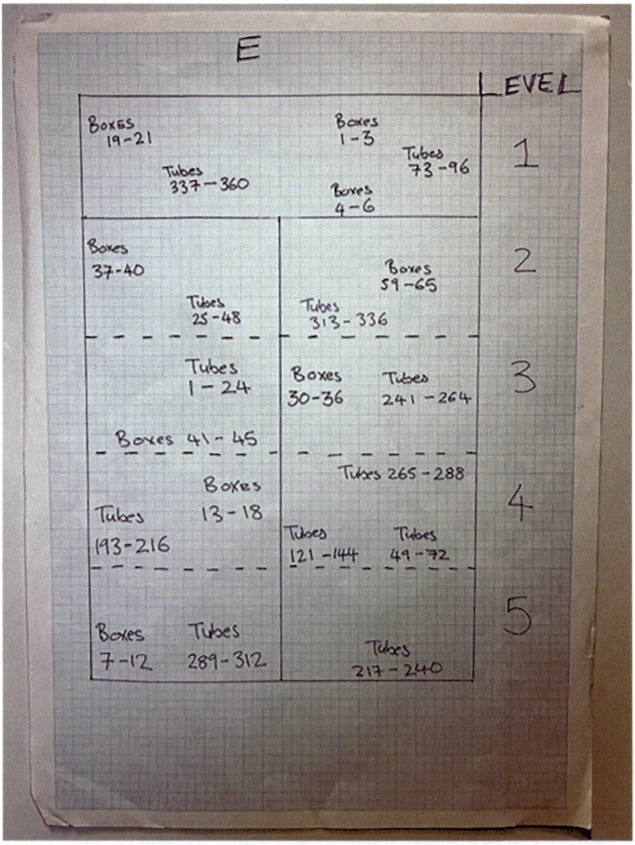

Figure 5.6 Freezer map used to show locations of all contents stored inside the Liebherr SFFvh5501 freezer. Each shelf and drawer were clearly shown to ensure that end users could quickly retrieve contents. Photo taken from an independent case study carried out at the Learning and Research Centre, University of Bristol. Case study commissioned by Calibre Scientific and Liebherr GB Ltd, testing carried out by Green Light Laboratories Ltd.

Table 5.8 Impact of having an inventory (organised access) compared to the absence of any inventory (unorganised access). Data taken from an independent case study carried out at the Learning and Research Centre, University of Bristol. Case study commissioned by Calibre Scientific and Liebherr GB Ltd, testing carried out by Green Light Laboratories Ltd.

Door opening data – Loaded freezers		
Manufacturer	Liebherr	Liebherr
Model	SFFvh5501	SFFsg5501
Net capacity (L)	257	394
Organised access data		
1st Door opening time	26 seconds	24 seconds
2nd Door opening time	21 seconds	53 seconds
2nd Door opening time	44 seconds	48 seconds
Total time	91 seconds	125 seconds
Peak recorded temperature during door openings		
Top compartment	−14.6 °C	−18.8 °C
Middle compartment	−20.6 °C	−18.6 °C
Middle compartment sample	−21.0 °C	−18.6 °C
Bottom compartment	−18.9 °C	−18.5 °C
Energy cost of total access duration	0.132 kW h	0.079 kW h
Unorganised access data		
1st door opening time	50 seconds	73 seconds
2nd door opening time	145 seconds	82 seconds
2nd door opening time	70 seconds	91 seconds
Total time	265 seconds	246 seconds
Peak recorded temperature during door openings		
Top compartment	−7.8 °C	−17.4 °C
Middle compartment	−18.6 °C	−17.8 °C
Middle compartment sample	−19.7 °C	−17.4 °C
Bottom compartment	−16.3 °C	18.3 °C
Energy cost of total access duration	0.212 kW h	0.153 kW h
Increase in total door opening time	**174 seconds (191%)**	**121 seconds (97%)**
Increase in total door opening energy costs	**0.080** kW h **(61%)**	**0.074** kW h **(94%)**

and third access steps of the day were carried out at ~12 pm and ~3 pm respectively.

The results of this case study are summarised in Table 5.10.

For me, the most impactful results are those of the organised access. Why? Well, this procedure involved using a basic inventory which was no more than a guide depicting who owned each shelf. This inventory took 5 minutes to produce and yet reduced door opening times by 50% and door opening recovery energy by 57%. For me it would be almost criminal for such a simple, quick action to be missed. Take 5 minutes, save 50%. Admittedly

Table 5.9 The three levels of organisation employed in the fridge inventory case study, the higher the level of organisation, the higher the amount of detail included in the associated inventory.

Access procedure	Description
Unorganised	Labelled contents are retrieved and then replaced with no inventory or shelf labelling to guide end users.
Organised	Labelled contents are retrieved and then replaced with the aid of a basic inventory affixed to the outer door indicating the ownership of each shelf space.
Highly organised	Labelled contents are retrieved and then replaced with the aid of a detailed inventory affixed to the outer door indicating the ownership of each shelf space and the exact location of all contents. In the chamber, each shelf is also labelled indicating the owner.

Table 5.10 Summary of fridge door opening times and recovery energy when using increasingly detailed inventories. Data taken from an independent case study carried out at the Learning and Research Centre, University of Bristol. Case study commissioned by Calibre Scientific and Liebherr GB Ltd, testing carried out by Green Light Laboratories Ltd.

Access procedure	Total door opening time	Door opening recovery energy
Unorganised access (U.A.)	498 seconds	0.081 kW h
Organised access (saving *vs.* U.A.)	247 seconds (50%)	0.035 kW h (57%)
Highly organisaed access (saving *vs.* U.A.)	166 seconds (67%)	0.031 kW h (62%)

this study assumes that a user without a guide has no knowledge of exactly where an item is located or needs to be placed, but there are many real-life situations where this is exactly the case. The other good reason for employing an inventory is avoiding wasting precious storage capacity. If somebody leaves do you have to hang onto their stuff? Free up your capacity just by knowing who owns what. Doing so will help you avoid buying extra items of cold storage, saving you space, energy, service and maintenance costs and money.

5.2.7 Cold Storage, The Bad

This baddie only applies to freezers or freezer compartments. The first one is the absence drawers and is closely followed by the drawer type employed when present. Having drawers is of huge benefit to you and your precious contents. Firstly, they make accessing your stuff easier, not reaching into the back of a freezer chamber, having to decant half of the contents just to retrieve that one thing saves you time. Secondly a drawer might save you time and money. Might? Yes. It depends on the drawer type we are talking about. The best drawer to go for would be a solid plastic drawer without any holes. Wire drawers hold your contents in place, although that said it is not uncommon for items to fall out of the gaps or dangle half-way out making pushing the drawer back into place a long-drawn-out affair. Also, wire drawers cannot really retain cold air when you carry out a door opening resulting in temperature rises akin to having no drawers at all. So, what is the difference for you if you compare using solid drawers to zero drawers? Well, when drawers are absent door opening temperature rises for a sample representative probe are over two times higher and air temperature rises are over three times higher. The energy to recover from these door openings also rises by 95% (Table 5.11).

Table 5.11 The impact of freezer drawers on temperature rises during a following door opening and the energy required to recover from door openings. Data taken from an independent case study carried out at the College of Health and Life Sciences, Aston University. Case study commissioned by Liebherr GB Ltd, testing carried out by Green Light Laboratories Ltd.

	Under bench freezer
Model	LGUex1500
Net capacity	129
kW h per day at −20 °C set point	0.912
Energy – W per L per day	7.07
Avg. sample temperature at −20 °C set point (lowest/highest)	−22.0 °C (−22.9 °C/−20.9 °C)
Avg. air temperature at −20 °C set point (lowest/highest)[pull down]	−22.3 °C (−24.0 °C/−20.2 °C) [71 min]
Drawers In – Door Openings – Sensor Drawer Pulled Out	
1 Minute Door Opening Performance Air	Rise of 2.8 °C to −18.6 °C
1 Minute door opening performance sample	Rise of 1.5 °C to −19.4 °C
2 Minute door opening performance air	Rise of 6.0 C to −16.4 °C
2 Minute door opening performance sample	Rise of 3.7 °C to −18.0 °C
3 Minute door opening performance air	Rise of 8.9 °C to −12.4 °C
3 Minute door opening performance sample	Rise of 7.2 °C to −14.6 °C
kW h per day at −20 °C set point + door openings	0.955
Energy – W per L per day with door openings	7.40
No drawers – door openings	
1 Minute door opening performance air	Rise of 13.3 °C to −6.9 °C
1 Minute door opening performance sample	Rise of 5.3 °C to −13.9 °C
2 Minute door opening performance air	Rise of 20.4 C to −1.1 °C
2 Minute door opening performance sample	Rise of 10.1 °C to −8.8 °C
3 Minute door opening performance air	Rise of 28.2 °C to 4.5 °C
3 Minute door opening performance sample	Rise of 14.8 °C to −4.9 °C
kW h per day at −20 °C set point + door openings	0.996
Energy – W per L per day with door openings	7.72

5.2.8 Cold Storage, The Ugly

The presence of ice in cold storage is unavoidable, particularly in areas which are relatively humid. However, like your fingernails, it is a good idea to minimise the growth of such things otherwise day to day activities may become difficult, time consuming and even painful. Icing progresses faster in freezers than it does in fridges. In the latter, icing may be rarer but can still develop from the rear of the chamber, slowly creeping forward, engulfing contents. In either chamber type, icing is an ugly thing bringing bad news. The bad news is:

1. Icing reduces the space available for storage.
2. Icing increases the time taken to retrieve contents. In turn this prolongs door opening times, meaning chamber and content temperature rises, increase in energy to recover from a door opening and a proportional increase in heat output into the lab.
3. Icing can damage drawers. Broken drawers are less effective in retaining content and cold air. This will increase door opening times, and temperature rises (see point 2 above).
4. Icing can impede proper door closure. This can have a couple of outcomes. The door will not seal as effectively, resulting in an increase of both chamber temperature and energy consumption (note that this can result in damage of your precious samples). The other outcome is that requiring more force or effort to close the door will damage the door handle and/or the door hinges.

There is a fifth point too. As icing can result in longer door opening times, the prolonged exposure to the warm air of the lab increases icing. Which in turn exacerbates the four points listed above. So, de-ice. Annually doing so should be frequent enough to keep these ugly outcomes at bay, and if you cannot find the time to do the entire unit, at the very least keep those seals clean!

5.2.9 ULT Freezers

Otherwise known as '−80's', ULT freezers are widely used for the stable, long-term storage of precious samples and products. Older units can consume >30 kW h per day when set to −80 °C, more modern incarnations will consume ≤10 kW h per day at the same set point. For context, the average European home consumes just 5.5 kW h of electricity per day (according to the EU's Eurostat). In the late 2000s energy efficient ULT freezers first appeared in the marketplace. For many people, including myself, this product range kicked off their journey into sustainability. It was recognised that with this great energy consumption, there could be a variety of ways to reduce and save, leading to sustainability gains. The causes of these differences in consumption were often striking and raising awareness of such things became a vocation and fascination. Over the last 15 years there has been considerable development and research in the field of ULT freezers. In a way, they have served as a prototype for product sustainability, dos and don'ts for manufacturers and end users. Many lessons learnt from this product range are applicable to other items of lab equipment.

5.2.10 ULT Freezers, The Good

The first good thing any ULT freezer owner should do is get hold of some racking. Racking comes with several wonderful benefits that more than justify the price of procuring this essential accessory. To demonstrate this, we have loaded a ULT freezer with 475 cryoboxes (Figure 5.7 and Table 5.12) and measured performance without and then with aluminium racking (Table 5.13).

The first benefit of racking is the reduction in temperature rises following a door opening. Racking reduces such temperature rises by up to 74%.

Figure 5.7 Left, the SU780XLE 79% full without racking. Right, the unit with racking installed and contents in the same locations. Each compartment apart from compartment 2, housed 75 cryoboxes. In compartment 2, 100 cryoboxes were housed. Compartment 2 housed all aluminium and cardboard cryoboxes which were each concentrated into 4 rows (front to back) of 5 cryoboxes. Data taken from an independent case study carried out at the Learning and Research Centre, University of Bristol. Case study commissioned by Calibre Scientific, testing carried out by Green Light Laboratories Ltd.

Table 5.12 Weights of ULT freezer contents, 79% full, without and with aluminium racking (PP = polypropylene, Al = aluminium). Data taken from an independent case study carried out at the Learning and Research Centre, University of Bristol. Case study commissioned by Calibre Scientific, testing carried out by Green Light Laboratories Ltd.

Stirling ultracold SU780XLE Contents data	Units	No racking Total weight (g)	Racked Total weight (g)
Aluminium racking units	30	0	11 4000
PP cryoboxes	435	67 516.35	67 516.35
Al. cryoboxes	20	4345.4	4345.4
Card cryobozes	20	1864.6	1864.6
PE bag	475	225	225
100 mL water/PE bag	475	47 500	47 500
Total Al weight		**4345.4**	**11 8345.4**
Total plastics weight		**67 741.35**	**67 741.35**
Total cardboard weight		**1864.6**	**1864.6**
Total water weight		**47 500**	**47 500**
Total weight stored		**121 451.35**	**235 451.35**

Furthermore, the energy required to recover from door opening is lower. The next benefit of racking is its time-saving capability. Racking keeps your contents organised and makes retrieval easier, especially when combined with a freezer map. How much easier? Well, when accessing the unit 3 times in a working day to retrieve and then return a total of 6 cryoboxes (and yes, the same 6 in both cases) door opening time was reduced by 33% with the door opening recovery energy falling by 56% (Table 5.14).

Racking also helps you to quickly transfer your contents from freezer to freezer (providing the freezer is the same model or compatible with the racking). The last benefit of using racking is temperature and time. In the event of a loss of power or if you need to relocate your ULT freezer, racking gives you a time cushion, holding your contents below the dreaded −50 °C threshold for longer (Table 5.15). Need an extra ~5–7.5 hours? Racking gives you that time due to its improved thermal capacity.

Now, as shown in Table 5.14, it would be silly not to mention the importance of having an inventory. As we have already covered this subject, we can be brief. In our case study evaluating a ULT freezer, an inventory will reduce the door opening time by 67% and door opening recovery energy by 70%. That 70% represents 2.214 kW h of electricity, enough to operate two ~300 L lab grade freezers at −20 °C for over 24 hours. That is just from one day of door openings.

Racking can either be made from aluminium or stainless steel. Now which one should you choose? Well, when we compare each type there's the expected pros and cons (Table 5.16). In terms of performance, we have compared both in the same freezer and had results that are not too dissimilar (Table 5.17). It must be noted that this testing was carried out without the ULT freezer being loaded.

Table 5.13 Impacts of racking upon door opening and temperature performance in a 79% loaded ULT freezer. Data taken from an independent case study carried out at the Learning and Research Centre, University of Bristol. Case study commissioned by Calibre Scientific, testing carried out by Green Light Laboratories Ltd.

Manufacturer	Stirling ultracold	
Model	SU780XLE	
Net capacity (L)	780	
Performance data	**79% Loaded No Racking**	**79% Loaded Racked**
kW h per day at −80 °C set point	7.602	7.654
Energy – W per L per day	9.75	9.81
C1 (Back) Mean air temp (Lowest/Highest)	−80.4 °C (−80.4 °C/ −80.4 °C)	−77.9 °C (−78.0 °C/ −77.9 °C)
C1 (Middlle) Mean air temp (Lowest/Highest)	−78.7 °C (−78.7 °C/ −78.7 °C)	−75.8 °C (−75.9 °C/ −75.8 °C)
C2 (Middlle) Mean air temp (Lowest/Highest)	−82.8 °C (−82.8 °C/ −82.7 °C)	−80.7 °C (−80.8 °C/ −80.7 °C)
C2 PP Cryobox Mean air temp (Lowest/Highest)	−84.4 °C (−84.4 °C/ −84.4 °C)	−81.4 °C (−81.5 °C/ −81.4 °C)
C2 Card Cryobox Mean air temp (Lowest/Highest)	−84.0 °C (−84.1 °C/ −84.0 °C)	−82.2 °C (−82.2 °C/ −82.2 °C)
C2 Al. Cryobox Mean air temp (Lowest/Highest)	−84.1 °C (−84.1 °C/ −84.0 °C)	−81.5 °C (−81.6 °C/ −81.5 °C)
C3 (Middlle) Mean air temp (Lowest/Highest)	−84.8 °C (−84.8 °C/ −84.7 °C)	−84.1 °C (−84.2 °C/ −84.1 °C)
C3 (Sample) Mean sample temp (Lowest/Highest)	−84.1 °C (−84.2 °C/ −84.0 °C)	−83.6 °C (−83.7 °C/ −83.6 °C)
C4 (Middlle) Mean air temp (Lowest/Highest)	−84.0 °C (−84.1 °C/ −83.9 °C)	−84.9 °C (−85.0 °C/ −84.8 °C)
C5 (Middlle) Mean air temp (Lowest/Highest)	−82.2 °C (−82.3 °C/ −82.1 °C)	−84.0 °C (−84.1 °C/ −83.9 °C)
C6 (Middlle) Mean air temp (Lowest/Highest)	−77.8 °C (−77.9 °C/ −77.8 °C)	−81.5 °C (−81.6 °C/ −81.3 °C)
C6 (Front) Mean air temp (Lowest/Highest)	−77.0 °C (−77.2 °C/ −77.0 °C)	−80.8 °C (−80.9 °C/ −80.7 °C)
Timed door opening data		
A. Single 60 second door opening		
C1 (Back) Start air temp (Peak/ Rise)[Recovery Time]	−80.4 °C (−77.9 °C/ 2.5 °C)[86 minutes]	−77.9 °C (−76.0 °C/ 1.9 °C)[−77.8 °C in 616 minutes]
C1 (Middlle) Start air temp (Peak/Rise)[Recovery Time]	−78.7 °C (−75.8 °C/ 2.9 °C) [−78.6 °C in 429 minutes]	−75.8 °C (−73.9 °C/ 0.8 °C)[−75.7 °C in 616 minutes]
C2 (Middlle) Start air temp (Peak/Rise)[Recovery Time]	−82.7 °C (−80.5 °C/ 2.2 °C)[0 minutes]	−80.7 °C (−79.2 °C/ 1.5 °C)[179 minutes]
C2 PP Cryobox start air temp (Peak/Rise)[Recovery Time]	−84.5 °C (−82.5 °C/ 2.0 °C) [0 minutes]	−81.4 °C (80.8 °C/ 0.6 °C)[0 minutes]

Table 5.13 (*Continued*)

Manufacturer	Stirling ultracold	
C2 Card cryobox start air temp (Peak/Rise)[Recovery Time]	−84.1 °C (−82.5 °C/1.6 °C) [0 minutes]	−82.2 °C (−81.4 °C/ 0.8 °C)[0 minutes]
C2 Al. Cryobox start air temp (Peak/Rise)[Recovery Time]	−84.0 °C (−82.3 °C/1.7 °C) [0 minutes]	−81.5 °C (−80.8 °C/ 0.7 °C)[0 minutes]
C3 (Middlle) Start air temp (Peak/Rise)[Recovery Time]	−84.7 °C (−81.7 °C/3.0 °C) [0 minutes]	−84.1 °C (−83.1 °C/ 1.0 °C)[0 minutes]
C3 (Sample) Start sample temp (Peak/Rise)[Recovery Time]	−84.0 °C (−81.6 °C/2.4 °C) [0 minutes]	−83.6 °C (−82.5 °C/ 0.9 °C)[0 minutes]
C4 (Middlle) Start air temp (Peak/Rise)[Recovery Time]	−83.9 °C (−80.6 °C/3.3 °C) [0 minutes]	−84.8 °C (−84.0 °C/ 0.8 °C)[0 minutes]
C5 (Middlle) Start air temp (Peak/Rise)[Recovery Time]	−82.1 °C (−80.0 °C/2.1 °C) [0 minutes]	−83.9 °C (−83.3 °C/ 0.6 °C)[0 minutes]
C6 (Middlle) Start air temp (Peak/Rise)[Recovery Time]	−78.4 °C (−77.0 °C/1.4 °C) [4 minutes]	−81.4 °C (−81.2 °C/ 0.2 °C)[0 minutes]
C6 (Front) Start air temp (Peak/Rise)[Recovery Time]	−77.0 °C (−76.9 °C/0.1 °C) [6 minutes]	−80.7 °C (−80.7 °C/ 0.0 °C)[0 minutes]
B. Single 90 second door opening		
C1 (Back) Start air temp (Peak/ Rise)[Recovery Time]	−80.4 °C (−76.7 °C/ 3.7 °C)[110 minutes]	−77.8 °C (−75.2 °C/ 2.6 °C)[−77.4 °C in 550 minutes]
C1 (Middlle) Start air temp (Peak/Rise)[Recovery Time]	−78.6 °C (−74.3 °C/2.3 °C) [−78.4 °C in 884 minutes]	−75.6 °C (−72.9 °C/ 0.8 °C)[−75.2 °C in 550 minutes]
C2 (Middlle) Start air temp (Peak/Rise)[Recovery Time]	−82.8 °C (−79.3 °C/ 3.5 °C)[59 minutes]	−80.3 °C (−78.1 °C/ 1.8 °C)[−79.5 °C in 550 minutes]
C2 PP Cryobox Start air temp (Peak/Rise)[Recovery Time]	−84.4 °C (−83.8 °C/0.6 °C) [0 minutes]	−81.1 °C (−80.0 °C/ 1.1 °C)[0 minutes]
C2 Card Cryobox start air temp (Peak/Rise)[Recovery Time]	−84.1 °C (−83.1 °C/ 1.0 °C)[0 minutes]	−81.9 °C (−80.7 °C/ 1.2 °C)[0 minutes]
C2 Al. Cryobox start air temp (Peak/Rise)[Recovery Time]	−84.1 °C (−81.4 °C/2.7 °C) [0 minutes]	−81.1 °C (−80.0 °C/ 1.1 °C)[0 minutes]
C3 (Middlle) Start air temp (Peak/Rise)[Recovery Time]	−84.9 °C (−80.2 °C/ 4.7 °C)[0 minutes]	−83.9 °C (−82.3 °C/ 1.6 °C)[0 minutes]
C3 (Sample) Start sample temp (Peak/Rise)[Recovery Time]	−84.2 °C (−80.4 °C/ 3.8 °C)[0 minutes]	−83.4 °C (−81.6 °C/ 1.8 °C)[0 minutes]
C4 (Middlle) Start air temp (Peak/Rise)[Recovery Time]	−84.3 °C (−79.0 °C/ 5.3 °C)[35 minutes]	−84.8 °C (−83.4 °C/ 1.4 °C)[0 minutes]
C5 (Middlle) Start air temp (Peak/Rise)[Recovery Time]	−82.7 °C (−79.1 °C/ 3.6 °C)[29 minutes]	−84.0 °C (−83.0 °C/ 1.0 °C)[0 minutes]
C6 (Middlle) Start air temp (Peak/Rise)[Recovery Time]	−78.4 °C (−78.2 °C/ 0.2 °C)[0 minutes]	−81.5 °C (−81.4 °C/ 0.1 °C)[0 minutes]
C6 (Front) Start Air Temp (Peak/Rise)[Recovery Time]	−77.8 °C (−77.8 °C/ 0.0 °C)[0 minutes]	−80.9 °C (−80.8 °C/ 0.1 °C)[0 minutes]
Door opening energy (A+B) (kW h)	**1.289**	**1.057**

Table 5.14 Inventory usage impacts upon door opening time and energy consumption. Data taken from an independent case study carried out at the Learning and Research Centre, University of Bristol. Case study commissioned by Calibre Scientific, testing carried out by Green Light Laboratories Ltd.

Manufacturer	Stirling ultracold		
Model	SU780XLE		
Performance data	**79% Loaded no racking**	**79% Loaded racked**	**79% Loaded racked**
Door opening data	**Organised access**	**Organised access**	**Unorganised access**
1st Door Opening Time	57 seconds	40 seconds	120 seconds
2nd Door opening time	164 seconds	105 seconds	271 seconds
2nd Door opening time	169 seconds	117 seconds	414 seconds
Total time	**390 seconds**	**262 seconds**	**805 seconds**
Door opening energy consumption	**2.139** kW h	**0.951** kW h	**3.165** kW h
Peak recorded temperature during door openings			
C1 (Back) Mean air temp	−76.8 °C	−74.1 °C	−67.3 °C
C1 (Middlle) Mean air temp	−73.8 °C	−71.2 °C	−63.4 °C
C2 (Middlle) Mean air temp	−76.6 °C	−76.9 °C	−71.4 °C
C2 PP Cryobox mean air temp	−82.9 °C	−78.5 °C	−74.0 °C
C2 Card cryobox mean air temp	−81.7 °C	−79.8 °C	−76.0 °C
C2 Al. Cryobox mean air temp	−80.3 °C	−78.9 °C	−74.3 °C
C3 (Middlle) Mean air temp	−79.9 °C	−81.3 °C	−76.9 °C
C3 (Sample) Mean sample temp	−79.8 °C	−80.6 °C	−75.9 °C
C4 (Middlle) Mean air temp	−77.6 °C	−83.1 °C	−79.7 °C
C5 (Middlle) Mean air temp	−77.2 °C	−82.6 °C	−80.9 °C
C6 (Middlle) Mean air temp	−77.0 °C	−80.2 °C	−79.0 °C
C6 (Front) Mean air temp	−76.0 °C	−79.7 °C	−78.4 °C

Beyond racking, the next good thing you can do with your ULT freezer is warm it up. Just a little bit is good, by 10 °C is great. Now there is plenty of talk regarding the energy saved by warming up your ULT freezer. The old headline quoted figure is 30–40% reduction in energy consumption, but that's very 2009. For the more modern units, from the 22 different models made in the last 5 years that we have tested, the energy saved by warming up from −80 °C to −70 °C ranges from 18.88% to 38.44%. These savings represent 1.148 to 5.270 kW h per day saved. You may ask, why such a range?

Table 5.15 Impact of racking upon warm up times from −80 °C set point to −50 °C threshold in ULT freezers with 79% loading. Data taken from an independent case study carried out at the Learning and Research Centre, University of Bristol. Case study commissioned by Calibre Scientific, testing carried out by Green Light Laboratories Ltd.

Manufacturer	Stirling ultracold		
Model	SU780XLE		
PERFORMANCE DATA	**79% Loaded No racking**	**79% Loaded racked**	**79% Loaded racked**
Door opening data	**Organised access**	**Organised access**	**Unorganised access**
1st Door opening time	57 seconds	40 seconds	120 seconds
2nd Door opening time	164 seconds	105 seconds	271 seconds
2nd Door opening time	169 seconds	117 seconds	414 seconds
Total time	**390 seconds**	**262 seconds**	**805 seconds**
Door opening energy consumption	**2.139 kW h**	**0.951 kW h**	**3.165 kW h**
Peak recorded temperature during door openings			
C1 (Back) Mean air temp	−76.8 °C	−74.1 °C	−67.3 °C
C1 (Middlle) Mean air temp	−73.8 °C	−71.2 °C	−63.4 °C
C2 (Middlle) Mean air temp	−76.6 °C	−76.9 °C	−71.4 °C
C2 PP Cryobox mean air temp	−82.9 °C	−78.5 °C	−74.0 °C
C2 Card cryobox mean air temp	−81.7 °C	−79.8 °C	−76.0 °C
C2 Al. Cryobox mean air temp	−80.3 °C	−78.9 °C	−74.3 °C
C3 (Middlle) Mean air temp	−79.9 °C	−81.3 °C	−76.9 °C
C3 (Sample) Mean sample temp	−79.8 °C	−80.6 °C	−75.9 °C
C4 (Middlle) Mean air temp	−77.6 °C	−83.1 °C	−79.7 °C
C5 (Middlle) Mean air temp	−77.2 °C	−82.6 °C	−80.9 °C
C6 (Middlle) Mean air temp	−77.0 °C	−80.2 °C	−79.0 °C
C6 (Front) Mean air temp	−76.0 °C	−79.7 °C	−78.4 °C

Well, I believe one reason is that some ULT freezers are already a tad warmer than −80 °C at that set point. It is a great way to save energy by not being as cold as what the display on the door reads. So, when units are warmed up to the −70 °C set point, they are not jumping up a whole 10 °C and therefore the energy savings aren't as great. Furthermore, the chamber temperatures measured at the −70 °C set point are sometimes warmer than −70 °C, meaning that door openings can result is temperatures peaking closer to and warmer than −50 °C. For this reason, warming up to 70 °C is only a good

Table 5.16 Aluminium or stainless-steel racking?

Property	Aluminium	Stainless steel
Price	Lower	Higher
Thermal conductivity	High – Cools down faster	Low – Cools down slower
Heat capacity	Low – Requires less energy to change temperature	High – Requires more energy to change temperature
Cooling efficiency	Efficient – Stays cold for a short time	Less Efficient – Stays cold for a longer time
Strength	Strong but prone to denting and deformation	Superior strength, resistant to deformation
Corrosion resistance	Susceptible to corrosion but forms a protective oxide layer	Highly resistant to corrosion due to chromium
Weight	Lightweight	Heavier

Table 5.17 ULT freezer performance with different racking metal types. Unit had zero loading. Data taken from an independent case study carried out at the School of Plant Sciences, University of Oxford. Case study commissioned by Eppendorf UK and Scientific Laboratories Supplies Ltd, testing carried out by Green Light Laboratories Ltd.

Measurement	Empty F570h ULT freezer	Aluminium Racked F570h ULT freezer	Steel racked F570h ULT freezer
Energy consumption	7.57 kW h	7.20 kW h	7.77 kW h
Compartment 1 Avg. temperature	−78.6 °C	−78.9 °C	−76.7 °C
Compartment 2 Avg. temperature	−79.9 °C	−79.3 °C	−78.6 °C
Compartment 3 Back Avg. temperature	−81.5 °C	−80.5 °C	−79.3 °C
Compartment 3 Middle Avg. temperature	−80.2 °C	−79.4 °C	−78.9 °C
Compartment 3 Front Avg. temperature	−80.4 °C	−79.5 °C	−78.6 °C
Compartment 4 Avg. temperature	−79.2 °C	−79.2 °C	−77.9 °C
Compartment 5 Avg. temperature	−76.7 °C	−77.4 °C	−75.3 °C
Compartment 1 warm up time to −50 °C	4 hours 19 minutes	8 hours 44 minutes	8 hours 11 minutes
Compartment 2 warm up time to −50 °C	5 hours 12 minutes	11 hours 4 minutes	11 hours 1 Minute
Compartment 3 back warm up time to −50 °C	5 hours 55 minutes	13 hours 11 minutes	13 hours 8 minutes
Compartment 3 middle warm up time to −50 °C	5 hours 54 minutes	13 hours 5 minutes	13 hours 30 minutes
Compartment 3 front warm up time to −50 °C	5 hours 49 minutes	13 hours 2 minutes	13 hours 15 minutes
Compartment 4 warm up time to −50 °C	6 hours 10 minutes	14 hours 23 minutes	14 hours 36 minutes
Compartment 5 Warm Up Time To −50 °C	6 hours 10 minutes	14 hours 24 minutes	14 hours 24 minutes

thing if one prerequisite is adhered to. Yep, you have guessed it. Rack your ULT freezer, then warm it up. That way you save energy and protect your contents.

5.2.11 ULT Freezers, The Bad

For me, there are many bads when it comes to ULT freezers. However, for reasons that will become apparent, I am going to go for just one bad which may be surprising. For me, the use of specific **modes** is a bad thing. There are numerous models out there that either have a mode that is energy savings or 'eco' and then a high-performance mode. The former saves energy whilst the latter has a better temperature performance (colder, better uniformity). For me, this is bad because why should an end user have to decide between them? The more sustainable units, in my opinion, would be those that are truly as cold as their set temperature, stable, uniform and capable of delivering this performance in an energy efficient manner. Sometimes, marketing materials for units that have modes will advertise specific performance data from distinct modes as one picture, whilst in reality you will have to choose. It is often in a small print, the tiny, miniscule, small print. Being forced to choose between these two aspects of performance is a bad thing. Unnecessary too.

5.2.12 ULT Freezers, The Ugly

Why not put bad practice in The Bad section? Well, because ULT freezers in bad practice generate some particularly ugly situations. So, what are these bad things you should avoid? They include:

- Blocked or dirty filters. Not only will this increase the energy consumption of the unit, but more modern units will simply turn themselves off to prevent their compressors from dying.
- Spacing. We all need room to breathe. ULT freezers are not different. Give them space all around and do not store items on top.
- Seal. Keep the seal free of obstructions and damage. If either of these circumstances occur the chamber temperatures will rise, as will the energy consumption.

There are other bad things to avoid; not having any racking or inventory. We could even decide to stick to operating at −80 °C instead of warming up. But we have covered these previously. So, these three will be ample. Ample enough to show that even if you were to do something good like warming up to −70 °C, being bad can lead to ugly results what negate any good that you may have done. How do we know this? That's right, we tested it too (Figure 5.8).

We did all those things you shouldn't do to a ULT freezer and then measured their impact upon running costs. In a nutshell, warming up to

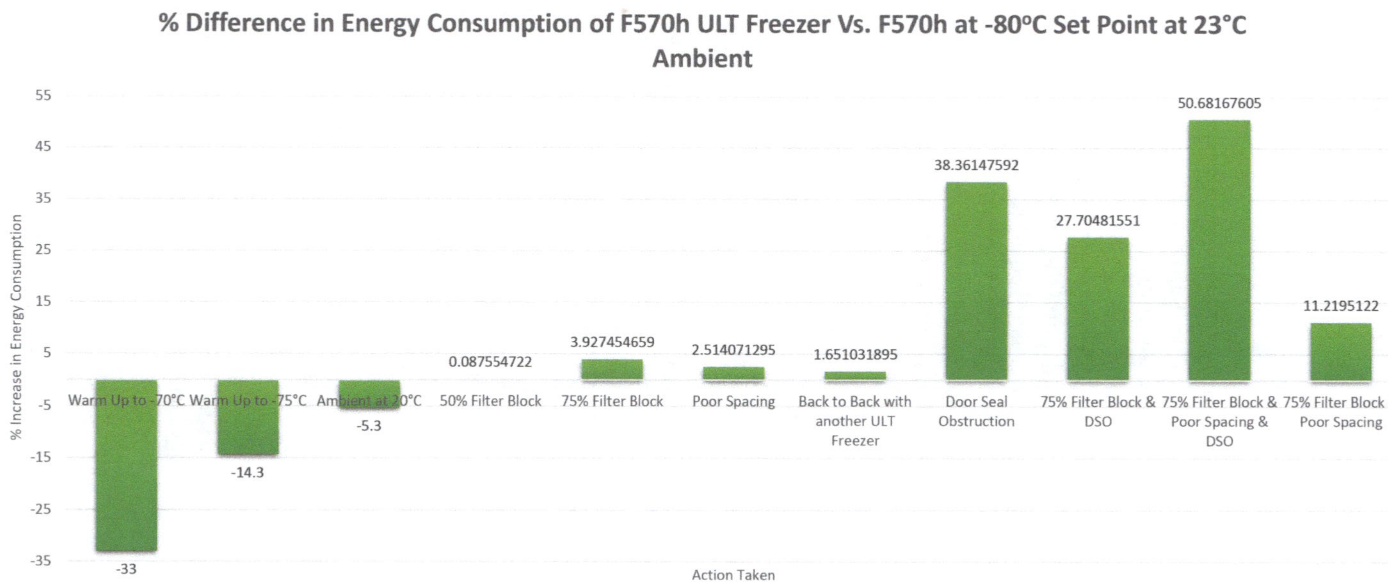

Figure 5.8 Impacts of good and bad practice upon ULT freezer energy consumption. Data taken from an independent case study carried out at the Learning and Research Centre, University of Bristol. Case study commissioned by Eppendorf UK and Scientific Laboratories Supplies Ltd, testing carried out by Green Light Laboratories Ltd.

−70 °C reduced electricity consumption by 33% whilst bad practice increased energy consumption by ≥50%. Now that's ugly and a complete turn off.

5.3 Turn Offs

Turning off should be a turn on. For a variety of reasons, the equipment that could be switched off is left on. It might be required at some point today, it is easier to start it up in the morning and then leave it on or it's just been forgotten about. During lab audits I often find a variety of equipment left on and not in use. My heart sinks that little bit more when I find such occurrences at 8 am and I am the first person in the lab. I have energy monitored thousands of items of equipment over the last 16 years. Listing those would be a gargantuan task, however, in Fig. 26 there is a sample of what you would waste by leaving some common items of lab equipment on. Annually, buildings can save thousands of pounds/dollars/euros on running costs whilst simultaneously slashing their scope 2 carbon emissions (Table 5.18).

Labelling such equipment with 'ok to turn off' stickers helps and a quick sweep of facilities before the end of the working day is a great way to cut wastage. Furthermore, a great tip is to check how long it takes for equipment to warm-up or cool-down to the standard operating temperatures used in

Table 5.18 Turn these items off when you can, or this is what you expect to waste and pay for. Data from independent testing carried out by Green Light Laboratories Ltd.

Equipment type	Examples of savings
Water bath	A 12 L unit filled with 10 L of water: 37 °C: 0.027 kW h h^{-1}/0.648 kW h per day. 65 °C: 0.090 kW h h^{-1}/2.16 kW h per day. 90 °C: 0.186 kW h h^{-1}/4.464 kW h per day
Block heater	A two block unit: 37 °C: 0.008 kW h h^{-1}/0.200 kW h per day. 65 °C: 0.021 kW h h^{-1}/0.492 kW h per day. 90 °C: 0.038 kW h h^{-1}/0 906 kW h per day
Laminar flow hood	1200 mm width. 0.400 kW h h^{-1}/9 6 kW h per day
Microbiological safety cabinet	1200 mm Width (recirculating): Older Design: 0.760 kW h h^{-1}/18.24 kW h per day. New Design: 0.113 kW h h^{-1}/2.712 kW h per day
Glassware drying cabinet	100 L set to 75 °C. Oder Design: 0.635 kW h h^{-1}/15.24 kW h per day New Design: 0.232 kW h h^{-1}/5.56 kW h per day
Rotary evaporator cooling	Water Cooled (20 min): 120 L h^{-1}/2880 L per day Chiller Cooled (single rotovap chiller): set at 6 °C: 0.266 kW h h^{-1}/6.384 kW h per day
Hotplate stirrer	Fitted with 500 mL round bottomed flask adaptor: 250 °C: 0.168 kW h h^{-1}/4.032 kW h per day

your laboratory. This means you can better plan your experiments throughout the day and equally know when you can turn things off.

5.4 Greenwashing and Purchasing

Another form of waste is buying the wrong thing. We all make mistakes but being misled is infuriating because by the time you realise it, you have already bought the item and plugged it in. Funding is hard-won and finite, so spending it in the wrong place, on the wrong thing is a waste. There are a multitude of reasons why the wrong item of equipment may be purchased, some of the most common reasons given are:

- 'We've always used this brand'
- 'The sales rep is great'
- 'It was the cheapest'
- 'They are all the same really'
- 'They said they were the most efficient'

However, for me, the worst factor in guiding poor procurement is greenwashing. The first form of greenwashing is the traditional form, original recipe, greenwashing. Greenwashing is the process by which companies give a false or misleading impression about the environmental benefits of their products. It involves making unfounded claims or unfair comparisons to mislead consumers into thinking that the products are eco-friendly or have a larger positive impact on the environment and running costs than they truly do. So why do companies greenwash? There are a few reasons but the overriding one is that **it works!** In my experience people are inherently good and so if they are going to buy something and they can choose an option that is both efficient and better for the planet, they'll go for it feeling that they have done a positive thing. Greenwashing increases the intent to buy.

I have been teaching about greenwashing for 10 years and raising awareness is a fantastic way to empower end users to better evaluate marketing materials and make the most informed procurement decisions. Procurement officers are fantastic negotiators, so let's help them by filtering out the less efficient, less sustainable, often greenwashed products. The result will be your organisation getting the best deal for the right product.

5.4.1 Traditional Greenwashing

There are a few tricks from the dark arts of marketing to put you in the right frame of mind. For many lab operators, this will be the first impression regarding this *amazing* product, and we all know how important first impressions are. Here are some things to look out for:

- **Framing** – This is often overlooked because it is subtle. From the get-go you need to believe in this document. It is facts, it is irrefutable. So, let's start by naming it appropriately. Names like FACTSHEET or DATASHEET are used to frame the contents. It is more official that way and you will be more invested in what is written. Always remember, it's still marketing literature.

- **Green Galore** – Greenwashing just would not be complete without the use of the green colour and iconography. You are already thinking of leaves and trees, aren't you? A bonus point if you saw them inside a flask. Green worlds also pop up with E being particularly popular: Eco, Efficient, Environmentally friendly.
- **Better Than What?** – '35% reduction in energy consumption'. Compared to what though? Quite often, these figures are often a result of the manufacturer comparing their new product to their old one, or carefully selected units from the marketplace. These models may not be the best ones available. Also, where did they get this information from? Sometimes, the models, manufacturers and sources may be anonymous. *Thanks.*
- **Conditions and Usage** – Differences in or undefined testing locations, ambient conditions, set point, program or cycle type can all be employed to present and exacerbate a difference in running costs or performance.
- **Tips of Icebergs** – Lets focus on this one feature or saving. Just the one. All the other performance and running cost factors may not see the light of day. There are often hidden trade-offs. Less energy used. Great! But maybe not if that is at the expense of performance, longevity, build quality, recyclability, warranty, service and maintenance costs. If such a factor is missing, you need to start investigating.

Let me be clear. You **can** have a great product that has a leaf icon on it. But be on the lookout for the marketing materials where you can see two or more of these greenwashing indicators present. I wish that this was the only warning I will be giving, but there's a second form of greenwashing that has been growing and evolving in recent years.

5.4.2 'Unintentional' Greenwashing

Greenwashing is a problem. Humans are problem solvers. However, sometimes our intended solutions do more harm than good. One of the most frequently asked questions is simply "How can I know which item is more sustainable?", which sadly in this moment in time, is not easily answered. Companies use too many variations in how they assess product "sustainability", and there are no verified programmes which would allow this. Current attempts to label, badge or certify lab equipment to eliminate greenwashing have often resulted in more greenwashing, as companies may confuse assessing their products for sustainability with actually *being* more sustainable.

If we are to eliminate greenwashing and achieve sustainable procurement, then quantified, transparent, and robust data is where it is at. To make informed decisions we need to compare products with a fine granularity of data for all relevant running cost factors. Simplification, wherever possible, is great providing quantification is still included. I am using the term 'unintentional' here because I am optimistic that by highlighting these mistakes we can avoid employing them with immediate effect. So be on the lookout for the following:

- **A, B, C is as easy as 1, 2, 3** – Scales look best on a snake. So, giving an item a rating or 'A' or a numerical mark without an understand of what this means has limited usage if it is not accompanied by transparent data. Furthermore, a scale is only truly useful if it encapsulates the data for all available products, old and new.
- **Omissions** – Like the tip of the iceberg, performance is at best barely mentioned if at all included. Energy efficiency is great but not very useful if I have no idea if it will provide the performance needed.
- **Awarded!** – This product has been awarded a sticker! Awarded or was this a transaction? The language often used in describing how labels are attained is concerning as it implies the product is 'green' or sustainable. This may not be the case.

Until these mistakes are ironed out, stickers, labels, and certifications may lend themselves to being third party marketing instead of an effective tool for guiding procurement.

This is not to be too despairing on systems, as in the end, something is needed to help users better understand what is the most sustainable. In the meantime, users should try to align in how they engage suppliers for information, to create a unified voice. Engage with suppliers on how they have made their claims. Some are genuinely trying to do good. With less generic types of equipment or consumables, users may have few options from which suppliers they are buying from, and so it really requires a dialogue on how products can better perform. The more specified a product becomes, the less suitable it likely is for a certification or standard which relies on generalities to permit comparison. When users unite in their requests, they can sometimes find suppliers who are making genuinely positive strides, but sadly it takes some work in this current day to differentiate the good from the bad, and the ugly.

5.4.3 Final Words

Thank you for reading my chapter. I am optimistic that there is something you can use to improve your equipment-related lab operations. My final tip is that you should ask for help. Quite often, we will jump straight into a task and just go for it. You do not *need* help, but you will benefit from it; speed, quality, motivation, and results are all better with a bit of help. My colleagues and I are working on numerous projects that will help you all out in the coming years, so keep in touch.

Best regards
Andy Evans

CHAPTER 6

Research Quality, Reproducibility[†]

BENOÎT P. NICOLET

Netherlands Cancer Institute, the Netherlands
Email: b.nicolet@nki.nl

6.1 Introduction

Imagine that you are a car manufacturer, and your goal is to reduce the carbon footprint of your car. For this, you would isolate that the goal of your car is to transport, as energetically efficiently as possible, passengers and goods from A to B. In other terms, you would make it as fuel-efficient as possible. Would you first go after how much plastic goes into the dashboard and streamline its manufacturing process? Maybe. But when you realize that most of the footprint of your product comes from the use of the car and not the manufacturing itself, you would probably focus on making any modifications that increase fuel efficiency.

If we now look at a laboratory and project the same thinking onto it, we would consider that a lab is a place where knowledge is gained. Increasing the carbon and resource efficiency of research undertaken, relates to how much knowledge is gained per unit of resource and ultimately carbon. Would you focus on how many tips you use in an experiment? Probably not after hearing that your research's footprint is complex, and plastic tips are only a small fraction of it. We tend to focus on aspects we can see (plastics, consumables). But most of a scientist's footprint is invisible (energy, travel, commute, supply chains). One would rather focus on making their time in

The Sustainability of Science: How Science Impacts the Environment, and What Can Be Done
Edited by Martin Farley

Published by the Royal Society of Chemistry, www.rsc.org

the laboratory count as much as possible, impacting all the above-mentioned aspects at once. Most importantly, one would focus on ensuring that the results acquired in the lab are available to others and that others could easily replicate your findings. In essence, scientists should focus on making their research count. This is in other words, making our results reproducible and accessible. In this chapter, I argue that there is plenty a scientist can do to increase the carbon efficiency of their research, and can start now.

6.2 Why Do We Research?

Scientific research aims to understand the laws of nature, improve life, cure diseases, uncover new technologies to advance humankind or provide knowledge onto which innovation can be built. This can stem, for example, from deep interest in a subject or wanting to crack complex puzzles. To achieve these overall gigantic goals, researchers slice up the task into small steps and experiments, to test ideas and slowly but surely advance knowledge of the unknown. These small steps and associated findings are then packaged into a scientific publication with a story-telling touch to communicate effectively with the rest of the scientific community, who can then build upon the reported findings. To further engage with peers, scientists will present and discuss findings and novel ideas at conferences, often travelling across continents to do so. Funding in academia then allows researchers to continue with more experimentation, and is awarded partly as a result of publications, novel ideas, collaborations (and maybe fame?). And with more funding, a field of research moves forward into new unknowns, and iterates.

6.3 The Footprint Pyramid of Research

I often like to picture the way we do scientific research as a system. Try to zoom out and picture in your mind the entirety of all the things a scientist needs to run an experiment. Often, we will picture the consumables we use (*e.g.* plastics, chemicals...) or the energy a certain machine uses. Try to add the energy needed to keep you warm/cool in the laboratory or office, to store your samples, the building itself, the travels you make to conferences and your commute, the coffee (to keep you shaking just enough while trying to perform precise tasks), the computer (and the making of it, most likely made in Asia) and the computer infrastructure and software (and the associated energy wasted complaining about it), the support staff and their precious work, *etc.* All this is actually needed to run an experiment. Now let's crank up the pace: picture in entirety all the things needed to make one publication... It is very hard to fully grasp how much stuff is needed to create one publication. Probably no one knows precisely. I tried to depict the main things that a scientist needs to produce one scientific publication in Figure 6.1. Think now of all the publications of a single lab over 20 years, or even an

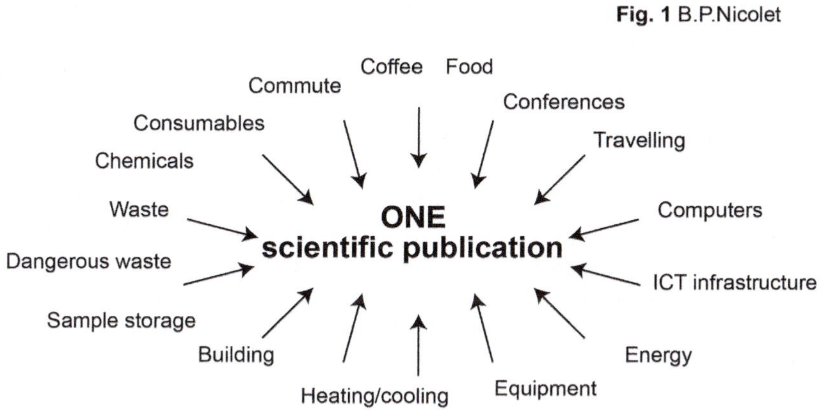

Figure 6.1 A cartoon representation of resources needed to create a scientific publication. This list is non-exhaustive.

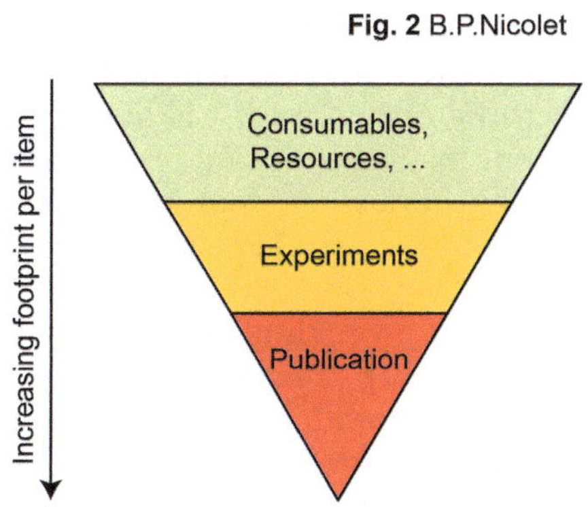

Figure 6.2 The funnel of scientific publication. This graphic represents the amount of resources going into several experiments resulting in one scientific publication. This graph assumes 100% success rate of experiments that will all be included in the final paper (which is not realistic).

institute, university… Humongous is the word that pops to my mind when picturing all this.

Next, try to picture the carbon and the environmental footprint associated with this. Bam! Humongous is here again. We tend to focus on the plastics and the consumables used in a lab. Now if you bring these into the big picture, it is only one or two aspects of the total footprint. If you now look at the level of a scientific publication, the few pipette tips or tubes that you saved/recycled/upcycled look pretty tiny in comparison to the many resources needed for the many experiments in a publication (remember that recycling still requires enormous amounts of energy). I think of this as a pyramid. There are many consumables/resources/things needed to make an experiment, and there are many experiments needed to make a publication (see Figure 6.2).

Following this logic, adding or leaving out a figure in a publication will have major repercussions on the need for further experiments and resources. Arguably, any addition/deletion of a figure in the paper has much more impact on the environmental/carbon footprint than cutting down on a few pipette tips or probably freezer temperature (not that they are not worth doing). In my discussions with colleagues and friends, I talk about this subject and often the reaction is in the line of "Oh... crap... yeah... maybe we can optimize our experiments and our papers! Maybe we should include fewer biological replicates (*e.g.* donors), so there is less need for experimentation...". I was also thinking of this in terms of my work in biomedical science, what if instead of 5 biological replicates I used 4, or 3 or 2?

Unfortunately, I need to flash back to why we do scientific research here: to ensure our findings can be reproduced anywhere with the same settings. This entails ensuring that the findings are true in more than 2 donors. Would I feel comfortable to conclude an entire paper where only 2 donors were used? Definitely not!! (Unless it is a medical case report).

For the sake of the argument, let us imagine I publish a paper using only 1 or 2 donors. By experience, I know that it is unlikely to capture the full variation of biological systems. In other words, if 1 or 2 patients that were given a new drug react in a certain way, it is not likely to inform me much about how a third patient may react to said drug. Furthermore, my findings are probably going to be difficult to reproduce elsewhere, or worse, the findings could just be due to statistical noise in the 1 or 2 donors that I used.

What would happen to my publication (assuming it would even get published)? I would likely present my findings at conferences, and other scientists would use my work as a starting point for their own research. Their future publications rely in part on my work and its robustness. What if indeed my findings were just noise in the data and not reproducible by anyone else who tries? What if it takes many years before someone picks up on my mistake? What happens when science does not self-correct? I can foresee that many papers would be published, citing my work as a foundational study for the research carried out. And it may be just noise as my findings were not robust... now let us bring back the humongous-ness of the carbon footprint for a second here. Can you imagine the resources, money, and carbon lost down the drain?

Let's play out a similar scenario but with 5 or 10 donors. You experiment, the first 2 donors look promising and then you bring a 3rd and a 4th and things start to look less good. You bring in another 6 donors, and actually you realize that the drug does nothing. Nada. The first (human) reaction is to triple-check for mistakes. But no. It all checks out. There is nothing. You make a figure, present this "fun story" in a lab meeting and decide to allocate more time to a different super cool project that you are excited about. There are no subsequent conferences, no tens of follow-up papers, and all the resources, money and carbon are then used for (probably) better research.

Now here, the devil's advocate comes in and says: "but that is idealistic... Science corrects itself and anyway the experiments are likely to be tried again in 3 other labs in the world. No negative findings will be in the public domain. No one will know about the failed experiments that you have nicely documented with 10 donors. Are you not exaggerating the gains in resources??". The devil's advocate would be right. But this goes beyond the scope of this chapter. Sharing negative findings, even only on a preprint server, can prevent quite a lot of other labs from repeating the same experiments and wasting resources. Alternatively negative findings can serve as a basis for new experimentation with unknown new parameters (this we know does not work, maybe with different parameters it does?). The "science corrects itself" concept will be covered later on in this chapter.

In the meantime, let's get dark for a minute. Maybe dark and sad. Let's push it and talk about fraud. What if a scientist/lab commits deliberate fraud? Other scientists will start their projects based on fraudulent data from published work. Maybe even set a new research line, or the start of a new field? Maybe it will take a few years before someone finds out that all the foundational work was bogus. Can you imagine the number of experiments performed for nothing? Picture again all the resources needed for experiments, and multiply this by the number of labs that started projects resting on the validity of (now known) fraudulent data? Hey... Here is humongous again.

There is a powerful way to visualize how fraud can impact other papers. González-Márquez *et al.*,[1] took 21 million abstracts from PubMed and ran a complex analysis (using Large Language Models) on the abstract of these papers to derive "semantics" of each paper in mathematical terms (called embeddings). These semantic terms were then used to map and cluster papers together, nicely reflecting fields of research. It is also a great way to realize how much jargon is used in our publications, allowing an algorithm to split apart abstracts by fields with only 100–300 words. The authors had the brilliant idea to also project flagged/retracted work onto this paper map (see Figure 6.3). Flagged papers represent publications that have received a letter of concern, a retraction of the work, or have been mentioned on Retraction Watch, a website that offers a platform to start a dialogue with authors that may have made mistakes in their publications (human error, or deliberate manipulation of data). You directly appreciate that the map is sprinkled by flagged papers (PubMed/Retraction Watch). To me, this tells me that fraudulent and retracted papers are unfortunately present in every field. This is worrying if once again we picture the "humongous-ness" of the footprint of all these papers influencing all these fields. Interestingly, it seems that some clusters of papers are enriched in flagged/fraudulent papers. This suggests that the language (and probably the author lists) is fairly powerful in grouping together research that is likely fraudulent. A well-meaning data scientists could even probably derive a confidence score in the likelihood of being a dodgy paper.

How can we prevent this? I am glad you asked. Good news: there are many things a scientist can do to prevent this. Let us first look at the core: the method.

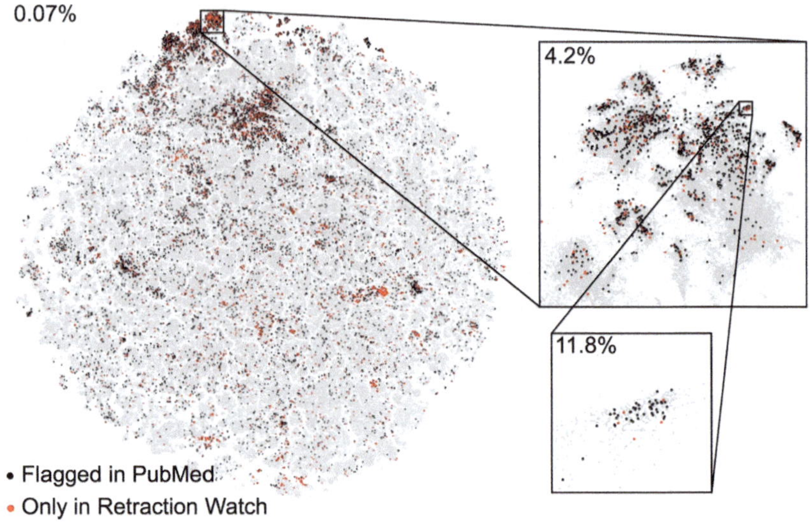

Figure 6.3 The landscape of biomedical research. Reproduced from ref. 1, https://doi.org/10.1016/j.patter.2024.100968, under the terms of the CC BY 4.0 license, https://creativecommons.org/licenses/by/4.0/.

6.4 The Scientific Method

Scientists are directly and indirectly trained to follow a certain methodology to ensure the quality of their findings. It is called the "Scientific method". It combines the strict ethics of experimentation, the acknowledgement of flaws or bias, the rigor of centuries of mathematics and statistics, and many more aspects. All this is aimed at ensuring the validity and quality of results stemming from scientific research and to resist human biases and egos. Of course, the scientific method evolved over centuries and continues to evolve with technological advances. The scientific method revolves around 4 steps: hypothesis – design – test – conclusion.

6.4.1 Hypothesis

Hypothesis driven investigation allows for the testing of key parameters and control factors that could interfere with the trustworthiness of the results. It must account for alternative hypotheses to ensure that only one aspect is tested. For instance, a study looking at the relationship between height and weight of humans, should probably not ignore age, sex, income, *etc.* At the same time, a hypothesis should not be about *trying* to prove something. It should rather *test* something. Not: "I want to prove that A is better" but: "I want to test whether A or B is better". The scientist creating the hypothesis must adopt a disinterested look at the hypothesis. Sleep over it. It is not about the emotion of the experiment, whether one will be right or not. It is about the test: is it A or B?

Probably the most important aspect of the hypothesis generation is to discuss it in detail with peers. Peers will always be less involved than the experimenter and will naturally have a more detached look at your hypothesis. Pay great attention to what is said about your hypothesis. Sometimes you receive some very grumpy feedback from a senior scientist. If you feel it is unfair, ask someone else's opinion. If the second opinion comes to the same conclusion as in the initial feedback, you should probably listen to it (a true scientist would of course ask for at least 3 opinions (to boast a $n \geq 3$)). When presenting a new idea and there is zero feedback (and the meeting is not just after lunch), be worried. Silence is probably the worst. There is no way to tell whether there is really nothing to say and the hypothesis is bulletproof, or if no-one feels like telling you what is wrong about your hypothesis. The only solution: ask! Some of my colleagues repeatedly make presentation slides with ideas they have. Firstly, it is to ensure they do not forget, and secondly it is to easily present the idea to colleagues for efficient feedback.

Why do I hammer on about the quality of hypotheses? If a hypothesis is well thought through, the experiment design will be easier, and the more likely it is that something meaningful will come out of the experiment. There is less chance of having missed an obvious confounding factor or alternative hypothesis. If there are fewer mistakes, the efficiency goes up, and as a result, the resource costs and carbon footprint go down. Bottom line: think about your idea well, talk about it with peers, make it as good and bulletproof as possible. This will reduce your footprint.

6.4.2 Experimental Design

Alright, time to test this hypothesis: let us design an experiment. The delicate design of an experiment will allow one to test that single element in your hypothesis, and only that element. I think we, scientists, do not spend enough time designing a great experiment. I must admit that I have not always been that sensitive about the importance of design and have resorted to hasty testing in the past. It always resulted in me having to redo experiments early in my PhD.

Let us consider particle physics for a minute. It takes years to decades to form a hypothesis and design machines and devices which can test that hypothesis. Think of the CERN's Large Hadron Collider (LHC), the Higgs boson hypothesis was formulated around 1964 and it took 40+ years to design and build a machine that could test whether the particle existed (note: the LHC was not only built for that, there were also many other super cool experiments in the making). The LHC cost around 7.5 billion Euros.[2] You want to make dead-sure that you think very – very – well about your hypothesis and experimental design before investing that kind of money and time.

Why don't we realize that experimentation in science is expensive, and that we should think things through very well before we start experimenting? And not only for multi-billion experimentation in physics.

I believe the best option is to bake-in the experimental design with an A–B testing procedure. In computer development for example, to test a new design of a website, developers will split the visitors of a website into an A and B group. Then they internally redirect group A to the "old" website, and group B to the new website, before monitoring what happens – are the customers/visitors dwelling longer in A or B? Are they buying more of your products? I find this is a neat way to be detached from a question: "I don't mind if the old or new design works best, tell me which one works best". The design of an A–B like experiment in scientific research – inherently – pushes the experimenter to have the required detached look at an experiment. In addition, when the results come in after executing the experimental design, the pain of something not working as the formulated hypothesis is reduced due to this detachment. I believe this is the basis of not getting side-tracked into dead-end research lines. Even better, A–B testing prevents the dangerous grey zone where one would wish to have A be better than the other condition (basis for *p*-hacking). Bottom line: use A–B testing in your research.

Another borrowed concept for experimental design to make them resilient: the minimum viable product. This concept was brought to my attention in the books "The $100 Startup"[3] and "the lean startup" (I don't know if the concept existed before these books).[4] Startup companies don't often have too much cash to burn on mistakes, they need to move fast and see if the current product works with customers. If a company spends 3 years designing a "perfect" product, producing it, and realizing that customers don't want it, it would have burnt through 3 years of expensive development for nothing. The company is dead.

In contrast, if a company develops the minimum product needed to test if a customer likes or dislikes the idea, it doesn't cost as much. It asks for minimal development team size, minimal effort and time. If it fails, the (financial) pain of failure is minimal. The company lives and can move on to try a different product. It saves a lot of time and resources. That is the concept of the minimum viable product.

Why am I raising this here? I believe that the minimum viable product concept is extremely useful in reducing the time and resources spent on (new) research lines. Imagine that you start a new line of research. There are many unknown factors – what method to use? Which cell line to use? If a new method needs to be developed, what test would convince me (and reviewer #2) that the concept works? What is the minimal time and effort to get there?

Let us say you observe that a certain gene/protein is important for cellular function and could be important for tumor cell growth. What is the shortest path to a test that would convince us of an effect of this gene/protein? Typically, one would ask around colleagues for some of their funkiest tumor-derived cell lines and collect 3–5, maybe 10, and then order an antibody or DNA primers to test the expression of the gene/protein. This would require a good month of set-up before you can confidently test if the gene/protein is expressed and how it relates to your hypothesis. This is great from a training point of view, but it involves a lot of investment and time. This is not efficient at all. One could also just make a tiny presentation about the hypothesis and then be advised to go on the DEPMAP project website which is a great large-scale effort to characterize thousands of cell lines of different cancer origins.[5,6] There in a few hours, you could look at the expression of the gene and protein of interest in thousands of cell lines and directly test if your hypothesis holds. A few hours! Talk about a minimum viable product!

Now consider if we translate this example into a carbon footprint. One month of lab experiments, new reagents, energy to culture cells and store samples, commute to and from the institute, ventilation and heating... That is a huge amount of carbon, ranging from hundreds to thousands of kg of CO_2 equivalent. If you bring in the minimum viable product scenario, it is one day at best. Even if you factor in the energy used to prompt the large datasets online, the carbon footprint is in the hundreds to thousands of grams of CO_2 equivalent.

From a scientific point of view, the minimum viable product described here is also arguably more sensitive, reproducible and precise than the alternative I presented above.

Final points on the design: the *a priori* and *a posteriori* design. I believe that what I am about to say will resonate with you. Early in my PhD, I spent days experimenting in cell culture labs. After a few months, I developed a feeling for what should be the next step. I would come into the lab in the morning with an idea to test, quickly draft an experiment on a post-it (yes, a post-it). I would get excited about the new idea and experiment and felt I could do it right away. Which I did. I would write down a few details I needed not to forget, but all the "routine" tasks were the same so why write them down? The thrill of testing something supercool and winging it was pleasant. I would acquire the data (let's say by flow cytometry), analyze the data, and then started writing a protocol. A few times, I realized that I had made a mistake in calculating the concentration of a drug or thought "I wish I had included this condition, oh well I'll include it in the next experiment". Just writing this I feel silly. Then I started looking at the impeccable protocols of a colleague (technician) who calculated all the drug concentrations and all the conditions to include, which were decided before even stepping into the lab. For this colleague there was no sweat in the lab from a fear of having forgotten something, it was a relaxed pleasant time because all was easy, it was written down like a recipe. There were rarely any calculation mistakes, there was never a thought of "oh I wish I had included that condition". All was well thought through. Once I understood that it was more efficient, I quickly adopted the approach of designing prior to going into the lab. The *a priori* design (technician style) was clearly superior and more efficient that the rookie style, *a posteriori* design (winging it on post-it(s)). Sadly though, I believe that this rookie style is not uncommon and is something that most will be familiar with in some manner.

Bringing back the efficiency aspect is important. Again, efficiency in the lab means fewer resources used and a smaller carbon footprint.

Bottom line: the efficient *a priori* design reduces the experimental footprint. There again, talk to your peers for their input. They will have good ideas that make your design better or may share a great experimental design template.

6.4.3 Conclusion

The conclusion of an experiment is very important as it allows the formulation of a new/revised hypothesis. We tested A and B, we concluded that A is better than B at something under these conditions. This crystalizes the next steps needed for the (potential) next hypothesis/experiment. In my experience, the conclusion is something that is often forgotten. It is present in publications, but mostly missing from everyday experimentation.

I have been teaching 'sustainability in the lab' for a few years at my university. I do this from BSc level onwards to PhD student level. In this course, I try to provide food for thought in a similar manner to this chapter. I do not really focus on plastics/energy/freezers but rather on scientific efficiency, as well as the impact of the quality of our findings on our carbon footprint. What has shocked me the most so far is the lack of statistical knowledge of some students. The students are not even the ones to be blamed... There are surprisingly few universities which are forcing statistics (stats) courses onto their students post BSc level. Stats is such a core aspect of a scientist's work, and it is not even compulsory to educate PhD students in it. I feel it may derive from the uncomfortable feeling of more senior scientists who are not versed in stats and are fearful to admit it. I must admit that it was 1 to 1.5 years into my PhD before I acknowledged that I knew surprisingly little about stats. Oh, Prism and SPSS, you are so easy to use. Damn you! Oh, wait no, it is maybe more appropriate to say "damn me" for not knowing enough and pretending that I did by using software that simplifies my life.

Why do I talk about stats in a book about sustainability? The answer, once again, is simple: efficiency! Efficient research means lower resource costs and smaller carbon footprint. But how is research more efficient with stats?

I believe that every human, versed in scientific reasoning or not is subject to cognitive biases. For instance, remember the "I want to prove that A is better than B"? That is biased. There are so many human cognitive biases, see for yourself:[7] ok cool but so what, we are biased, big deal, what does it have to do with sustainability? Once again, if we listen to our biases and leave them unchecked, we will want to see what we wish to see in our experiment. We will dismiss the outliers in our results as being just the odd one out and will publish findings which are in fact simply our biased "analysis" of an experiment. This could set a whole field of research in the wrong direction. There are countless examples in different fields of research where this happened.[8] By setting a whole field in the wrong direction we use crazy amounts of resources and ultimately carbon footprint for nothing but noise generation. That is very far from the efficient research which I think we should strive for.

One way to prevent this from happening is stats. Stats provide an objective assessment of a finding. They offer a level of certainty that a finding is not due to chance or biased interpretation. In a way it also enforces the testing aspect ("is A different than B"). There are many tests which can be run but one of the foundational one is the Student's t-test proposed by William Gosset in 1908 and pushed forward by Ronald Fisher.[9] Gosset worked for Guinness (one of my favorite beers) and wanted an unbiased estimate whether the chemical properties of a batch of barley/stout were different to those used in previous batches. This was a small sample size problem. One cannot create many more batches than the ones being tested. So, to remedy to this small sample size problem, he assessed the probability that the batch was different. (It is sort of similar to simulating the expected parameters for many virtual batches). This was then used later to carry a probability that, given a known distribution of probability (of the virtual batches), would be above or below a threshold.[10] The probability distribution was published under the name "Student",[11] as Guinness only allowed the publication of the findings under a false name, with no mention of beer or Guinness (probably to protect their interests, while still allowing academic freedom).

When testing two sets of data (with some specific assumptions), the p-value (p is for probability) gives a probability that the differences observed are outside of expectation by chance given the data distribution. This is a super simplified definition that will give goose bumps to statisticians reading this. Sorry. Although less layman-appropriate, the more correct definition is: "the p-value is the probability of obtaining test results at least as extreme as the result actually observed, under the assumption that the null hypothesis is correct."[10] The fun part is that p-values are not an absolute value which is comparable across fields or experiments. It is relative to the data. Its threshold for what is considered "true difference" varies between fields. It is important to note that this threshold is *arbitrary*! In biology, for small sample testing, p-value < 0.05 is the golden standard. That means if we draw many times at random data with the same distribution as the samples, one expects that there is less than 1 in 20 draws where the sampled data would not match the actual data (1 in 20 *i.e.* 95% is somewhat high but not bulletproof). In physics for instance the p-value threshold used is very different, the threshold that had to be met for the Higgs boson particle was 3×10^{-7}, (~1 in 3.5 million draws showing similar data distribution). Also, p-values are corrected for sample size and number of tests taking place. If considering testing the effect of a treatment at the scale of the genome (~30 000 genes) an uncorrected p-value threshold of 0.05 (1 in 20) would mean that by chance 1500 genes would be considered different. There are many ways to correct for multiple testing. Dig in: ref. 12.

Appropriately applied stats are useful to unbiasedly test whether two or more conditions are different beyond what one would expect by chance. Alright that was a lot of simplification and stats stuff. Stats are useful to reduce the bias that we humans would otherwise apply to our data.

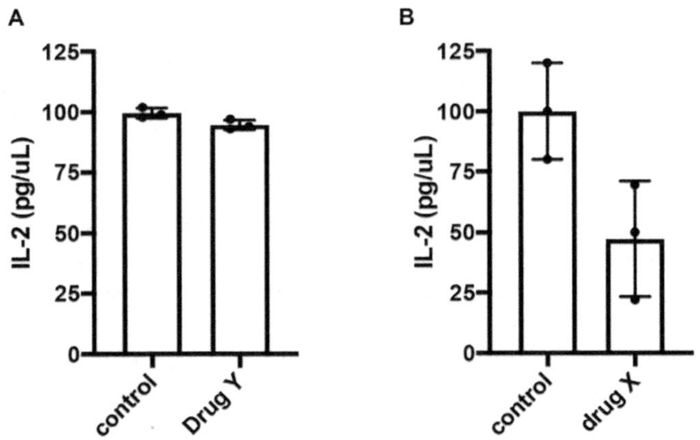

Figure 6.4 *p*-values are not a measure of the strength of biological effect!! The two graphs above are made up and present two sets of data (supposedly interleukine-2 concentration) whose spread and values give in both cases a *p*-value of 0.042 (below the 0.05 threshold). Graph (A) has a mean difference of 5, and (B) of 50 units. Student's *t*-test.

Now for the sad part. Stats is not an oracle; it is a tool. That means that one can misuse it intentionally (*e.g.* *p*-hacking, see fraud part above) or unintentionally. In my experience, the latter comes from a lack of knowledge of basic stats. For example, the *p*-value is often mistaken as a measure of how strong the difference is between conditions and equates to biological effect. It is not. I often trick students when showing a custom designed (made up) bar graph (see Figure 6.4). I made graphs that have different data distributions but have both a *p*-value of 0.042 (woohoo it is significantly different! <0.05). Whenever I have asked colleagues or students with a show of hands who believed A or B to have a lower *p*-value, there is first hesitation (I suspect that it is partly because students sense the trap coming and because they genuinely do not know), then B is selected the clear winner. But there is no winner. They have the same *p*-value. The only thing that changes is the mean difference and the data distribution. *p*-values are not a measure of biological effect! Print it and paste it at the back of your toilet's door.

One of the worst things I have heard was a combination of the *p*-value misconception above, and an illusion of how stats work. The person was using different tests on data (repeated Student's *t*-test, Welsch test, paired/non-paired version of these...) and was picking the one with the highest *p*-value. Not the lowest (*p*-hacking), but the highest. The reason behind was that one should pick the most conservative test, hereby ensuring that any difference was "really true". Completely false, of course! All depends on your data (how many datapoints, what is the distribution (normal?), and the data type (categorical of continuous)). And again: *p*-values are not a measure of biological effect!

One way to ensure that you feel comfortable standing behind your stats is to get comfortable with them. Like, really comfortable. You can follow the stats course of your university or institute, get these from online platforms such as EDX (edx.org) or Coursera (coursera.org), or watch the amazing StatQuest videos by Josh Starmer (youtube.com/@statquest). I have followed all the above-mentioned, but my favorite is still StatQuest! It is simplifying the core message down to the basics and stepwise increasing in complexity. Very well done. EDX and Coursera are more alike university lectures (and are also great!). You can audit the courses for free (with a time-limit), or pay and get a diploma/certificate. These courses will get you more comfortable with stats.

Bottom line: stats are important to remove some bias in our conclusion on our data. Once you have a nice crispy graph that does/doesn't show any difference, it is time to draw conclusions. Write these down or make a slide deck. Next step is again to engage feedback from peers regarding your conclusion. Is it not too strong? Are you missing something? Could an alternative hypothesis that you had not considered explain part of the results?

What is the next step? A new hypothesis! You observed patterns in your data and now you want to test the next step. And the loop is looping: start again at the "hypothesis" stage!

6.5 Best Practice in Experimentation

Now you are all aware of the importance of efficient research in applying the scientific method. There are further aspects that can make your research efficient in terms of reducing resource use and carbon footprint. Some are just straight forward like data storage and safety, others are 'soft' such as honesty and integrity.

Let's start with data storage. I vividly remember my PhD supervisor mentioning a case of an article published in a prestigious journal, where the raw data had been stored on a single laptop. The laptop got lost/stolen and only the data that had been shared with co-authors was recoverable (the high level, analyzed data). There was still a chance that this was a covering of fraud, but I deliberately choose the naïve option that indeed the data was just stupidly only saved on one laptop. In terms of lost resources, it is not just a laptop or a bunch of gigabytes of data, but all the weeks/months/years of experimentation needed to generate all this data. All the commuting, the coffee, the lab-hours, the freezer storage, the heating/cooling and ventilation... humongous comes again to mind.

How to prevent this? First do not store data locally on a laptop... These machines are fairly sensitive to rather mundane things such as coffee... Use a cloud/server backup system. I am sure your institute has this in place. If not, you can have "3rd party" products such as Dropbox, OneDrive... (this is not an endorsement). Note that some providers are not compliant with patient and sensitive data. Ask your data management officer for help with patient data storage). Some people will also back up things on external hard drive disks (HDD). It is better than nothing, but the redundancy to data loss is minimal. While cloud/server storage is in a water-tight, electrical surge-protected, safely located, and have back-up in an identically safe space somewhere else, external HDD are often stored just next to the laptop/computer where the other copy of the data is. Think of fire damage to the building and both your data copies are gone. Or just a clumsy coffee accident. It is not a safe backup if 20 mL of hot water can obliterate it. In the case of cloud or server storage, several copies of your data are still available. Just gain access to your data with a fresh laptop and you can continue where you left off.

Same goes with data safety. Malignant parties always try to find ways to infect computer systems. One way to prevent this is to have encrypted, redundant (several copies in different locations), backed-up data storage. Backed up in this context means that there are versions of the data that are stored every day for the past 30-days or more. So, if someone implemented malware two weeks ago and activated it yesterday, you have not lost all your data. You can ask your data manager to revert to the last known uninfected back up and get most of your data back. Please do not consider back up as "a copy of my data that I put on an external HDD but not on any other computer and if the HDD drops on the floor my last four years of data are evaporated". Data management and safety may sound boring to you, but it is super important especially because all our science is data. In Europe, many grant-givers are requiring a data management policy when applying for grants. I am not aware of any assessment of the actual policy being implemented, but regardless it is important for your work, your resource efficiency, and your carbon footprint. Bottom line: have a harsh look at how your research data is stored!

While we are on data, lets continue with metadata. When running experiments we generate tons of data. Think of a typical flow cytometry experiment with ten to twenty parameters for one hundred thousand to millions of cells. These are literally tens of millions of data points. It is critical not to have to redo, and subsequently store data of experiments. Flow cytometry itself is fairly cheap, but having to redo an experiment is expensive. In particular, the generation of cells is probably more expensive than the measurement itself. This is why it is critical to have the annotation (metadata) of the sample, the protocol, and any details relating to the sample generation (and analysis) together with the raw and analyzed data. Three years down the line, if the metadata are not attached to the raw/analysis data, no-one will remember how these were generated. What a waste of time and money. Worse, it could lead to misremembering what the protocol/preparation was and mislabeling the samples. This data could then be mistakenly introduced in papers and lead to unintentionally misleading data.

Related to the centralization of metadata is the logistics involved in multicenter/university testing. It is very hard to coordinate research collaboration. There are different agendas, different research cultures, personalities, preferences, *etc.* In the face of collaboration and association of metadata with raw/analyzed data, it is important to again ensure that these are in a safe, secured, and centralized location. If someone in lab A wants to replicate something from lab B, and doesn't manage, how do we expect a completely different lab to reproduce the data? (More on reproducibility later on).

Once again this is very bad for research efficiency, and as a result very bad for the carbon footprint associated with it. Bottom line: put the protocol, data and analysis in the same folder/location!

One last point about concluding research findings and papers. We are humans, with bias, preference, and ambition. We want a good career and want to impress at conferences and win over editorial boards. This is natural. But this can lead to conclusions being crispier than they scientifically should, *i.e.* they present findings in a more black-or-white way than they should be, whereas reality is complex. I do not have an answer to this dilemma, but much of the work done in biology is replicated 3–5× with different donors/animals/cell types. I have always wondered if it is fair to conclude how data from 3× European blood donors can truly support a conclusion applicable to all of humankind. No one says in conclusions that we expect the findings to be applicable to the entire humankind, but we imply it. Is it fair? One part of me says "no", the other part says, "we have to start somewhere and can't sample everyone".

When one writes an overconfident conclusion, the damage seems minimal. Down the line though, when a paper is published, many people may read it and take your word for it. This will likely become foundational knowledge in the field. Now, if the conclusion doesn't stand in replication attempts, the damage becomes much greater, both to the scientific accuracy and the trust in the researcher, once again resulting in a large unnecessary carbon footprint.

Food for thought.

6.6 Data Sharing

Big data in biology and science are more ubiquitous than ever. For instance, in genomics we went from measuring the mRNA expression of single genes by targeted probes (*e.g.* primer sets) to measuring the entirety of mRNA expressed (the transcriptome). Even more impactful, we needed 20+ painstaking years to sequence most of – one – human genome in the early 2000s while we can now get hundreds of genomes at greater sequencing depths under two days (and at ~10 000th of the cost of the original genome project). These technological advances have allowed insights on many topics such as how cancer develops, the root cause of many genetic diseases, and genome-wide association studies (GWAS).

Nowadays, RNA sequencing or mass spectrometry measurements are routinely carried out in biology and immunology. The technologies are cheaper, but they remain somewhat expensive and mostly still tricky to perform. All this data has been generated following a protocol and prepared in a specific way with many critical steps. This requires a lot of resources and quite some financial investment. Typical (single cell) RNA-sequencing or mass spectrometry-based proteomics still represent thousands to tens of thousands EUR of financial investment and a lot of specialist (*i.e.* expensive) machines.

To ensure that these resources are well spent, the resulting data should accomplish the biggest impact. Keeping them secretly locked in a (virtual) drawer would be the worst outcome. The ideal scenario is to share this data at publication, openly without restriction and easy to find. This can be achieved by following FAIR principles throughout the experimental process (see https://www.go-fair.org/fair-principles/ and ref. 13). "F" stands for findable, which means that the data should not be hidden somewhere in a private database. "A" stands for accessible, which entails that the data should be clearly stating how to access the data. "I" stands for interoperable, which means that the data can be integrated with others including a detailed protocol on how the data were processed. Finally, "R" stands for reusable, which entails that the data are very well annotated with rich metadata describing the experiment ensuring that another researcher can reproduce the findings. This last R is probably the most important as it is critical for the impact of the data I mentioned above.

There are unfathomable amounts of data that are shared (see Figure 6.5). For example, the Sequence Reads Archive (SRA) of the National Center for Biotechnology Information (NCBI) has close to 10^{17} nucleotides stored in its database. And it is only one of the largest repositories (there is also the European nucleotides archive (ENA) ebi.ac.uk/ena, which interoperates with SRA). Yet, for these near 34 Petabyte (3.4×10^{16} bytes) to be useful, they should be accessible. Only 69% of these are openly accessible. Generally, the non-open access data require an authorization. For instance, the data shared

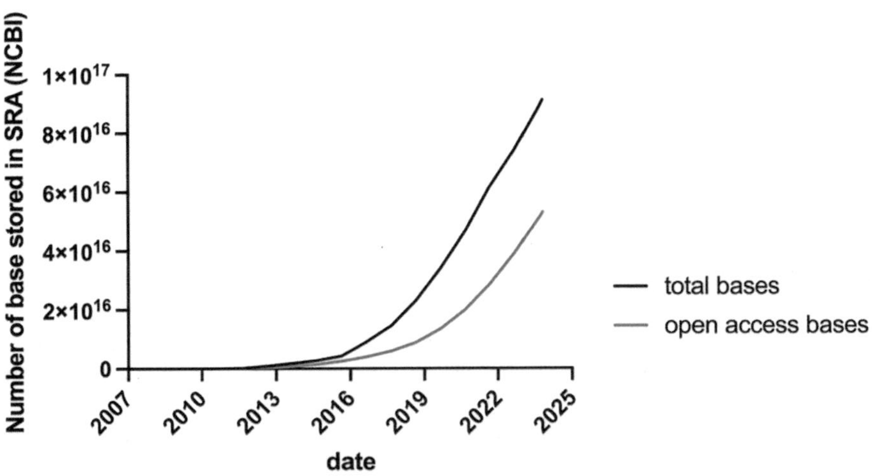

Figure 6.5 Bases (nucleotides) stored in the Sequence Reads Archive (SRA) of the National Center for Biotechnology Information (NCBI). The data were queried at https://www.ncbi.nlm.nih.gov/Traces/sra/sra_stat.cgi (records until 25-02-2024).

via the European Genome-phenome Archive (EGA, ega-archive.org) which contains some sensitive information about donors, often put the responsibility of giving access to the raw data onto the original study's authors. That means that if the authors are unresponsive (something that never, ever, ever happens in academia (hmm or maybe all the time?)), or simply moved to a different institute, the prospective scientist seeking the original data can wait an eternity to obtain access. From personal experience, only one of my many attempts to request such data was successful (this is not EGA specific). To me the need to check with original authors for approval equates to "can be obtained upon motivated request to the last author". That is inappropriate with publicly funded research. It is important to share data, and to share them well. Another aspect that is suboptimal is the reusability of shared data. For instance, Puntambekar *et al.*, demonstrated that most single cell RNA-sequencing published is shared without metadata.[14] So basically useless, as one cannot guess which sample is which or how it was prepared. Only 20% of studies shared such metadata. I consider this being 80% of grant money, time, resources and carbon footprint wasted! Let us push it further: where does grant money come from? From donations, taxpayers, and tuition fees in most cases. So, it means that of the donation to [insert favorite medically related charity], 80% has been wasted in the case of these single cell RNA-sequencing data. That is very inefficient and just unacceptable.

It is up to all of us to do our best to openly publish our data. I have been adamant about this in my work. I am far from perfect but have tried my best to share with FAIR principles all the data I have produced. I believe in this FAIR sharing as it not only allows data to be shared for reproducibility purposes, but it reduces the need for new data to be generated. During my PhD work, I have routinely augmented my work with the reuse of data from published work. This is great as it gives a level of certainty, where the data from two different labs come to the same observation. Additionally, I have used existing data to quickly test my hypothesis. For example, it was easier for me to test the expression of a set of genes of a tumor cell line from published work than to design an experiment to culture, optimize, and test. This real-life example saved ~two weeks of work (one afternoon *versus* two weeks). Even crazier and more useful, I used an entire set of data with over one hundred samples to test the expression of some strange new RNA (at the time). The data were very well annotated when shared and offered full reusability in my work. I ended up publishing a paper (with additional thorough lab validation) in record time (one year twenty-three days from idea to publication) and saving my supervisor several thousand Euros in sequencing and preparation costs. This shows that it is possible to successfully reuse data from others if these are FAIR-ly shared. (Do not forget to cite the beautiful original work!) Single cell RNA sequencing can also be a source of data for quick hypothesis testing. One can acquire several datasets from different studies (and labs) which can be mined for new hypotheses. Let us say you have a new favorite gene, and you want to test its expression in blood cells of Covid patients. There you have tens of datasets to mine for this gene and in under a week you can test your hypothesis. This was also super useful in my PhD studies. It is quite a satisfying feeling to contemplate the savings in resources, time, and resulting carbon footprint stemming from these projects. I invite you to get comfortable with (big) data analysis or collaborate with people who can analyze such datasets with you. Your footprint will be lower, your testing will be faster, and your supervisor/financial officer will be happier.

One last aspect of data sharing in our data-heavy scientific era: the code. Let's say you have done an amazing job in generating, annotating, and FAIR-ly sharing data of your study, but you omit to share your code for analysis. In many cases, it is just as bad as not sharing the data at all. For example, you may compute a specific score/output from algorithms that is critical if someone else wants to use your technology/method. Without the code, that becomes impossible. It is super important to share the code alongside the data and metadata to give the maximum impact to your work.

Bottom line of all these last parts: Share your data code and metadata complying to FAIR. It is not rocket science, but it is being an integrated scientist in the 21st century. Sure, many people will not do it. But, "There is nothing noble in being superior to your fellow man; true nobility is being superior to your former self." (Ernest Hemingway).

6.7 Self-correction

Let us zoom out from the scale of a single experiment or a single publication. Let us try to mentally picture a field of research. Tens, hundreds, or thousands of papers collectively contributing to the understanding of a disease, possible cure or the basis for drug development... Such a great amount of work. So many human hours put into it. So many resources and carbon emissions. Can you visualize all the conferencing done in your specific field/subfield? Can you start to tally-up the grant money altogether? Humongous again...

What if it was for nothing? There are several events in academia where a field took a turn towards one research line because of seemingly promising pilot studies, only to end up publishing hundreds of papers and realizing after twenty years that it was for nothing. I investigated one such event together with experts in sustainability and reproducibility in science.[15] We wanted to estimate the carbon footprint of the failure of a field to self-correct despite un-refutable repetitive negative findings. The original 1996 paper proposed and somewhat supported a hypothesis, that was then questioned from 2002, then showed to be non-reproducible in meta-analysis and larger studies in 2005 to 2009. That should have been it. But it was not, as most of the papers associated with these non-reproducible findings (~70%) were published after 2010. This is a clear example of a scientific research field that should have self-corrected but did not. It resulted in an estimated 30 000 tons of CO_2 equivalent that were not needed. The costs involved in this mishap are likely to be in the tens of millions of Euros. Devil's advocate would note here that it is too easy to say so in hindsight. There is some truth to that. Yet, it is useful to try not to repeat the same mistakes.

What can we do? I think it starts with good scientific conduct: following the scientific method coming from observation, sharing of findings with peers for critical input on the design, execution, and conclusions. Take part in a journal club, and read articles (beyond the abstract)! Also, there is merit in becoming "reviewer #2" and looking pessimistically at our own findings (see experimental design). Quite often, it seems that the core problem is the seductive narrative around the findings while completely ignoring an alternative hypothesis. "Reviewer #2" mindset can help in this regard. Quite some work has gone into addressing the reproducibility crisis in science.[16] We know quite a lot about it. I invite you to read in more detail the relevant pieces of work. At worst it will provide you with food for thought.

6.8 Final Thoughts

We are all human. We will at times fail in achieving our highest standard and quality of research. It does not mean though that we should not try our best to do so. Start today in becoming a better researcher by getting comfortable with statistics, in applying the full cycle of scientific method: observe, hypothesize, design, test, analyze and conclude (and repeat it if promising). Once you have generated beautifully annotated datasets, provide these openly and FAIR-ly. This will make your work even more useful to others by augmenting their research (and increase your citations), will make your research more reproducible, and will help your field to quickly self-correct if needed. Altogether, these factors will make our scientific research more efficient in resources and again, ultimately lower in carbon footprint. Let us start today.

Bottom lines of the chapter:

- Annotate your experiments well.
- Share your raw data and how you analyzed these in a FAIR way.
- Get comfortable with stats! Stop focusing on *p*-values.
- Apply the scientific method: observe > hypothesize > test > analyze > conclude > (repeat the loop if needed).
- Be critical of your own work, as you would with someone else's.
- Aspire to be an excellent scientist and make your research count!

References

1. R. González-Márquez, L. Schmidt, B. M. Schmidt, P. Berens and D. Kobak, The landscape of biomedical research, *Patterns*, 2024, 5, 100968.
2. Large Hadron Collider, Wikipedia, The Free Encyclopedia, https://en.wikipedia.org/wiki/Large_Hadron_Collider.

3. C. Guillebeau, *The $100 Startup: Reinvent the Way You Make a Living, Do What You Love, and Create a New Future*, Crown/Archetype, New York, 2012.
4. E. Ries, *The Lean Startup: How Today's Entrepreneurs Use Continuous Innovation to Create Radically Successful Businesses*, Crown Business, 2011.
5. Depmap.org, *Depmap*, https://depmap.org/portal/.
6. A. Tsherniak, F. Vazquez, P. G. Montgomery, B. A. Weir, G. Kryukov, G. S. Cowley, S. Gill, W. F. Harrington, S. Pantel, J. M. Krill-Burger, R. M. Meyers, L. Ali, A. Goodale, Y. Lee, G. Jiang, J. Hsiao, W. F. J. Gerath, S. Howell, E. Merkel, M. Ghandi, L. A. Garraway, D. E. Root, T. R. Golub, J. S. Boehm and W. C. Hahn, Defining a Cancer Dependency Map, *Cell*, 2017, **170**, 564–576.e16.
7. List of cognitive biases, Wikipedia, The Free Encyclopedia, https://en.wikipedia.org/wiki/List_of_cognitive_biases.
8. Replication crisis, Wikipedia, The Free Encyclopedia, https://en.wikipedia.org/wiki/Replication_crisis.
9. Student's *t*-test, Wikipedia, The Free Encyclopedia, https://en.wikipedia.org/wiki/Student%27s_t-test.
10. *P*-value, *Wikipedia,* The Free Encyclopedia, https://en.wikipedia.org/wiki/P-value.
11. Student, The Probable Error of a Mean, *Biometrika*, 1908, **6**, 1.
12. Multiple comparisons problem, Wikipedia, The Free Encyclopedia, https://en.wikipedia.org/wiki/Multiple_comparisons_problem.
13. M. D. Wilkinson, M. Dumontier, I. J. Aalbersberg, G. Appleton, M. Axton, A. Baak, N. Blomberg, J.-W. Boiten, L. B. Da Silva Santos, P. E. Bourne, J. Bouwman, A. J. Brookes, T. Clark, M. Crosas, I. Dillo, O. Dumon, S. Edmunds, C. T. Evelo, R. Finkers, A. Gonzalez-Beltran, A. J. G. Gray, P. Groth, C. Goble, J. S. Grethe, J. Heringa, P. A. C. 'T. Hoen, R. Hooft, T. Kuhn, R. Kok, J. Kok, S. J. Lusher, M. E. Martone, A. Mons, A. L. Packer, B. Persson, P. Rocca-Serra, M. Roos, R. Van Schaik, S.-A. Sansone, E. Schultes, T. Sengstag, T. Slater, G. Strawn, M. A. Swertz, M. Thompson, J. Van Der Lei, E. Van Mulligen, J. Velterop, A. Waagmeester, P. Wittenburg, K. Wolstencroft, J. Zhao and B. Mons, The FAIR Guiding Principles for scientific data management and stewardship, *Sci. Data*, 2016, **3**, 160018.
14. S. Puntambekar, J. R. Hesselberth, K. A. Riemondy and R. Fu, Cell-level metadata are indispensable for documenting single-cell sequencing datasets, *PLoS Biol.*, 2021, **19**, e3001077.
15. M. Farley, M. R. Munafò, A. Lewis and B. P. Nicolet, The carbon footprint of science when it fails to self-correct, preprint, 2025, DOI: 10.1101/2025.04.18.649468.
16. M. R. Munafò, B. A. Nosek, D. V. M. Bishop, K. S. Button, C. D. Chambers, N. Percie Du Sert, U. Simonsohn, E.-J. Wagenmakers, J. J. Ware and J. P. A. Ioannidis, A manifesto for reproducible science, *Nat. Hum. Behav.*, 2017, **1**, 0021.

CHAPTER 7

Equity and Justice in the Global Scientific Enterprise†

STAR SCOTT

University of Georgia, Athens, Georgia, USA
Email: SustainableStarScott@gmail.com

Author's note to readers: When writing a chapter on equity and justice, it is imperative to use terminology that acknowledges linguistic evolution and is modern, accurate, appropriate, inoffensive, inclusive, and culturally sensitive. Many racial descriptors historically used are rooted in systemic oppression, and are inaccurate, derogatory or centered around Whiteness. For example, the term "minority", which was once widely-used, is no longer acceptable, as it inaccurately implies Black and Brown people are fewer in quantity than White people or lesser in quality.[1] Inclusive language guidelines and standards published by well-known institutions reflect societal growth and promote more equitable terminology when reporting on topics of race or ethnicity.[2,3] The author has made every attempt to adhere to the *AMA Manual of Style* when referencing race and ethnicity, however she also acknowledges her personal, social and cultural experiences, and her potential for unrealized implicit biases. To this point, the author extends her desire for continued growth, as well as her regret for any oversight which normalizes or centers her own experiences or perspective. Finally, many sources of data are still reported using terminology which does not meet updated standards, such as governmental or census data. In these instances, the author adheres to terminology used by the data source for the purpose of reporting accuracy.

The Sustainability of Science: How Science Impacts the Environment, and What Can Be Done
Edited by Martin Farley

Published by the Royal Society of Chemistry, www.rsc.org

7.1 Introduction

The immense contribution of the global scientific enterprise to the wellbeing of the planet cannot be overstated. Science contributes vastly to global advancements: From conservation to public health to technology, science serves the interests and needs of society, the environment, economic growth and productivity. Science makes remarkable contributions to the planet and to people. Without it, we would lack vaccines and life-saving medicines, electricity, the internet, the periodic table of elements, an understanding of evolution and methods for measuring planetary health and change, just to name a few. Most scientists are driven by a desire to make positive contributions through their work. They are solution seekers; often committing their lives to improving planetary or social wellbeing. Yet, for all the positive contributions and intentions of the scientific community, the enterprise has long operated as though these positive contributions negate its operational footprint and exempt it from the need to mitigate negative impacts associated with scientific research operations. But science has a carbon footprint, as well as many other environmental and social impacts, many of which are contributing to the exact problems scientists are working so hard to solve. In this manner, the operation and intention of the global scientific enterprise become paradoxical. For example, the operations of public health scientists working to cure lung diseases, such as asthma, contribute to global emissions which are well-documented to drive the disease.[4] With this paradox in mind, scientific and research operations must look inward at their own processes and strive to mitigate the negative impacts of their work.

This chapter serves as an invitation to redefine what it means to *"do good science"*. While this term has always represented the concepts of rigor and reproducibility, and perhaps the ability to secure adequate funding with which to carry out research, we must now include the concept of sustainability into this definition and rapidly incorporate methods which offset the associated adverse environmental and social impacts of scientific operations. To continue to serve the greater good and *"do good science"*, sustainability must be fully integrated into scientific processes and best practices. This includes looking beyond in-house operations to understand the impacts of the full value chain of science. How are scientific consumables created and what are the impacts of transporting them around the globe? Who are the workers creating these goods and are they paid a fair and living wage? What are the factory conditions in which they work? Is there oversight and governance ensuring labor laws are upheld? Did locals have community representation in decisions to bring industry to their area? Has the extraction of raw materials or creation of goods impacted their local ecosystems or public health? Has factory waste water impacted local waterways? Have factory emissions impacted air quality? Has the rate of illness or disease increased since the arrival of industry in the community? How is scientific waste managed when it leaves research facilities? Where does waste go when it leaves the lab? Is it incinerated or landfilled? What is the history of the landfill, including citations and infractions? What are the community sociodemographics at the point of waste generation and do they echo the sociodemographics at the point of disposal? These are all guiding questions which help expand one's understanding of impacts along the continuum of the value chain of science.

Scientists are often acutely or sometimes myopically focused on their subject matter and for good reason—it benefits their work to dive deep and know as much as they can regarding their subject of interest. While this approach is quite beneficial to *conducting* science, it does not naturally lend to the above suggestion of expanding one's purview to include the whole value chain of science; in fact, it inadvertently works against this approach. Thus, efforts to expand one's understanding and comprehension of the massive and complex scientific value chain must be intentional and thoughtfully approached.

7.2 Framing Equity and Justice Within Sustainability

According to the 1987 Brundtland Report, as well as Elkington's Triple Bottom Line (TBL) framework, the three foundational pillars of sustainability are *social*, *economic* and *environmental*[5,6] (see Figure 7.1). These are also referred to as the *Three P's of Sustainability:* People, planet and profits. Historically, sustainable action has focused almost exclusively on the environment and how to make those efforts economically viable. This is especially true when reviewing sustainable science or Green Labs efforts. Yet, in order to have true sustainability—to design balanced, sustainable systems and processes—*people* must also be considered and included. Otherwise, sustainability efforts are incomplete. The three pillars of sustainability are inextricably interconnected and cannot be isolated from one another.

Another means by which we root our interconnectivity and understanding of the role of equity within sustainability is through the United Nations 17 Sustainable Development Goals (SDG) (see Figure 7.2). As stated by the

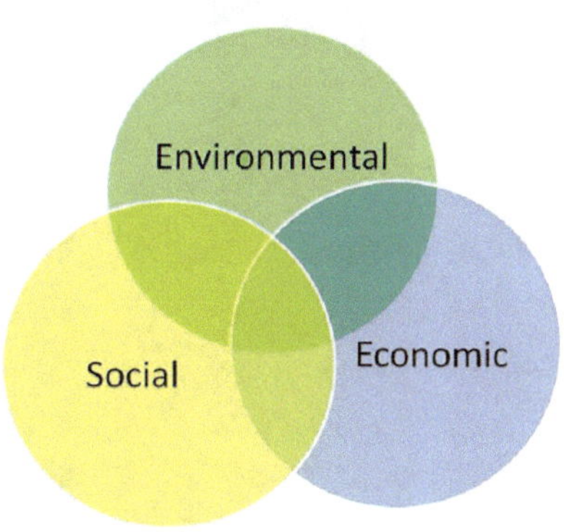

Figure 7.1 The Three Pillars of Sustainability: Social, economic and environmental. Also referred to as the "Three P's": People, planet and profit. Reproduced from ref. 193.

Figure 7.2 The United Nations 17 Sustainable Development Goals. Reproduced from ref. 7 with permission from the United Nations. The content of this publication has not been approved by the United Nations and does not reflect the views of the United Nations or its officials or Member States.

United Nations, these goals, "are an urgent call for action by all countries – developed and developing – in a global partnership. They recognize that ending poverty and other deprivations must go hand-in-hand with strategies that improve health and education, reduce inequality, and spur economic growth – all while tackling climate change and working to preserve our oceans and forests".[7] Each goal falls within at least one of three pillars of sustainability: Social, economic or environmental.

In order to have lasting, sustainable impact, these goals must come together simultaneously and in balance with one other. We are unable to sustain environmental goals without ensuring social and economic goals are met and supported, and *vice versa*. Another way to visualize the SDGs is as a multi-layered cake, *The SDGs Wedding Cake*,[8] demonstrates how economies and societies are embedded parts of the biosphere (see Figure 7.3). In this case, sustainability is demonstrated in each balanced "bite" of the cake, moving away from a sectorial approach and emphasizing the need for an integrated implementation of the 17 individual goals. Global challenges do not happen in isolation and neither do solutions. When one pillar of sustainability is impacted, those impacts ripple through the other pillars. Environmental disasters do not just impact an ecosystem, but also have great effect on people and the economy.

For example, it is well-established that rising atmospheric carbon dioxide levels drive climate change and global warming, thus changing weather patterns and resulting in more severe storms.[9] More severe storms impact not only the environment in the form of floods, fires, droughts, hurricanes and more, but also impact people *directly* as injuries or death, loss of property or housing, and *indirectly* by affecting water supplies, food or resource availability. All of these impacts have great effect on the economy, costing countries billions in damage costs and disaster management, and illuminating our interconnectedness.[10] Though the term *sustainability* by definition includes consideration for the environment and for people, current colloquial language does not often include people or equity when referencing the term. Thus, the author of this chapter will reference social considerations explicitly until which time colloquialism evolves to use the term within the context of the 3 pillars and its full definition.

Interconnectedness is a founding principle of ecology,[11] acknowledging that all living things in an ecosystem are connected through networks of relationships. Organisms and processes depend on one another for survival and when one is impacted, a ripple effect can be felt throughout the entire system. This same concept of interconnectedness can also be seen in most spiritual systems and philosophies throughout history.[12,13] Interconnectedness can be observed in the scientific sector, as well. Science is connected to the world through our operations, discoveries and advancements and is deeply interconnected to the well-being of the planet and society. The tether of interconnectedness is an implicit driver of scientific inquiry, illuminating processes and connections. Through science we are able to demonstrate how a lack of sustainable action in one place often results in negative outcomes

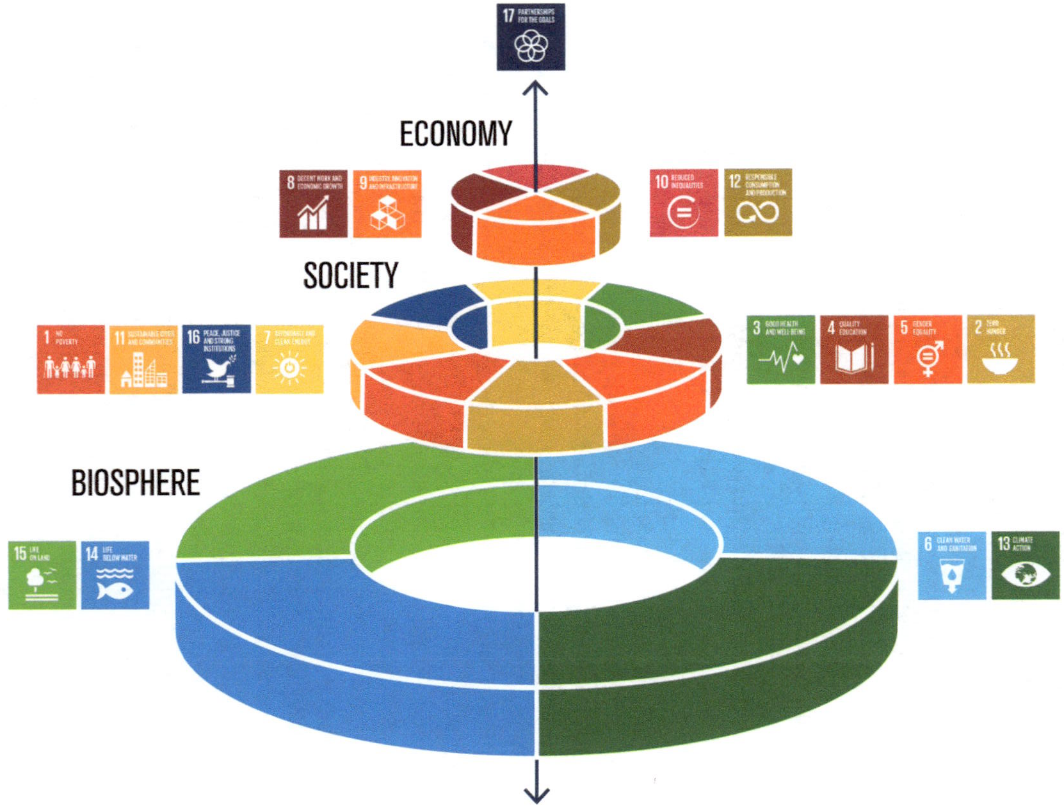

Figure 7.3 The SDGs Wedding Cake demonstrates how economies and societies are embedded parts of the biosphere. Reproduced from ref. 8, https://www.stockholmresilience.org/research/research-news/2016-06-14-the-sdgs-wedding-cake.html, under the terms of the CC BY 3.0 license, https://creativecommons.org/licenses/by-nd/3.0/.

for communities in other places. For example, many coastal communities are negatively impacted by ocean plastics and pollution coming from other parts of the world. This may impact local fishing operations and/or quality of harvests and the community suffers despite the fact they did not cause the pollution. In situations like these, we use the foundational principle of interconnectedness and the methodology described below to understand that justice and equity are lacking.

The methodology for exploring equity within a system is simple: Ask questions! Where did this come from? Who made this? Where does it go after use, who is impacted and can we trust this information? While a wide range of questions are helpful, the two *most* important guiding questions in the framework for identifying where greater equity is needed are: *Who benefits?* and *Who bears the burden?* This is referred to as the *Benefit/Burden Framework.* For any given operation, process or practice, one can ask these questions and if someone other than the process driver is bearing the burden, then justice has not yet been achieved and is needed. When someone outside of the scientific enterprise is bearing the burden of scientific operations, then equity and justice are lacking and effort is required by the scientific community to review and reform the current approach. The Benefit/Burden Framework can be applied to any scenario or any part of a process; from extraction to manufacturing to transportation to use to disposal and more. It is simple, objective, and illuminating, and can be a reliable guide for justice work. Use it as the powerful tool it is to help reveal bias, inequities, and blind spots. But also, and more importantly, to reveal better and more equitable products, processes, and methodology. Positive change need not be difficult or costly, as the majority of progress occurs in simply changing the way we think about science. When one begins to ask these two simple questions, a whole new paradigm of scientific approach begins to unfurl almost effortlessly. This new approach allows for easier expansion of thought to include those outside of science within the scope of scientific design. In short, it helps to remove the unintended elitism of science and move us towards processes which benefit the *Whole.* When scientific processes are aligned with global interconnection, it naturally leads to more sustainable and equitable outcomes; a win–win scenario for all.

According to Morton and Fasching-Varner, equity is, "the notion of being fair and impartial as an individual engages with an organization or system, particularly systems of grievance. *Equity* is often conflated with the term *equality* (meaning sameness). In fact, true equity implies that an individual may need to experience or receive something different (not equal) in order to maintain fairness and access".[14] Justice is defined as, "both a process and a goal. The goal of social justice is full and equal participation of all groups in a society that is mutually shaped to meet their needs".[15] It is also important to note different types and definitions of justice are utilized in different sectors.[16] While judicial law recognizes the four different types of justice as distributive, procedural, retributive, and restorative, this chapter adheres to the definitions shared above, used within the sustainability and human rights sectors. The author will frequently use the blanket term *justice*, as it is now broadly accepted that environmental justice, social justice, racial justice and spatial justice are intersectional and thus, each reflected within that term.

7.3 The *Equity and Justice in Science Framework* and Categories for Sustainable Action and Inquiry

The framework used for exploring topics of equity in science is the *Equity and Justice in Science Framework (EJSF),* the foundation for this chapter (see Table 7.1). This standardized framework provides a structure with which to organize inquiries of equity within the global scientific research enterprise. It allows for more consistent organization and shared language, which in turn support collective efforts toward greater equity by allowing data to be more easily amassed and shared by researchers worldwide.

The EJSF is comprised of three primary categories for inquiry; *Upstream, In-lab,* and *Downstream*. Understanding these three categories will allow for better organization and understanding of equity and sustainability efforts within the global scientific enterprise, as well as lend to improved program planning and implementation. Many equity and sustainability initiatives will touch or impact all three categories, though usually will have a dominant impact on one category.

Processes occurring outside of the lab and prior to the implementation of scientific research are considered *Upstream* actions. *Upstream* is a broad category, referencing processes, actions and efforts occurring within the scientific value chain which often intersect with procurement and policy. This can include environmental and social impacts of product creation, development and manufacturing, raw materials extraction, fuels used in manufacturing and transportation of goods, policy at various levels (manufacturing, governmental, institutional) and much more.

In-lab refers to actions or efforts directly impacting scientific laboratories or the researchers who work in them. Implementing resource conservation often

Table 7.1 *The Equity and Justice in Science Framework* provides a structure with which to organize inquiries of equity with the global scientific research enterprise.

Category for Inquiry	Description	Example
Upstream	Processes or actions occurring outside of the lab and prior to the implementation of scientific research	Procurement, policy, product creation, raw materials extraction, transportation of goods
In-lab	Actions, efforts or processes directly impacting scientific laboratories and/or researchers	Laboratory resource use, waste management/diversion, equity in publishing and promotion, invisible work
Downstream	Outcomes and impacts happening post-scientific processes	Waste disposal, end of life disposition, effects of scientific resource use

occurs at this level, such as temperature-tuning an ultra-low freezer, adding timers to power-down equipment when not in use, or establishing waste diversion efforts. *In-lab* also references processes impacting scientists and researchers; from quality of physical lab space to publishing, promotion, and more. When we reference topics that directly affect researchers or the space they work in, this is usually occurring at the *In-lab* level.

Downstream references outcomes and impacts happening post-scientific processes. This can include waste management of hazardous and non-hazardous materials leaving a research facility, end of life disposition for equipment and materials, and effects of scientific resource use. *Downstream* is a term often used to describe unintended or involuntary outcomes of a process. Though this term is neutral by definition, it often has a negative undertone as *Downstream* impacts are often discovered when processes negatively impact communities or environments outside of scientific research spaces.

7.4 Upstream Equity and Justice Considerations Within the Global Scientific Enterprise

There are many aspects of *Upstream* operations that impact equity and justice. From the harvesting of raw materials to shipping practices to a manufacturer's sustainability commitment—these matters have far-reaching consequences, not only for the scientific community purchasing the goods, but even more so for communities and ecosystems in areas where goods are sourced or produced. Scope 3 emissions, or value chain emissions, include purchased goods and services and account for 75% of companies' greenhouse gas (GHG) emissions on average,[17] thus having tremendous global impact on climate change, biodiversity and more (see Figure 7.4).

Scientists and research institutions can directly influence these emissions and related impacts *via* their procurement and purchasing power, as well through simply purchasing fewer items when feasible. When researchers give purchasing preference to products created in equitable and sustainable ways, it directly influences the market and helps drive culture shifts, leading to positive change. Researchers have an opportunity to minimize the suffering of communities and individuals globally, simply by choosing sustainably and equitably created products. Research institutions must use their collective purchasing influence to drive supply chain improvements by also withdrawing support for products created in inequitable ways. Historically, scientific supply chains have been long, multi-tiered, and opaque, lacking the transparency of sourcing information needed in order for consumers to make informed decisions. However, a movement is currently underway within the scientific sector to bring greater transparency to scientific procurement, thus enabling more informed purchasing. These efforts fall under the umbrella of sustainable procurement—a strategic approach to purchasing goods and services which minimizes negative social, environmental and economic impacts. Some foundational elements of sustainable

Figure 7.4 Supply Chain Iceberg Slide demonstrating more than 90% of an organization's GHG emissions are often embedded in their supply chains. Reproduced with permission from the Sustainable Purchasing Leadership Council, Copyright 2022.

procurement are environmental assessment, social justice, fair trade, economic viability, and ethical governance. International special focus groups and non-profits are helping to lead the way in promoting the engagement of sustainable procurement through efforts creating international sustainable procurement standards, such as the *International Organization for Standardization's (ISO) Sustainable Procurement Guidance (ISO 20400)*[18,19] and the *Sustainable Purchasing Leadership Council's Sustainable Procurement Guidance for Laboratory Supplies and Services.*[20]

New methodology, procurement system improvements, and integrative software platforms are helping to transform procurement standards by offering various filters and rating systems designed to better support sustainability. For example, many procurement platforms now allow for filtered searches of companies to identify those which may be small businesses or owned by females or people of color.[21] Additionally, there are a growing number of platforms seeking to allow purchasers to sort products according to fair labor practices, or social or environmental certifications. Although opportunities to identify products created using fair labor are currently very limited within the scientific sector, it remains of utmost importance to seek out fair labor products when available and to continue to request supply chain transparency from suppliers. The more demand there is for sustainably and equitably created products, the more manufacturers will strive to improve their processes in order to meet these demands and retain the business of their customers. It is also worth mentioning that while many improvements have been made within procurement platforms, most purchasers still struggle to access comparable and transparent data which would inform more ethical purchasing. More must to be done to address this.

Finally, it must be acknowledged that the legitimacy of existing fair trade certifications (which exist in other sectors) remains under question by human rights groups, who report on-the-ground conditions which dramatically counter the fair labor claims made by many of these certifications.[22] Thus, it serves to remain skeptical while continuing to demand transparency.

7.4.1 Raw Materials Extraction

When reviewing or researching *Upstream* considerations, raw material extraction is one of the first processes to consider. Raw material extraction is the harvesting of foundational materials needed to create a product and usually requires removing natural resources from the Earth. This is one of the first opportunities for sustainable and equitable action when creating a product and is often an indicator of a company's overall commitment to sustainability. The extraction of virgin raw materials is often exploitative to nature or local communities. Purchasers should instead seek to support regenerative and renewable processes and avoid those which exhaust natural resources.[23] According to the 2024 Global Resources Outlook Report, the extraction of raw natural resources is responsible for over 55% of global warming impacts, 40% of air pollution impacts and over 90% of global water stress and land use related biodiversity loss. Global raw material extraction is expected to increase by 60% by 2060, from 100 to 160 billion metric tons, having devastating consequences on the environment, the climate, and ultimately, people.[24] It is imperative these trends are reversed, in order to avoid extreme weather events which will have costly financial, environmental and social impacts.

7.4.1.1 Raw Materials Extraction for Plastics Production

The use of single-use disposable plastics in science is more pervasive than ever before and is commonplace in daily scientific operations, having displaced many more sustainable (albeit historically more laborsome) processes. Urbina *et al.* estimated in 2015 that the bioscientific enterprise generates over 5.5 million metric tons of plastic waste annually, the combined tonnage of 67 cruise ships.[25] As single-use plastics use has continued to skyrocket and global annual plastic waste has more than doubled over the past two decades,[26] this estimate is now likely much higher. Plastic, and particularly single-use plastic, is one of the most disruptive materials on the planet, causing devastating environmental and public health impacts, poisoning oceans, animals and humans. It is ubiquitous in every place on Earth – in the breast milk of nursing mothers,[27] the testes of dogs and men,[28] the deepest ocean ecosystems in the Mariana trench,[29] rain in the pristine Pyrenees mountains,[30] even in human brains[31] – plastics are everywhere and are one of the greatest ecological and public health threats of the 21st century. The use of single-use plastics is certainly one of the most paradoxical aspects of scientific operations, allowing for more efficient research operations while simultaneously contributing to a massive and growing global crisis which requires urgent mitigation.

Current raw materials extraction processes for plastics are inequitable and an environmental and social injustice. More than 90% of plastics are created from fossil fuels,[32] specifically crude oil and natural gas, *via* increasingly disruptive extraction methods. The most common method of extracting natural gas is through hydraulic fracturing, commonly referred to as fracking. Fracking extracts oil and natural gas from deep below the Earth's surface *via* various drilling techniques, often in conjunction with the injection of gases (such as natural gas, carbon dioxide or nitrogen), steam, or chemicals to thin the oil and allow for easier extraction. Fracking is often poorly regulated and poses significant health hazards for communities living in proximity to its operations.[33] In the US, fracking operations experience limited oversight and with the exception of the use of diesel, are exempt from the Safe Drinking Water Act.[34] The impacts on local communities are considered a public health crisis by many, especially in the US – the largest global producer and consumer of natural gas.[35] Fracking releases hazardous air pollutants, including hydrocarbons and volatile organic compounds, known to cause respiratory illnesses and other diseases *via* the contamination of air and local water sources.[33,36] 55 Chemicals associated with fracking are known or suspected carcinogens, including 20 that specifically increase the risk of leukemia and lymphoma.[37] These operations are disproportionately located in rural communities with sociodemographic indicators which include age (children and elderly), education level, poverty level and race.[38] Low income and communities of color are most impacted by fracking operations, with an increased risk of cancer the closer one lives to fracking sites.[39] Additionally, locals employed at fracking facilities face serious injury and death, with an increased death rate quadruple that of the fatal injury rate of all United States (US) workers.[34] These communities were not given the opportunity to deny these facilities, nor do they have the resources or agency to move to another, safer location. Populations with higher socioeconomic status are known to prevent these types of facilities in their own communities *via* NIMBY (Not in My Backyard)[40] protests, utilizing their financial resources and political influence to sway development. Unfortunately, communities facing economic disparities or historically marginalized communities often lack these opportunities, thus bearing the burden of these harmful extraction processes and little to none of the benefits.

The use of plastics created from fossil fuels drives demand for the inequitable processes described above, and scientific plastics are no exception. Though the intent for single-use plastics used in scientific research is often to serve the greater good by improving research processes, the outcome remains the same; science reaps the benefit of utilizing these materials while communities outside of scientific operations bear the burden. Thus, it is imperative that researchers and scientific procurement utilize scientific data to guide internal processes and materials selection. If scientists are not willing to make accommodations in processes based on scientific data, then it cannot be expected that anyone else would. Scientists must champion these opportunities to "walk the walk" in order to help lead the public, government and industry in directions that are healthy and sustainable for the planet and people. While it is not feasible to eradicate single-use plastics from scientific research completely, steps must be taken to minimize their use and unintended negative impacts. While the sterility provided by single-use plastics is demanded by many scientific processes, it is not demanded by *all* processes and this is where efforts to reduce lab plastics should be focused. Innovative technology now exists to support scientists' efforts to move away from single-use plastics, and back to glass and other renewable processes of the past, such as microplates, Petri dishs and pipette tip cleaning systems guaranteeing levels of sterility suitable even for high-sensitivity assays and DNA/RNA analysis.[41] These novel developments allow researchers to use reusable materials without the time burden and fear of contamination experienced by researchers in the past. Finally, the *disposal* of lab plastics is also rife with injustices, and will be discussed further in Section 7.6.

7.4.2 Manufacturing and Factory Processes

There are many factors impacting equity and justice at the manufacturing and factory level. From the fuel used in manufacturing processes to forced or unethical labor to the degree of community representation in decisions regarding where factories are built, the creation and operations of manufacturing facilities have tremendous environmental and social impacts. Though most countries have environmental and social policy guiding manufacturing, they are often poorly regulated and enforced. Additionally, in some countries, labor laws may be directly undermined by migration and national security policy.[42] Companies located in countries with more stringent environmental and social regulations often move factory and manufacturing operations overseas to countries with fewer regulations in order to save money.[43] Countries with fewer regulations also tend to have weaker enforcement capacity, often due to a lack of officials to maintain appropriate oversight.[42] The capitalistic drive for creating more while spending less is a fundamental underlying challenge that must be addressed, as it encourages companies to skirt social and environmental protections in order to save money in manufacturing processes.

7.4.2.1 Fuel Used in Manufacturing

Manufacturing accounts for 54% of the world's total energy use.[44] While the primary fuels used in manufacturing vary by location and the type of products created, most utilize carbon-intensive fuels, such as natural gas and diesel. Manufacturing contributes an overall 1/5 of total global emissions;[44] these are emissions from both fossil fuel combustion and industrial processes,[45] such as creating feedstocks. There is great opportunity for decarbonization and reduction of emissions within the manufacturing sector simply by replacing on-premises fossil fuel sources with low carbon energy sources, recovered heating, heat pumps, electric heating and switching to renewable energy sources.[45] As stated in Section 7.4.1.1, the extraction processes for these fuels are inequitable and unjust, disproportionately impacting lower socioeconomic status communities and communities of color.

Additionally, many regions such as Malaysia and Indonesia use fuels touted to be more sustainable, such as palm kernel oil. These claims are countered by many experts and are surrounded by great debate. Palm kernel oil is used in the production of laboratory products, such as nitrile gloves in Malaysia. While in theory it may be a renewable and sustainable fuel source, environmental impact studies counter these claims with studies reviewing the associated deforestation, loss of biodiversity, increase in GHG emissions, and human rights violations, including forced labor and child labor.[46] It is common practice to convert tropical forests to monoculture palm kernel agriculture operations, thus dramatically and negatively impacting forest ecosystems and leading to a significant loss of species.[47] In concept, palm kernel is often considered a renewable biomass,[48,49] but in reality, the associated land use changes lead to net greenhouse gas emissions *via* the clearing of tropical forests for palm kernel agriculture, swamp peatland conversion, and Palm Oil Mill Effluent (POME) processing. These practices fall under the category of *conflict palm oil*. Fire is often used in the clearing of these lands, leading to air pollution, which negatively impacts local communities. Additionally, surface and ground water is commonly contaminated *via* runoff from land clearing processes, POME and chemicals used in agriculture.[46] This is ecologically devastating to important keystone ecosystems, such as the Leuser Ecosystem, home to many critically endangered and keystone species.

Additionally, the palm oil industry is known to have many negative social impacts. While child labor is the most well-known social concern of the industry, there are often other negative impacts for palm oil communities. Limited protections for mothers, low breastfeeding rates, lack of childcare, poor maternal healthcare and a lack of education for workers' children pose significant threats to health and developmental wellbeing in these communities.[50,51]

While data demonstrating fuels used in a product's manufacturing are not commonly shared by manufacturers and suppliers, it is a major *Upstream* consideration for which purchasers require greater transparency in order to optimize sustainable purchasing efforts. Lab purchasers and procurement managers must inquire about manufacturing conditions and processes. The more frequently these questions are asked, coupled with consumer demand for greater information, the more likely this important information will be shared.

7.4.2.2 Factory Resource Use

Resources used in manufacturing have an indelible impact on a product's overall sustainability. The quality and sourcing of materials can vary greatly;

products can be created from raw materials extracted *via* disruptive processes or from sustainable, recovered, or recycled materials. Products created using recovered or recycled materials tend to have lower global warming potential (GWP),[52] lower associated production costs, and more secure supply chains by avoiding the need for virgin materials.[53] They also support a circular economy (more on this in Section 7.6.5).

Factories owned or utilized by multi-million or billion dollar corporations often engage in inequitable resource use, taking an inordinate amount of local resources for use in manufacturing operations.[54–56] Companies commonly utilize community or regional resources for private benefit and financial gain, to the detriment of the wellbeing of local communities. This often leaves local communities in shortages or without the resources they need,[57] especially considering many of these communities were water insecure prior to the presence of industry.[56] According to the United Nations Convention to Combat Desertification (UNCCD) *Drought in Numbers* report, by 2030 an estimated 700 million people will be at risk of being displaced by drought and by 2040, an estimated one in four children will live in areas with extreme water shortages.[58] Additionally, women and children are most heavily impacted by water scarcity, with detrimental impacts on wellness, safety, education and sanitation. Furthermore, 72% of women and 9% of girls are responsible for collecting water, using as much as 40% of their total caloric intake transporting it.[58]

Though manufacturing resource use is rarely transparent information shared with consumers, scientists must remain aware these inequitable power differentials exist and avoid purchasing from companies negatively impacting local communities when data exist to allow informed decisions.

7.4.2.3 Factory Waste Streams

The methods by which manufacturing facilities manage waste are also an indicator of a product's overall sustainability. Waste from manufacturing has significant equity and justice implications. In fact, many companies in countries with strong environmental protection laws export manufacturing to other countries in efforts to reduce costs associated with adhering to regulations. As one might expect, this can be especially problematic for local communities and ecosystems in proximity to these facilities.

Though scientific post-consumer waste leaving research facilities is something to consider under Section 7.6, manufacturer waste is an *Upstream* consideration, as it should be considered prior to purchasing lab supplies and equipment. Consumers, including scientists, have historically lacked visibility of factory conditions and operations, making it nearly impossible to include as criteria in purchasing decisions. However, in recent years, scientists, procurement professionals and other consumers have begun demanding greater transparency from suppliers regarding factory processes by asking questions and giving preference to companies demonstrating a willingness to be transparent. These questions include: Where is the factory located? Does the factory engage in circular economy initiatives or materials recovery? What is the factory waste management plan including waste diversion efforts, programs or commitment, such as *Zero Waste* initiatives? Do they recycle? How much waste do they generate and what is their diversion rate? Does factory waste impact the community?

When scientists purchase products created sustainably and ethically, they are helping their own science and institutions to be more sustainable, as well as helping to drive more sustainable market shifts through their collective purchasing power. The methods by which a company manages their waste says a lot about their overall commitment to sustainability, as waste is the lesser-seen and often hidden side of operations, giving insight into company values when they think no one is watching. A company's willingness to be transparent regarding internal practices is a significant step in the right direction, as well as an indicator of their willingness to grow towards greater sustainability. Preference should be given to companies willing to be transparent about their processes over those who are not.

For some products, it is important to ask about specific factory waste streams, such as hazardous waste or waste water runoff. In nitrile glove and battery factories, waste water runoff is of significant concern and comes with a long history of environmental violations. In these cases, it is important to ask where these products are being manufactured, as country and regional environmental protection laws are of great significance.

Finally, it is important to understand if a company has made any commitments (either voluntarily or based on local regulation) regarding *Extended Producer Responsibility* (EPR). EPR refers to environmental policy that assigns greater responsibility to producers for their product over its entire lifecycle and specifically at a product's end-of-life (EOL). EPR is especially useful in mitigating waste management issues—when the burden of waste management for products is placed back onto producers, it drives more innovative and efficient processes in the product design and manufacturing, often inspiring manufacturers to engineer out hard to manage materials.[59] EPR legislation has expanded globally, starting in the European Union (EU) and followed by many others, including the US.[60] Examples of existing EPR legislation include take-back programs for batteries and electronics, recycling of packaging materials, and other efforts to remove the burden of EOL disposition of products from the consumer.

7.4.3 Labor

The labor used in the creation of laboratory consumables and equipment is one of the most important *Upstream* considerations during the procurement process. Many items are created using forced or unethical labor practices, directly impacting marginalized communities and workers. If the cost of a product is significantly lower than its competitors, this should prompt purchasers to inquire about the labor used in its creation, as this can often be an indicator of unfair labor practices.[61] More than 50 million people are currently living in modern slavery with over 28 million of those in forced labor. Modern slavery impacts every country, with about half of all forced labor occurring in lower-middle or upper-middle income countries and driven by demand from higher-income countries.[62] Nearly two thirds of all forced labor cases are connected to global supply chains, specifically in the lowest tiers of the supply chain; raw materials extraction and production.[63] Scientific products and equipment are a contributing factor to this global crisis of exploitation and the global research enterprise must prioritize purchasing products and goods created ethically using fair labor.

According to the Global Slavery Index, the US is by far the biggest importer of at-risk products, that is products at-risk of being created with forced or unethical labor. In 2023, the US imported more than $169 billion in products considered at-risk of modern slavery. This is more than the next highest five countries combined (Japan $53.1B, Germany $44B, United Kingdom $26B, India $23.6B, South Korea $20.2B).[62]

The International Labour Organization (ILO) sets forth international standards to "promote opportunities for women and men to obtain decent and productive work, in conditions of freedom, equity, security and dignity"[64] and serves as a unifying agency to educate and guide governments and protect workers. The five main principles of the ILO are the freedom of association and the right of collective bargaining, elimination of forced or compulsory labor, abolition of child labor, elimination of discrimination at work, and a safe and healthy working environment. These principles are at the heart of human rights and supply chain efforts and serve to promote equity and social justice. They are also the basis for the Ethical Trading Initiative's (ETI) Base Code, an internationally recognized code of labor practices.[65]

Governments' responses globally to prevent forced labor in supply chains have increased in recent years, though greater action is still urgently needed. Until 2021, the US was the only country in the world with legislation prohibiting the importation of goods created using forced labor per the Tariff Act of 1930.[62] In 2021, Canada amended the Customs Tariff Act of 1997 to prevent the importation of goods made with forced labor and in 2022, the EU adopted a similar proposal.

Additionally, more countries are embracing mandatory Human Rights Due Diligence (mHRDD) legislation, requiring companies and their supply chains to ensure humans and the environment are not harmed during the creation of their products.[66] The United Nations Guiding Principles on Business and Human Rights (UNGPs) is considered the authority on Human Rights Due Diligence and are the building blocks for mHRDD laws. Countries are also taking action to prevent the importation of products from known forced labor regions, such as the US' 2021 *Uyghur Forced Labor Prevention Act*. In the UK, EU, US, Australia, Canada and others, legislation has also been passed to allow for *Magnitsky* sanctions against those involved with human rights violations.[67] These sanctions allow for restricting travel and freezing the assets of human rights offenders.

There is a dire need for greater transparency of labor conditions within supply chains, and scientific supply chains are no exception. Due to this lack of visibility, scientists and other scientific purchasers lack, for the most part, the ability to make informed choices in their purchasing to no fault of their own. While avoiding forced labor and human rights violations is absolutely a priority, it just is not feasible for scientists to individually research every consumable or piece of scientific equipment entering one's lab, while also meeting the demands of their research. Scientists and scientific purchasers must demand radical supply chain transparency of labor conditions within the scientific supply sector; not only to minimize harm, but also to uphold institutional commitments and policy prohibiting the procurement of goods created using forced labor. Scientific suppliers will not willingly forfeit the business of the global research enterprise and will be driven to meet these demands in order to maintain their sales. Scientists can advocate for greater equity in their supply chains in a variety of ways; from asking distributors to identify products created using fair labor and to share information about companies' social commitments, to refusing to purchase conflict products—that is, products created using forced or inequitable labor. The purchasing power of the global research enterprise cannot be overlooked or overstated—collectively, scientists have the ability to impact significant and lasting change within and beyond their sector—simply through their purchasing alone.

Recently, some scientific suppliers have launched efforts to allow for greater transparency of products housed in their inventory, and their efforts are to be applauded. However, these efforts reflect only a miniscule percentage of their catalogue, and must be urgently upscaled in order to have valuable impact. A few manufacturers have also implemented traceability of product origin into their entire product line—primarily those in sectors at highest risk for forced labor or product fraud in the supply chain. One method implemented is *via* QR codes on individual products, allowing purchasers to scan to see a product's origin, testing and other information.[68] Unfortunately, labor conditions are not included in this information. Despite these novel advancements, the majority of the scientific supply chain remains opaque. Urgent action is needed to achieve full transparency of manufactured items through policy and greater regulation at the institution, state and federal levels. Purchasers should remain skeptical regarding claims of traceability and transparency using stock keeping units (SKUs). Currently, there is no international standard for SKUs, as their purpose is for companies' internal inventory management. Within a company, products created in separate factories, even in separate countries, with wildly varying conditions, can potentially share the same SKU. Thus, they do not usually offer traceability.

7.4.3.1 Forced Labor in the Scientific Supply Chain—Focus on Nitrile Gloves

One of the most commonly used consumables in science and health care is the nitrile glove (see Figure 7.5). Nitrile gloves are critical personal

Figure 7.5 Photo of nitrile gloves in production.

protective equipment (PPE) which keep users safe from hazards, such as hazardous chemicals or pathogens, while simultaneously allowing for great dexterity and sterility. The global market for nitrile gloves was valued at $15.18 billion USD in 2023 and is projected to grow to $24.85 billion USD by 2028.[69] In June 2020, during the Covid-19 pandemic, it was estimated 65 billion gloves were being used monthly globally.[70,71] In 2023, over 307 billion gloves were used globally and use is expected to surge to 450 billion gloves used annually globally by 2027.[72] Malaysia is the global leader in the manufacturing of nitrile gloves, accounting for half of global exports with the US leading demand, followed by the EU.[73] Malaysia supplies an estimated 75% of all nitrile gloves imported to the US.[74]

Though there are many aspects of the nitrile glove lifecycle which are concerning from a sustainability perspective—the raw materials extracted, the detrimental impacts on the local communities, the fuel used in manufacturing (commonly controversial palm oil), the impact of nitrile glove waste at end-of-life disposal—perhaps the most concerning is the inequitable, and often forced, labor which creates them. The entire nitrile glove value chain is rife with harmful environmental and social impacts. This is extremely problematic for the global scientific enterprise, as disposable nitrile gloves are an essential staple for scientific operations and often cannot be replaced by another currently existing technology or consumable. In some cases, the use of nitrile gloves can be reduced or engineered out of processes through careful review. Scientists and medical professionals alike are accustomed to using PPE at all times, even when not scientifically or medically required. In these cases, wearers should audit their use to determine if nitrile gloves are required based on a hazard assessment. For example, many scientists wear nitrile gloves when rinsing clean glassware. Gloves are not scientifically required for this task and the wearer could instead use reuseable gloves or no gloves at all. However, in cases where gloves are required based on hazard, the scientific community must identify and engage effective and transparent pathways by which to obtain conflict-free nitrile gloves.

More than 15.6 million people comprise Malaysia's labor force, including an estimated 3.5 to 5.5 million foreign migrants[42] primarily from Indonesia, Bangladesh, and Nepal. Many of these migrant workers are victims of human trafficking and forced labor,[75] having paid high recruitment fees for the opportunity to work—up to $5000 USD—leaving them trapped in debt bondage once they arrive in country. These high recruitment fees, coupled with the high cost of factory food and the inability to leave the factory property, traps workers for years in abusive working conditions as they work off their initial debt. In many cases, their passports are confiscated by their employers, they are not permitted to leave the factory and are forced to stay in an over-crowded dorm on site. These conditions are rampant in Malaysia's nitrile glove industry, which sprung up in the 1980s to meet demands from the US and Europe during the AIDS crisis.

The prevalence of forced labor in Malaysia's nitrile glove industry is so pervasive and problematic that US Customs and Border Protection (USCBP) has issued six Withhold Release Orders (WROs) over the last 5 years to Malaysian glove companies barring the importation of these goods to the US,[76] citing forced labor and human rights abuses. Currently, USCBP has two active WROs against foreign glove companies, one in Malaysia and one in China.[77] Additionally, a class action lawsuit is currently pending in the US District Court against the two largest glove providers to the US scientific and medical sectors. This lawsuit alleges violations of the Trafficking Victims Protection Reauthorization Act (TVPRA), including forced labor, human trafficking, and verbal and physical abuse.[78]

7.4.3.2 Efforts to Implement the Sustainable and Equitable Procurement of Nitrile Gloves

With the procurement limitations described above, including insufficient tools and a lack of supply chain transparency, scientists and research institutions are left to traverse sustainable and ethical nitrile glove purchasing on their own. The University of Georgia (UGA) Sustainable Science and Green Labs program (the program which this author founded) has taken efforts to create a nitrile glove "safe list" to prioritize the procurement of nitrile gloves created using fair labor in their research enterprise. Similar to most research institutions, the most commonly used laboratory consumable at UGA is nitrile gloves. Also similar to most research institutions, scientists and staff at UGA are deeply concerned to learn this common lab item often involves human rights violations in its supply chain.

Unfortunately, there is no fair labor certification within the nitrile glove sector and companies' claims, Code of Conduct policies, and sustainability commitments and reports are of little value in determining the quality of labor used in their production, as nearly *every* nitrile glove company cited with WROs, including those currently involved in the class action lawsuit, had fair labor statements or commitments at the time forced labor and other human rights violations occurred. These public-facing statements can often be a means of greenwashing (including equity washing), and are not suitable guides for determining equitable operations within a supply chain. The only thing more detrimental than not considering equity and human rights within a supply chain is greenwashing the suffering of individuals experiencing forced or inequitable labor conditions.

Clear, objective criteria were required to guide the process of creating a nitrile glove safe list for the equitable procurement of nitrile gloves. The original proposed criteria included data which could be easily integrated into procurement systems, considerations of company and factory certifications, manufacturing location, and the existence of credible allegations of forced labor and/or human rights violations.

The ability to have data integrated easily into procurement systems is required in order to meet the needs of a large research institution with decentralized purchasing, meaning many different labs purchasing for themselves. Procurement systems routinely auto-populate data provided by manufacturers and distributors, saving time and money for institutions and allowing for greater system flexibility, such as the ability to use purchasing filters at the institutional or user level. Auto populated data usually consists of information such as specs, materials and origins, as self-reported by the manufacturer. Integrating equity data is far more challenging, as currently there is no universally required supply chain transparency standard. Additionally, even those companies striving for supply chain transparency may not be aware of human rights violations happening further up the supply chain due to long and multi-tiered supply chain opacity. Thus, self-reported equity data not rooted in external standards or certifications is subjective and inadequate for procurement system integration.

There are several global social accountability certifications which reflect the fair treatment of workers, such as Social Accountability International's (SAI) SA8000 standard which implements controls across 9 different focus areas with the goal of fair treatment of workers.[79] This certification involves a self-audit, certification body review, and two initial external audits, and results in a three-year certification period. During the certification period, factories can expect twice-yearly unannounced onsite surveillance monitoring. It is important to note that all actions taken by the manufacturer are voluntary and reviewer reports are confidential to the client. Companies earning SA8000 certification often promote this achievement in public-facing documents, and with good reason as this is currently the most rigorous factory fair labor certification in existence. However, companies often use language implying SA8000 certification is a company-wide certification and it is not—it is a factory-level certification. It is of utmost importance to inquire how many factories a company has, how many of those are SA8000 certified, and what percentage of total production output they represent. It is possible, as well as not entirely uncommon, for a company to have SA8000 listed in their Supplier Code of Conduct *and* also have forced labor in their operations. For example, a company may have 7 factories in 4 different countries and SA8000 certifications at only two locations. The two locations with SA8000 certification would be considered to have equitable labor, but the other 5 factories would remain unknown. Thus, in order for SA8000 to be an impactful identifier and objective criteria to use in procurement systems, it would need to be attached to specific inventory items which were created at a SA8000 certified factory, and not attached to an entire product line. This concept is not yet reflected in specific products, and thus would need to be further developed and adopted to be an adequate identifier used in procurement platforms. Finally, many companies claim they have an "SA8000 equivalent certification", but while ethos and review areas may be similar, the auditing process is much weaker, often involving no surprise or follow-up audits and fewer site visits overall. The criteria by which certifications are achieved are only as important as the degree to which they are audited and proven.

Currently, only one company-wide certification standard for social and environmental performance, transparency and accountability exists, called a B-Corp certification.[80] This rigorous certification process measures a company's social and environmental performance through the B Impact Assessment and is considered to be an international standard for corporate sustainability commitments. However, this certification would also be insufficient to use as an objective indicator of overall labor from a supply chain due to several limitations: companies with human rights abuses are not automatically disqualified from certification, and companies scoring poorly in one category can compensate by scoring higher in other categories. For example, low scores in categories covering overseas operations can be compensated for *via* programs benefitting white-collar workers in higher income countries. Thus, while B-Corp certification overall is something for companies to aspire to, the mere presence of the certification itself does not guarantee a fair labor supply chain.[81]

Manufacturing location is another potential criterion reviewed by UGA Sustainable Science and Green Labs for the creation of a nitrile glove safe list. Though most countries, including Malaysia, have labor laws preventing forced labor, not all have the resources to enforce them. Additionally, Malaysia's labor laws are directly undermined by their migration laws, leaving the majority of their nitrile glove work force vulnerable and without protection from established labor laws.[82] Work visas are issued for employment with a specific company and workers who do not comply with abusive demands are at threat of deportation. Additionally, if a worker is fired and still in the country, they run the risk of being undocumented and arrested. Globally, migrant workers have a three times higher risk of being in forced labor than non-migrant workers.[62] With these concerns in mind, research was conducted to determine if other countries had similar loopholes and if a country's income had bearing on the degree of forced labor in its work force. Though most countries do not have such detrimental loopholes, according to the Global Slavery Index 2023, upper-middle and high-income countries are where 52% of forced labor occurs.[62] This is an important distinction to make, as lower-income countries have long been scapegoated as primary locations where forced labor occurs and higher-income countries have been assumed to have little or no forced labor. Thus, while a country's income and location can be an indicator of how likely forced labor may be, it is also not able to be used as objective criteria in determining labor practices or risk of human rights abuses at this time.

Credible allegations of forced labor and human rights violations were another category of criteria reviewed for integration into procurement systems with the goal of creating a nitrile glove safe list. While objective sources such as USCBP WROs, verified complaints by ILO, Global Slavery Index and others, or lawsuits in which companies have been found liable can collectively and objectively confirm a company's liability of forced labor or other human rights violations, they cannot objectively rule out forced labor for companies they have not researched or are not yet on their radar. Thus, while these methods may be beneficial for ruling out previous human rights offenders, they cannot affirm fair labor in companies which are not on these lists. Thus, this criterion also cannot be currently used to determine a safe list for nitrile glove providers.

While the UGA Sustainable Science and Green Labs program continues to champion efforts towards more equitable nitrile glove procurement through ongoing research, unfortunately none of the above proposed criteria in its current state is effective for objectively determining a nitrile glove fair labor safe list. While each has benefits to help indicate a presence of forced labor, they remain unsatisfactory tools for clear declarations of fair labor and are unable to be objective, determining factors of fair labor within procurement systems at this time. One possible solution and policy recommendation is for a fair labor certification to be created specifically for nitrile gloves, funded by the industry itself and developed through a reputable and independent third-party certification group with vigorous ongoing on-site audits, as well as a requirement of transparency and traceability. Coinciding this effort, nitrile gloves purchasers must commit to only purchasing gloves with a universal nitrile fair labor certification and use their unified and collective influence to help transform the industry. However, as previously mentioned, the integrity of fair labor certifications which currently exist in other sectors remain in question by human rights experts and in many cases, are often considered *equity washing*. Thus, in the event a fair labor certification is developed for the nitrile glove industry, it must be created in close collaboration with human rights experts and watch groups and include a vigorous and ongoing auditing process.

7.4.4 Transportation

Transportation of goods is another important *Upstream* consideration; how far does a product travel to reach its point of use? The further a product travels, the higher its emissions and carbon footprint and the more it negatively impacts communities around the world through climate change, air and water pollution and environmental destruction. While the benefits of buying local are well-known, this is often not possible when purchasing scientific equipment and consumables. Therefore, transportation considerations often involve *minimizing* negative impacts, rather than avoiding them all together. When reviewing two comparable products, consider which is manufactured closest to your scientific operations and balance this with further sustainability factors.

7.4.5 End-of-life Disposition Requirements

While most end-of-life (EOL) disposition and disposal requirements are *downstream* considerations, anticipating EOL impacts should also be considered during the procurement process. EOL refers to what happens to equipment or products when they are no longer of use. Products may require special disposal dictated by the product itself or by region-specific processes for disposal, such as incineration or landfill. If a scientific supplier offers a take-back program for products at EOL, this is usually socially and environmentally superior to traditional disposal methods. Many companies are moving away from linear economy models of "take-make-waste" and moving towards circular economy models, where materials are recovered and reused by manufacturers[83] (more on this in Section 7.6.5). Scientific purchasers should give preference to products with manufacturer-funded take-back programs or other established and proven waste diversion programs, as well as to companies exhibiting Extended Producer Responsibility (EPR), meaning those companies who exhibit greater responsibility of product stewardship thus reducing the burden of EOL disposition on consumers.[84] EOL disposition is a justice and equity concern as the prevalence of many products in landfills, especially those containing heavy metals or chemicals, increases environmental pollution which disproportionally impacts lower socioeconomic status rural communities and communities of color. Additionally, the negative environmental and social impacts of raw material extraction are reduced for products created using recovered materials.

7.5 In-lab Equity and Justice Considerations Within the Global Scientific Enterprise

It is not only important to consider how the global scientific enterprise impacts environments, communities and individuals outside of laboratories, but also within them. Despite the embedded progressive nature of science itself, there are still many inequities within the field which require collective action to resolve. From calls to decolonize the academy[85–87] to efforts resolving wage and publishing disparities, the scientific community is being beckoned forward to bring the same objective and equitable approach to its intracommunal equity and justice as is brought to the process of science itself. In order for the field to embody its full potential, scientists must strive to bring equity to internal operations, processes, and culture. Prioritizing an equitable culture leads to better scientific outcomes, including more novel contributions and greater innovation rates.[88] Creating an equitable enterprise not only allows for equal opportunities for underrepresented groups, but also improves the overall quality of scientific research.

The inclusivity of science remains an issue; from gendered language in syllabi undermining the competency of women and normalizing masculine belief systems and behaviors[89] to miscrediting discoveries of underrepresented groups,[90] many gender and race disparities persist. These disparities must be equalized and equity brought to scientific operations in order for the academy to continue to progress. Just as a weak link in a chain leads to mechanical failure, so too does inequity within the sector weaken its overall contribution and potential outcomes. When scientific inquiry is broad and inclusive of novel approaches, scientific outcomes improve.[91,92] Novel input comes from a diversity of voices, cultures and people having lived different life experiences and holding different perspectives. It is through these varying perspectives that unique insights are introduced. These differences of experiences not only strengthen the scientific community but also scientific research, which leads one to question, "If inclusivity improves our science, why has it not yet been prioritized?". In part, this is due to the often myopic focus of research interests, with researchers focused primarily (and often solely) on their subject matter and lacking the bandwidth to prioritize systems change which would require additional time and effort to implement. Additionally, many researchers may desire and be supportive of systems change, but are not clear on tangible actions they can take to do so.

Finally, the majority of scientific researchers, White males, have long benefitted from inequitable and exclusionary practices. This positions those in the majority, who are also the primary demographic in seats of power and responsible for making institutional decisions which drive cultural norms, to remain unaware of both the privilege they have experienced, as well as the degree of inequity and exclusion experienced by underrepresented counterparts. When an inequitable process serves a demographic, an implicit (subconscious) bias is created, providing an incentive to maintain the status quo and lack an intrinsic incentive for change. Additionally, those who are keenly aware of the need for greater inclusivity often lack the power or position to mandate such changes and must drive systemic changes *via* grassroots efforts. These efforts add to further inequities and negative experiences of marginalized researchers, as the time required for advocacy detracts from research time, as well as leads to more predominant experiences of exclusion and loneliness.[93]

7.5.1 Decolonizing the Academy

In recent years, there has been a movement calling for the decolonization of everything from art to media to education to science. Within higher education there have been calls to decolonize the academy, pedagogy, journals and even syllabi.[85–87,94] Decolonization is the unlearning and dismantling of deleterious teachings and concepts associated with colonization; primarily racist concepts of White European superiority over Black, Indigenous, People of Color (BIPOC). Decolonization requires demolishing racist systems of power while conducting honest, unbiased reviews in order to reveal and challenge inequitable systems. Decolonizing the academy can take many forms, such as decolonizing curricula, reimagining syllabi and classroom dynamics, rethinking pedagogy and exploring alternative methods to bring students to mastery.[85] Decolonizing curricula makes learning more inclusive and helps students to approach inquiry with a broader perspective and be able to more easily identify inequitable systems of power. Many institutions have taken formal action to decolonize, such as Keele University which created the *Keele Manifesto* for *Decolonizing the Curriculum (DTC)* in order to conduct an unbiased and honest exploration of the current state of colonization at the institution and to create and share resources, including actionable methods to decolonize curricula, syllabi and more.[86] These actions span far beyond adding indigenous or diverse contributors to syllabi or hiring more faculty of color, which can often be performative or tokenism. Instead, decolonization requires a conscious effort to change the way we approach knowledge *itself*—to change the way we approach learning through challenging assumptions that western intellectual traditions are superior or universal. It is also about recognizing the connection between knowledge and colonial hierarchies, status quos and long-existing systems of power, and then openly and mindfully challenging them.

7.5.2 Increasing Diversity in Science

Increasing diversity within the scientific community is not only about equity and justice; it also improves scientific outcomes. Research in the private business sector shows diversity correlates with better financial performance.[95] In a study examining public companies in Canada, Latin America, the UK and the US, companies in the top quartile for racial and ethnic diversity are 35% more likely to have financial returns above their respective national industry medians and companies in the top quartile for gender diversity are 15% more likely to have financial returns above their respective national industry medians.[95] Demonstrating the benefits of diversity within science may not be as easily quantifiable as financial returns in the business world, but can be clearly seen in citation counts,[96] as well as through the ability to solve more challenging problems *via* a diversity of life and research experiences and thus, perspectives.[97,98] Additionally, more diverse teams have more creativity and higher levels of scientific innovation.[99] Diversity maximizes our scientific potential, and yet the diversity of science remains at issue with a long way to go.

According to the *National Center for Science and Engineering Statistics (NCSES)*, men comprise nearly 75% of all science and engineering positions in the US and nearly 65% are occupied by those who identify as White individuals.[100] Women and Black, Indigenous, People of Color (BIPOC) have a very different experience in scientific jobs than do their White male counterparts, with expectations to contribute more,[101] while being paid less,[102–104] and having fewer opportunities for funding, publishing and promotion.[105–108] Data published in the 2020 *United Nations Development Programme Gender Social Norms Index* shows that nearly 90% of both men *and* women hold a bias against women,[109] and while this data is not specific to the scientific community, it does reflect global biases and prejudices which permeate nearly all sectors of society, including science. What we study and how we think is directly related to our real-world experiences and culture, thus our science is also influenced. Women in science, and particularly women of color, experience the highest levels of workplace harassment and hostility, often leading to the avoidance of professional events due to feeling unsafe, resulting in a loss of career opportunities.[110]

While change is typically incremental and takes time, there has been *no* significant improvement in the diversity of geosciences PhDs in the US in more than forty years.[111] Additionally, underrepresented groups account for only 3.8% of tenured faculty amongst the top 100 earth science departments[112] and faculty diversity in higher education remains less than 6%, also having not changed in decades.[113] Academic institutions must acknowledge the ineffectiveness of current diversity initiatives and engage in data-driven strategies requiring the participation of *all* faculty members—not only underrepresented faculty—in efforts to achieve greater diversity and inclusivity.[113] (More on this in Section 7.5.3). In order to create more diverse PhD candidates and thus, more diverse faculty, emphasis should also be placed on creating more master's degree to PhD pathways, as well as pathways better serving the specific personal interests of those from underrepresented groups who tend to be more motivated by altruistic intentions than their majority peers.[111]

7.5.3 Invisible Work, The Minority Tax, The Matilda Effect and Imposter Syndrome

Over 35 years ago, sociologist Arlene Kaplan Daniels coined the term *Invisible Work* to describe work being done by women which was unrecognized, unpaid, devalued and unregulated.[101] At the time, the term "women's work" was still commonly used to refer to essential tasks performed by women which kept households running, children bathed and fed, and the office party well-organized. This unpaid work, which was essential to the very foundations of society, was not valued in the same manner as paid labor, despite its relative importance. The term "invisible work" illuminated the stark contrast society placed on paid *versus* unpaid labor, including volunteering. Decades later, the term has been reembraced across academia and fields such as human resources, science and medicine, to recognize critical contributions women and people of color make in their careers which are not reflected in job descriptions, nor considered in advancement opportunities. These contributions take time and energy to make, yet do not count when assessing job performance or career status. These tasks are largely considered "extra" by employers, yet have significant impact on how well an institution or department functions. Examples of *Invisible Work* are providing emotional support to colleagues and other efforts to create a supportive work culture, resolving conflicts, mentoring, office organization and carrying more administrative duty regardless of title, communicating and coordinating on behalf of one's team, nurturing teams and investing in stakeholder relationships, to name a few.[114] In the scientific community, *Invisible Work* is seen in underrepresented demographics taking on more mentoring duties and serving on Diversity, Equity, and Inclusion (DEI) committees. This is largely in part due to departments requesting underrepresented researchers serve in these capacities, as well as from an altruistic and intrinsic desire to mentor and support students and upcoming faculty who share in the unique pressures and life experiences of being part of an underrepresented group. As *Invisible Work* falls disproportionally to non-majority individuals, it is of utmost importance institutions bring awareness to the detrimental impacts it can have on researcher satisfaction, faculty retention, workplace culture and more. The first steps in resolving the burden of *Invisible Work* are to recognize the labor and acknowledge harm.[114] On a departmental level, researchers should audit the amount of emotional labor they are carrying in order for leadership to redistribute it across the team or give fiscal value to those actions.

One specific type of *Invisible Work* seen frequently throughout the sciences is often referred to as the *Minority Tax*. Underrepresented groups in science are often expected to contribute more than their well-represented colleagues, with expectations to help achieve diversity and inclusivity at both the departmental and institutional levels. As with *Invisible Work*, the *Minority Tax* often includes mentoring trainees, serving on committees, and acting as a voice for underrepresented subgroups. These extra responsibilities require a significant investment of time and energy, which ultimately detracts from research time and scientific pursuit.[113,115] Additionally, these extra responsibilities actually impede academic promotion, as they are uncompensated and not considered promotion-earning activities, thus unmeaningful during tenure reviews. Rodriguez *et al.* suggests these disparities can be eliminated by valuing diversity efforts fairly, adjusting the assignment of responsibilities accordingly, assigning promotional value to diversity efforts, encouraging positive curiosity when encountering "otherness" and recognizing differences as a benefit to our institutions and processes.[115]

The *Matilda Effect* describes a common happening in which the value of contributions made by female researchers is diminished, overlooked, unacknowledged or attributed to a male colleague.[116] Within scientific research teams, women are significantly less likely to be credited with authorship than men.[104] There is an undeniable gender gap in attribution across most scientific fields and career stages, as well as a lower likelihood that a female scientist's contributions will be recognized.[104] This undervaluing of a woman's scientific contribution has direct impacts to her career; from being denied recognition and authorship of discoveries, such as Rosalind Franklin who should have been credited alongside Crick and Watson for the discovery of the structure of DNA, to having fewer opportunities for prizes, recruitment and tenure. Additionally, women are often expected to contribute substantially more than men to be recognized for their scientific contributions.[104] Researchers can combat *the Matilda Effect* by ensuring appropriate recognition is given to female teammates and speaking up if they observe a colleague's contributions are being disregarded or absorbed by a male colleague. This third-party voice is even more important when a power differential exists between the female colleague and the male colleague claiming their work.

Many women and other underrepresented groups in science experience *Imposter Syndrome*; a feeling of unworthiness or incompetency, regardless of their accomplishments and successes. They doubt their success, often feel they have deceived someone in order to have the status they hold or have "lucked" into their position or opportunities. *Imposter Syndrome* often stems from systems biases, experiences of workplace harassment, underrepresentation of one's gender or race, and the hyper-competitiveness of academia.[117] Even amongst faculty leadership, cultural undercurrents exist which undermine status and professional expertise in female leaders, driving further feelings of *Imposter Syndrome*. These undercurrents include gender inequities surrounding leadership qualities. For example, the same leadership qualities that are heralded in men, such as assertiveness, are seen as derogatory in women, causing women to feel less accepted and even more harshly scrutinized. Institutions can better support scientists who are suffering from *Imposter Syndrome* by improving researcher diversity and providing more professional development and mentoring opportunities.[97,117]

7.5.4 Wage Disparities

Despite the many technological advancements of the global scientific enterprise, there are aspects of the scientific community which have been slow to advance, such as wage equality. Female scientists are commonly paid less than their male counterparts[102,103,118] and the gender pay gap is 1.5 times wider in academia than in private-sector industry.[119] Some are concerned about the impact this could have on academic diversity, as this certainly does not incentivize the retention of female scientists. Female PhD recipients in STEM entering into a permanent job position after graduation experience a median salary gap of $18 000 in the US and $16 000 in Canada.[118] *The World Economic Forum (WEF) 2018 Global Gender Gap Report* states it will take 108 years to close the global gender gap and 202 years to achieve economic gender parity,[120] and though these numbers are not specific to science, it does paint an undeniable picture about the state of global wage disparities and the need to be conscientious and deliberate in efforts to resolve these inequities.

7.5.5 Hiring Disparities

In Section 7.5.2, we explored the need for greater diversity amongst scientific investigators, which is directly related to hiring and promotional gaps. Despite advancements in many fields, women in are still less likely to receive tenure than male counterparts[121] and within life sciences fields, a significant promotion gap remains for women.[122] One critical point of note is that the lack of diversity amongst scientific investigators is not due to a lack of qualified trainees within underrepresented categories, but rather due to a broken system of hiring, retention and promotion.[123] This system of hiring can be improved through various interventions, such as Bhalia's *Checklist to improve equity during faculty hiring*, actively advertising *via* databases for underrepresented groups of scientists, redefining excellence and merit, and conducting in-person interviews.[123] Career outcomes for Black/African American and White researchers start to deviate at the postdoctoral stage.[124] This is when mentoring scientists matters the most; research shows that mentoring at this career point leads to increased grants, publications and likelihood of receiving tenure.[124]

Hiring disparities are common in academic and scientific institutions, leading to a Diversity–Innovation paradox in which demographically underrepresented students have higher innovation rates than students in the majority, and yet their contributions are frequently disregarded, resulting in fewer career appointments. Women and other underrepresented scientists contribute higher rates of innovation than White male colleagues, but with much lower uptake, thus requiring them to innovate at higher rates in order to have comparable career opportunities.[88] As stated by Hofstra *et al.*, "Diversity breeds innovation, yet underrepresented groups that diversify organizations have less successful careers within them".[88]

7.5.6 Funding Disparities

Hiring disparities are not the only inequalities scientists are facing. In recent years, many studies have revealed disparities among grant recipients from major funding bodies. The National Science Foundation (NSF), a prominent funder in the US, was found to have prevalent racial disparities amongst grantees, and White Principal Investigators (PIs) were found to be funded at consistently higher rates than non-White peers; specifically Asian, Black/African American, Native Hawaiian, and Pacific Islander PIs.[107] Similar disparities have been identified in US National Institutes of Health (NIH) awards demonstrating Black applicants are 10% less likely than White applicants to receive funding.[125] These disparities can be explained in part by the number of citations a researcher's work receives, as Black/African American PIs have similar publishing numbers to White counterparts, yet fewer citations.[124] A researcher's point of focus may also play a key role in their likelihood to be awarded funding, as underrepresented researchers often focus on research topics geared at meeting the needs of underrepresented communities. Grant reviewers may not understand the importance or significance of these research topics as compared to topics focused on majority populations, and therefore, are less likely to award funding.[124,126]

In 2020, 22% of NIH funded recipients identified as Asian, <2% as Black, <5% as Hispanic and 65% as White. PIs holding more than one research grant simultaneously are considered super principal investigators (SPI). Women are 34% less likely to be a SPI than men, and Black PIs are 40% less likely to be a SPI than White PIs.[106] Black female PIs are 74% less likely to be a SPI than are White males.[106] Though the number of women PIs funded by the NIH continues to rise, they are still well below parity with nearly 62% of NIH awards going to men.[106,127]

On a global level, funding disparities are widespread, with the US, Germany, the UK, China, Japan and France leading in global research and development (R&D) funding.[108] Scientists in lower-income countries experience less in-country funding opportunities, as well as a lower probability of being awarded international funding, due to coming from less-robust research cultures. This not only limits great scientists to no fault of their own, but also their potential contribution to science and to the planet. Funding bodies must strive to bring greater equity to international funding opportunities with the knowledge that it will improve the overall body of science through the diversity of ideas.

The pervasiveness of race and gender disparities across funding bodies demonstrates how widespread this issue truly is. The impact of disparities in funding spans far beyond a single grant award, having cascading inequitable effects on a researcher's career. Policy and process-oriented solutions have been proposed by various research groups to minimize and resolve these disparities, including improving data transparency, increasing funding accountability for equity efforts, examining the culture of meritocracy, and reimagining scientific funding.[107,125,128,129]

7.5.7 Building a Healthier and Antiracist Research Lab

Due to the inequities shared in this chapter, as well as the highly competitive culture of scientific research and academia, scientists are often stressed or facing mental health challenges, such as anxiety or depression.[130] Even in the healthiest of research teams, pressures abound: the pressure to publish, competition for scientific positions, maintaining a work/life balance in a culture that promotes workaholics, and more. PIs and lab leaders can directly influence and minimize these stressors through a conscientious effort to build a healthier and non-racist lab. Maestre outlines ten rules towards this effort: Promote the well-being of lab members, let people set their own schedules, express gratitude, treat lab members as teammates not subordinates, collaborate, remember every lab member is unique, respect holidays and encourage work/life balance, give credit where it is due, destigmatize failure and promote the professional development of lab members.[131]

Part of creating a healthier research team is creating a safer environment. Challenging explicit racism and bias is, of course, extremely important for lab leaders. However, it is just as important to challenge implicit bias in order to build team resiliency and shift inequities in scientific culture. To accomplish this, PIs should focus on creating an antiracist lab. To clarify, a nonracist lab is one which attempts to take a neutral stance regarding racism, which inherently has no neutrality. An antiracist lab is one that actively supports antiracist policies and initiatives, as well as strives to challenge and dismantle implicit biases within a supportive team. Chaudhary and Berhe offer ten simple rules for building an antiracist lab: Lead informed discussions about antiracism in your lab regularly, address racism in your lab and field safety guidelines, publish papers and write grants with BIPOC colleagues, evaluate mentoring practices, amplify the voices of BIPOC scientists, support BIPOC in efforts to organize, intentionally recruit BIPOC students and staff, adopt a dynamic research agenda, advocate for racially diverse leadership in science, and hold the powerful accountable.[132]

7.6 Downstream Equity and Justice Considerations Within the Global Scientific Enterprise

In order to comprehensively understand the impacts of the global scientific research enterprise on equity and justice, we must also look *downstream* to understand how scientific processes can impact communities and environments both near and far from research institutions. *Downstream* impacts are the effects that occur after scientific operations are conducted, including waste management and emissions. Most scientists have little awareness of how their scientific waste is managed and what impacts it may have outside of their lab. In fact, due to the lack of visibility, most waste is not ever thought of again after it is thrown *away*. Yet, there is no "away"—when waste leaves research labs, it goes *somewhere* and has an impact on the ecosystem and the people in those places. Thus, *downstream* operations should be analyzed to better understand the impact of scientific waste on equity and justice, and also how operations can be better managed to minimize negative effects. When the Benefit/Burden Framework (Section 7.2) is applied to scientific waste management practices, it is quickly and easily seen that researchers and scientific institutions carry almost all the benefit associated with the generation of waste streams and very little of the associated burden. Long after the research is over, and scientists have moved on to other projects, communities are often left dealing with the detrimental impacts of scientific waste.[133,134] There are many opportunities available at both the lab and institutional levels to positively impact laboratory waste streams, many of which can be implemented *via* grassroots efforts and then scaled up with additional resources and institutional buy-in, such as micro scaling experiments, practicing green chemistry, or creating robust laboratory waste diversion programs. Each of these examples minimizes waste produced during scientific processes, and this reduction of waste results in less burden on outside communities and more equitable waste management.

7.6.1 Laboratory Waste Streams

Laboratory waste streams differ based on a number of factors; the type of research being conducted, the sustainability culture of the institution, the resources available to the lab, and the specific point within the project timeline. While teaching labs tend to generate an anticipated amount of waste during an academic semester, waste from research labs tends to be less predictable and for-profit labs tend to generate the same waste streams consistently, as dictated by their operational focus. It can be challenging to anticipate the volume or variety of waste streams leaving a specific research space, and also to extrapolate lab-specific waste audit data to estimate institutional or sector-wide lab-specific waste metrics. Not only are all labs different, but the waste streams within a single lab are often ephemeral and change throughout the year, based on field seasons, research focus or as required by environmental or biological cycles. Still, some common waste streams have been identified and form the foundation of laboratory waste management and diversion practices (see Figure 7.6).

Management of these waste streams also varies, as diversion and recycling opportunities are highly location and region dependent, relying on services available in one's area. For example, while a material may be recyclable, that does not indicate it can be recycled in *all* areas or by all waste processing facilities, and may require specialized recycling opportunities not found locally in order to actually be diverted.

More and more research institutions are developing lab-specific waste diversion programs including laboratory recycling, composting, and rehoming initiatives.[135–138] These programs not only help institutions avoid costly waste disposal fees, (*i.e.*, landfill tipping or incineration fees), but also reduce the burden of waste on outside communities and environments.

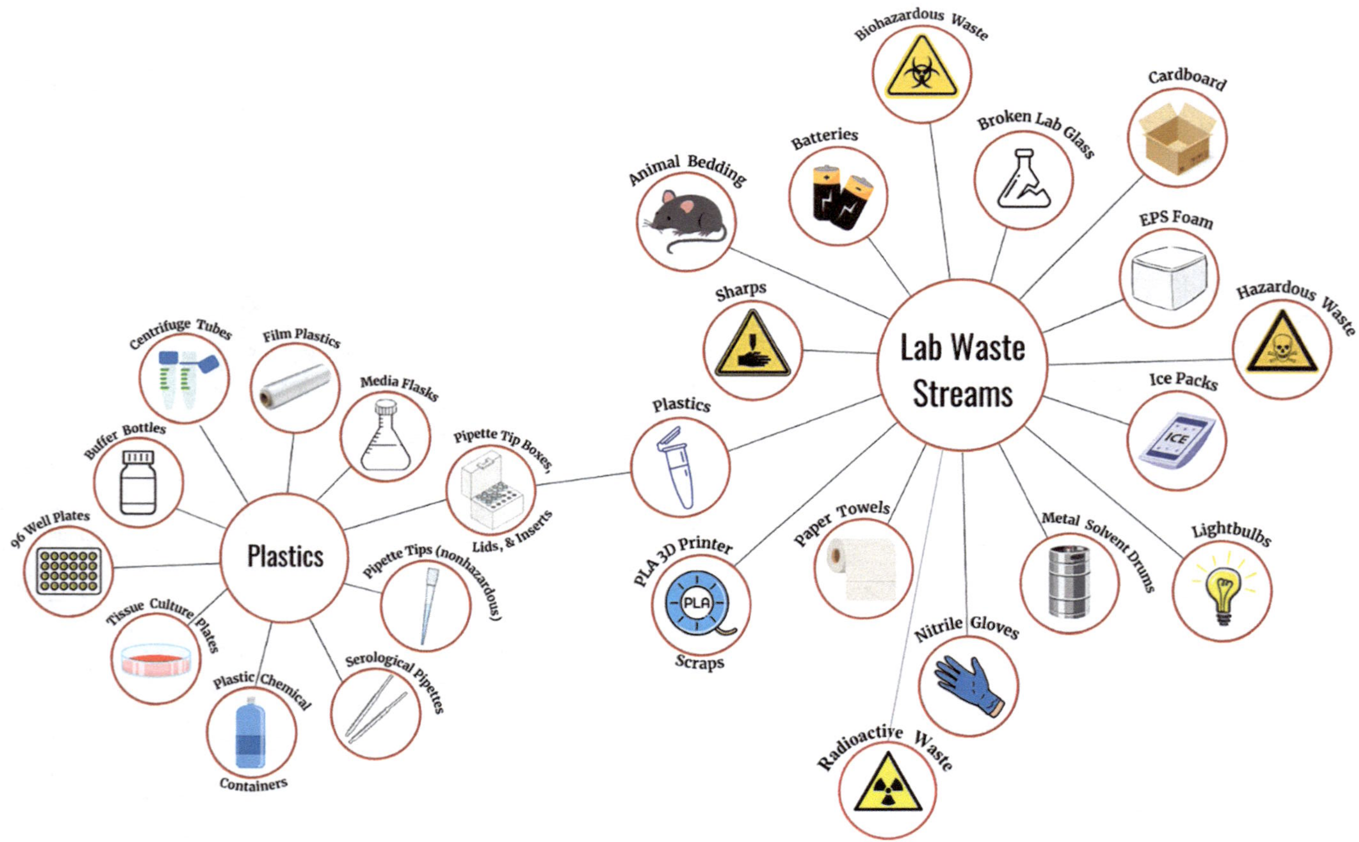

Figure 7.6 A schematic diagram of common laboratory waste streams at the University of Georgia. Reproduced with permission from UGA Green Labs, Copyright 2022.

7.6.2 Scientific Non-hazardous Waste Impacts

Non-hazardous laboratory waste falls under the category of municipal solid waste (MSW) and includes cardboard, paper, glass, plastic, plastic film, metal, organic material and more. These materials, though not hazardous, can still pose a threat to the environment and public health[139] if not appropriately managed through effective recycling, recovering, minimization, or composting. Thus, management of MSW is both a socioeconomic and environmental issue. It is well-documented that landfilled or incinerated waste from higher socioeconomic status areas, such as areas where many research facilities are located, often ends up in lower socioeconomic status areas (or countries), disproportionately impacting underserved communities.[140,141]

85% of waste incinerators or waste-to-energy (WtE) facilities in the US are located in Environmental Justice (EJ) communities,[141,142] meaning communities of color or lower socioeconomic status communities. Incinerators are a substantial contributor to the climate crisis through the release of greenhouse gases, emitting more GHG emissions per unit of electricity produced than any other power source.[143] They also contribute significantly to toxic air emissions, including dioxins, sulfur oxides, heavy metals and particulate matter[143] and are associated with an increased frequency of adverse health outcomes, including asthma, miscarriage and heart disease. Studies examining the health impacts of incinerators have had mixed conclusions,[144,145] due in part to the review of incinerator emissions in isolation and not considering the compounded effects of high volumes of truck traffic and related emissions, an increase in other polluting industries around incinerator sites and common pollution already existing in EJ communities.[143] These compounded exposures lead to significantly greater health risks.

Solid waste landfills have similar impact and site placement as incinerators, as they are also disproportionately located in EJ communities, and are the 3rd largest contributor in the US to anthropogenic GHG emissions due to their release of methane and CO_2.[146] Air quality in communities next to landfills is compromised, often with serious contamination[147] and living in proximity to solid waste landfills significantly increases one's risk of respiratory diseases and other health concerns.[139,148] The breakdown of waste in landfills results in the production of gases and leachate; leachate which can contaminate ground water, as well as surface water through holes in the landfill liner.[149]

Though the processes shared above for managing MSW are not exclusive to scientific waste management, scientific waste streams are also not exempt from contributing to global waste management challenges and impacts. Thus, scientists must remain educated and aware of the larger processes and global challenges to which scientific operations contribute. The enterprise must acknowledge these impacts and prioritize waste diversion initiatives which reduce the burden of scientific waste on outside communities and ecosystems.

7.6.3 Scientific Plastic Waste Impacts

Though the majority of plastic laboratory waste is considered MSW and shares the related impacts noted in the previous section, it also has profound and unique environmental and social impacts associated with its use and disposal, thus dictating the need for its own subsection. In addition to the extraction and manufacturing concerns of plastic noted in Section 7.4.1.1, the *downstream* impacts of plastics are also extremely problematic.[150]

As Wojnowska-Baryła *et al.* stated, "End-of-life does not equate to the end of the impact of plastic waste".[133] Petroleum-based plastics persist for hundreds of years beyond their point of use, posing a significant threat to the environment, flora, fauna and humans. Landfilled plastics experience long-term degradation leading to a structural breakdown of polymers into tiny particles known as microplastics (MP) and nanoplastics (NP). MPs are extremely problematic for the environment and climate, as their small particle size allows for them to easily spread very long distances *via* wind and water. Landfills are significant sources of MPs, due to plastics deposited within, as well as uncontained plastic litter blowing out of dumps. Landfills are the primary source of MPs in groundwater with over 90% of MPs in ground water derived from buried plastics.[133,151] MPs enter groundwater through landfill leachate, gases, surface runoff, and open dumping and effluent, thus demonstrating landfills aren't the final sinks for plastics, but rather generators of microplastics.[152] This is extremely problematic as MPs are not considered under landfill pollutant regulations in *any* country.[133]

MPs and NPs pose significant concerns for human and environmental health. Petroleum-based plastics contain a large variety of chemical additives, such as bisphenol A (BPA), polybrominated diphenyl ethers, and phthalates—all known endocrine disruptors.[133] These endocrine disruptors bioaccumulate in food chains and have significant adverse health impacts on mammalian endocrine systems, impacting reproduction, hormone regulation, cancer, immune function and more.[153,154] Additionally, the small particle size of N/MPs allow for greater specific surface area and absorption capacity, permitting co-contaminants to readily bind to them more easily. Many co-contaminants present greater hazards than the N/MPs themselves, such as heavy metals, antibiotics, bacteria, fungi and persistent organic contaminants. MPs act as vectors when co-contaminated with metals and antibiotics, resulting in the spread of multidrug-resistant human pathogens over long distances.[155]

As N/MPs spread, they impact the health of ecosystems and communities, disproportionately impacting oceans and marginalized communities. With over eight million metric tons of plastic entering the oceans annually,[156] plastic pollution resides at the center of many human rights and environmental justice cases, driving the creation and use of the term, "plastic justice". Communities most impacted by plastic pollution are usually the least responsible for it, bearing the burden of it but having shared in none of the benefits. They also have the least capacity to avoid its detrimental impacts.[154] Indigenous communities are most at risk, with cultural practices often centered around country foods, such as fish. Lower socioeconomic status and coastal communities are also disproportionately impacted, depending on wild seafood for their diet. These communities face the greatest risks from ocean plastics due to no fault of their own, as the pollution in their food chain does not come from their community, but rather from outside sources. MPs are ingested by small marine invertebrates, which are in turn ingested by a wide variety of sea life, leading to the bioaccumulation of plastics and co-contaminants up the food chain reaching highest concentration levels in top predators (*i.e.*, humans).[157] Some fish species occupying higher trophic levels have methylmercury levels up to a million times higher than surrounding waters.[158] Thus, methylmercury in fish is of significant concern, as mercury exposure is a known global health issue.[159]

While recycling plastic is a more sustainable and equitable process than landfilling, it is not a perfect solution and also has associated social and environmental injustices. The reality of global plastics recycling is often unknown by those producing plastic waste, including scientific researchers. Only 9% of all plastics ever made have been recycled[160] and many countries lack the internal infrastructure needed to manage and recycle their own materials. Historically, plastic waste generated in high-income countries has been exported to low or middle-income countries for recycling.[161] Thus the majority of plastics exported for recycling are leaving countries with more developed waste management systems and entering countries with less developed, and often insufficient, waste management infrastructure. Countries importing plastic materials are often blamed for mismanaged plastics with no consideration of who produced the waste or what country exported it. The problem is so pervasive that experts have suggested the global management of plastic waste would be greatly improved if waste producers maintained liability of exported material, similar to requirements designated in hazardous waste laws, ensuring it's proper management.[162] Despite having more developed waste management systems, many high-income countries lack the recycling infrastructure to recycle their own plastic waste.[163–165] For years, the majority of plastic recyclables from the Netherlands, US, Australia, Germany, UK and others were exported to China, but this stopped with the implementation of China's National Sword policy in 2018, a ban on the importation of plastic waste. This major change in global plastic waste management has left many countries struggling to manage their plastic waste streams. Many higher-income countries are now exporting waste to Indonesia, Malaysia, Thailand and Vietnam, where it remains significantly mismanaged.[156] It is estimated that over 80% of imported plastic waste in Indonesia is dumped or burned in the open, releasing hazardous pollutants and impacting the health of local communities.[163] For this reason and others, human rights and plastic justice groups have called for a halt of the exportation of plastic waste to countries lacking adequate infrastructure for the appropriate handling and recycling of materials. Outdated regulations and policies also contribute to current recycling limitations; in the US, for example, regulations governing solid waste management were codified nearly five decades ago, under the Resource Conservation and Recovery Act (RCRA).[166]

The recycling market is in constant flux; plastics accepted at recycling facilities may be purchased by recycling buyers one day and be undesired the next, leaving facilities to either store or dispose of collected materials. The recycling industry has historically engaged in mechanical recycling, a process in which plastics are cleaned, shredded, melted and made into new plastics. This process allows for recycled material to be used in place of virgin plastics, without the need for additional fossil fuel extraction for feedstocks. This process degrades the mechanical properties of plastic materials and thus, is most commonly used in down-cycled (less valuable) products.[167] For this reason, virgin plastic is often less expensive and of higher quality than recycled plastics.

Many plastics are now being recycled *via* chemical or so-called "advanced" recycling processes, such as pyrolysis, which converts plastic waste back to feedstocks and in theory, is able to produce high-quality virgin-like plastics again. However, this process rarely produces new plastic, with 86 to 99% of chemically recycled waste becoming fossil fuels or solvents, such as benzene or styrene.[168] In this sense, chemical recycling plants are predominately incinerating waste and the term "recycling" may be misleading. Additionally, many studies indicate chemical recycling has a *significantly* higher global warming potential than mechanical recycling or landfilling plastics.[167–171] Experts have identified chemical recycling to be an ineffective and unsustainable form of plastic waste management due to its energy intensive process and reliance on fossil fuels.[172] It must be noted that conflicting data exists regarding the global warming impact of chemical recycling due largely to differences in impact methods and consideration of recyclate quality.[167,173,174] It is recommended to consider funding sources for conflicting studies, as many were funded by special interest groups standing to directly benefit from specific outcomes. Additionally, the US plastics industry has invested over $30 million since 2023 in pro-chemical recycling advertisements.[175]

Surprisingly, few studies have examined the comprehensive toxicity and hazards of chemical recycling necessitating a more thorough inquiry and understanding of operational hazards, as well as the toxicity of products, byproducts and waste products.[172] It is known, however, that chemical recycling plants generate significant hazardous waste and air pollutants, many of which are known carcinogens or neurotoxicants.[176] The majority of chemical recycling facilities are located in communities of color and/or low socioeconomic status communities,[176] echoing the injustices of previous sections of this chapter and having a disproportionate impact on the environmental and public health in these communities.

The final option for plastic waste management at EOL is *via* Municipal Solid Waste Incineration (MSWI) or WtE. Though chemical recycling facilities are currently performing as incineration plants, as noted above, this section is focused on traditional MSWI or WtE plant processing of plastic waste. Many western and northern European countries, as well as Asian countries predominantly utilize MSWI, while other countries such as the US primarily utilize landfills. In the US, only 65 WtE facilities currently exist, supplying less than 1% of the country's energy, with only one new facility created in the entire country since 1995.[177] In the EU, there are over 500 WtE facilities accounting for 2.4% of the total energy supply with their use rapidly growing.[178]

Due to the recovery of electricity in newer MSWI facilities and more favorable GHG emissions, it is often touted as a more sustainable process than landfilling. Some studies counter this point, stating landfills have lower GHG emissions than incineration,[169] though these results are likely due to assumed levels of methane recapture in study calculations. When taking into account actual (non-theoretical) landfill gas collection, MSWI is suggested to be the GHG emissions preferable option.[179] Furthermore, energy produced by WtE plants offsets tons of CO_2 emissions that would otherwise be generated through the burning of fossil fuels; another sustainable "win". However, credible questions remain about the methods used for the analysis of CO_2 emitted by MSWI and some data suggests the carbon intensity of energy produced *via* MSWI could be double that of the average electricity grid intensity.[180]

Though MSWI is generally accepted to be sustainably superior to landfilling, it is still not an ideal option; creating social and environmental concerns due to its adverse impact on air and water quality, public health and overarching sustainability and circular economy goals and strategies. Many experts and advocacy groups argue that MSWI works against overall sustainability goals and efforts, as much of the MSW being incinerated are materials which could be recycled, composted or reused; thus working against environmental objectives of circular economy commitments and delaying the adoption of less carbon intensive power infrastructure, such as solar or wind renewable energies.[181]

WtE facilities are significant generators of toxic ash, gases containing particulate matter, heavy metals, dioxins and more. Long-term exposure to these by-products is associated with increase rates of cancer, reproductive issues and respiratory diseases.[182] Though comprehensive research exists regarding the exposure to these health hazards individually, the direct impact of WtE facilities or MSWI on public health, and specifically nearby communities, is sparce and requires comprehensive assessment and further inquiry. It is widely accepted, however, that older WtE facilities pose greater threat to public and environmental health and newer facilities are equipped with better technology and filtration systems, thus reducing risk. Additionally, WtE plants have constant truck traffic of vehicles delivering waste, thus compounding emissions and pollution from the facility itself. To reiterate, 85% of WtE facilities in the US are located in communities of color or low-income communities,[141,142] disproportionately impacting historically marginalized populations, meaning that decisions about waste management made in the lab can directly impact the wellbeing of these communities.

7.6.4 Scientific Hazardous Waste Impacts

While scientific waste streams may appear similar to household waste streams in many ways, managing them requires greater oversight and consideration due to Federal, State and Institutional regulations guiding laboratory operations and chemical safety. Many countries have regulations for the safe operation of research labs, including specific guidance for the safe handling of chemical, biological or radiological hazards, such as *The Occupational Safety and Health Administration's (OSHA) Laboratory Standard 29 CFR 1910.1450.*[183] Additionally, most countries also have regulations guiding the disposal of hazardous waste and materials, such as the *Resource Conservation Recovery Act (RCRA)* which requires hazardous waste in the US to be managed from *cradle-to-grave* and deposited in either a Subtitle C landfill or be incinerated at a special facility.[166] *Cradle-to-grave* liability means generators of hazardous waste are accountable for the responsible management and disposal of hazardous waste *in perpetua*, meaning decades or even a hundred years after it was generated. *RCRA* also provides guidance for the appropriate disposal of empty chemical containers of acutely toxic (or P-listed) hazardous waste. While many chemical containers can be landfilled or recycled and are not regulated, P-listed chemical containers are. Scientists are required to know their relative regulations and take careful measures to apply them to the in-lab management of laboratory waste. But, once hazardous waste leaves the laboratory, scientists are finished thinking about it and other teams take over from there. While non-hazardous lab waste streams are often managed like household waste, hazardous waste leaving a lab requires acute tracking. Most institutions generating hazardous waste contract with a third-party vendor, who sub-contracts with a roster of *Final Treatment, Storage, and Disposal Facilities (TSDF)*, including Subtitle C landfills and *Hazardous Waste Incineration (HWI)* facilities. Thus, while institutions may be aware of where their hazardous waste ends up, they are often not making those direct decisions. *This is a key point to emphasize regarding improving the equity of laboratory waste management.* Scientists and contract managers should ask where their waste goes when it leaves their institution and demand the equitable management and disposal of research-related waste.

Many of the global systems and processes guiding the disposal of hazardous waste are antiquated, inequitable, and urgently require updating. Hazardous waste leaving higher education and research institutions in the US is primarily generated in high-socioeconomic status White majority communities, and is disposed of in low-socioeconomic status communities of color.[184] Hazardous waste facilities are located almost exclusively in low-socioeconomic status rural communities; communities who had little to no representation in decisions regarding the facility placement and often lack the resources needed to effectively prevent such placement. In fact, in the US, race is the most significant predictor of proximity to hazardous waste treatment facilities.[185–187] This is in stark contrast to majority White and/or higher-socioeconomic status communities who often successfully prohibit the placement of hazardous waste facilities through effective and well-resourced NIMBY campaigns.

Most US *Carnegie* classified R-1 research institutions generate around 100 000 lbs of RCRA hazardous waste annually;[184] waste that is a by-product of scientific research operations. Most researchers fail to understand the full implications of their hazardous waste generation, functioning under the assumption that legally managed hazardous waste equates to equitably managed hazardous waste. While it is true most hazardous waste leaving research facilities is appropriately managed according to regulations and legalities, that does not indicate it is being disposed of equitably. This is why greater researcher advocacy is needed to encourage institutions to expand institutional policy with goals and requirements for equitable waste management and not settle for only legal waste management.

For US scientists and researchers, the *Environmental Protection Agency's (EPA) EJ SCREEN* app can be used as a tool to help determine how equitable an institution's hazardous waste management is. Researchers should inquire through their institutional hazardous waste team, usually part of Environmental Health and Safety (EH&S), to identify the name and location of TSDF(s) utilized for institutional hazardous waste through the tracking of waste manifests. Most waste vendors use one or two primary TSDFs for any given period of time, based on various factors such as distance from point of generation and cost of deposition. Once the TSDFs have been identified, researchers are encouraged to use *EJ SCREEN* to compare community sociodemographic indicators of the location where waste was generated to the community sociodemographic indicators where waste is disposed. This one exercise is often adequate to objectively identify inequities in the management of institutional hazardous waste. If waste is generated in a higher-socioeconomic status or White majority community and is disposed of in a lower-socioeconomic status or racial and ethnic minoritized community, this is inequitable and requires reformation. Reforming this process may be as simple as sharing these data with the institutional stakeholder who manages the contract and setting goals to include improved considerations for equity in the next contract renewal. However, it may also require petitioning one's administration to make changes in policy which prioritize equity within institutional hazardous waste management, thus allowing these factors to be prioritized over low-bid or other cost-driven incentives.

7.6.5 Circular Economy of Labs and Managing Equitable Laboratory Waste Streams

Scientists and research institutions have the ability to positively impact their laboratory waste streams to become more equitable and reduce injustices. While some actions such as changing hazardous waste vendors and EOL destinations for waste may require an institutional approach, many actions improving the equity of laboratory waste management can be easily implemented by scientists themselves. This empowers individuals or research teams to mitigate the unintended inequitable impacts of their scientific waste generation and cultivate more equitable processes into their science. This also expands the concept of "doing good science" into a more collective, unified approach, echoing the interconnected application of the *UN SGDs*.

A circular economy is a model of production and consumption which keeps materials in use and extends the life cycle of products. The goal of a circular economy is to reduce waste through sharing, reusing, repairing, refurbishing and recycling materials for as long as possible.[188] Various organizations promote engagement in circular economy efforts, such as the *Ellen Macarthur Foundation*[188] and *the International Institute for Sustainable Laboratories (I2SL) Circular Economy for Labs Community of Practice.*[189] Circular economy efforts are a win-win for manufacturers as well as for consumers and the environment. Many businesses have adapted their business models to include a circular economy, as recovering and reusing materials helps to secure their supply chain and saves money by avoiding the harvesting of raw materials. According to the *United Nations Global Resource Outlook Report 2024,* it is not only possible to reduce resource use while growing the economy, but will simultaneously protect the environment, reduce inequality and improve well-being.[24]

Initiatives supporting a circular economy should be prioritized within the scientific community, as keeping materials in use and diverting materials reduces the amount of waste leaving a research facility. This in turn minimizes negative impacts of waste management on public health and ecosystems. It also reduces the need for raw material extractions, which as stated in Section 7.4.1, poses significant risk to surrounding ecosystems and communities. Examples of initiatives supporting a circular economy in labs include Green Chemistry (GC), materials rehoming, repurposing, or recycling, and sustainable EOL equipment disposition, to name a few.

Implementing GC practices contribute to a circular economy and reduce hazardous waste in a number of ways, such as using less toxic chemicals, using catalysts instead of stoichiometric reagents, and promoting the use of renewable feedstocks.[190]

Materials rehoming can happen *via* grassroots efforts at an individual lab level or at a program or institutional level through the sharing of unwanted lab glass, consumables, and equipment. When a researcher retires or changes their processes, they are often left with an abundance of unwanted or unneeded materials. Rehoming these materials can serve internal or external stakeholders, depending on institutional policy, by diverting waste from landfills and improving educational equity through the resourcing of future scientists. Some rehoming efforts serve underfunded schools which may otherwise lack the resources needed to purchase expensive lab glass.[191] This can provide students the opportunity for hands-on experience; allowing each student to conduct an experiment rather than observing.

Finally, international non-profits, such as *Seeding Labs*,[192] also support a circular economy while improving scientific equity through the refurbishing and rehoming of scientific equipment. Equipment donated by private companies and research institutions is provided to scientists in developing countries, thus providing critical resources to students and researchers who may otherwise lack access to such equipment. This improves global scientific equity and contributes to more diverse scientific perspectives while also keeping equipment in use, saving EOL disposition fees for institutions and keeping items out of landfills.

7.7 Conclusion

The global scientific research enterprise carries nearly all the benefits of scientific operations with very little of the associated burden. It is time to move beyond regulatory requirements to ensure sustainability and equity are firmly embedded into scientific operations and design, not only for the benefit of the world, but also for the benefit and advancement of *science* itself. We must shift scientific culture to routinely include topics of equity and justice for all aspects of our research and operations and make every attempt possible to incorporate these considerations. Finally, scientists and sustainability professionals alike must work harder to fully embody the three pillars of sustainability so that when *sustainability* is mentioned, it immediately includes equity, justice and social considerations alongside topics of the environment and no longer requires a separate term or discussion.

Abbreviations

BIPOC	Black, indigenous and people of color
BPA	Bisphenol A
DEI	Diversity equity and inclusion
DTC	Decolonizing the curriculum
EH&S	Environment, health and safety
EJ	Environmental justice
EJSF	Equity and justice in science framework
EOL	End-of-life
EPA	Environmental protection agency
EPR	Extended producer responsibility
GC	Green chemistry
GHG	Greenhouse gas
GWP	Global warming potential
GWP	Global warming potential
HWI	Hazardous waste incineration
I2SL	International Institute for Sustainable Laboratories
ILO	International Labour Organization
mHRDD	Mandatory human rights due diligence
MP	Microplastic

MSW	Municipal solid waste
MSWI	Municipal solid waste incineration
NCSES	National Center for Science and Engineering Statistics
NIH	National Institutes of Health
NIMBY	Not in my backyard
NP	Nanoplastic
NSF	National Science Foundation
OSHA	Occupational Safety and Health Administration
PI	Principal investigator
POME	Palm oil mill effluent
PPE	Personal protective equipment
RCRA	Resource Conservation Recovery Act
SAI	Social accountability international
SDGs	Sustainable development goals
SKU	Stock keeping unit
SPI	Super principal investigator
SPLC	Sustainable Purchasing Leadership Council
TBL	Triple bottom line
TSDF	Treatment, storage and disposal facility
UGA	University of Georgia
USCBP	United States Customs and Border Protection
WEF	World economic forum
WRO	Withhold release order
WtE	Waste to energy

Acknowledgements

This chapter represents a union of two halves of Self; the scientist who is driven by data, and the advocate who is driven by love. I would like to express my deepest gratitude to Dr Kathy Ramirez-Aguilar who encouraged me to no longer separate the two and to embrace my passion for equity in science. I would also like to express my deepest gratitude to Visionary and editor Martin Farley for his endless patience, encouragement and support. Thank you to my beloved partner, my incredible family who taught me at an early age the joy found in caring for others, and my patient friends for all your support during this process and beyond. Finally, I humbly bow in gratitude to all the teachers, elders, Earthwalkers, brave stewards and protectors of this beautiful planet. I remain deeply inspired by, and in awe of, your wisdom and courage.

References

1. C. Black, J. P. Cerdeña and E. V. Spearman-McCarthy, I am not your minority, *Lancet Reg Health Am.*, 2023, **19**, 100464.

2. T. Frey and R. Young, Chapter 11: Correct and Preferred Usage. 2024 (Updated) 2020, in AMA Manual of Style: A Guide for Authors and Editors 11th edn, Oxford University Press. Available from: https://www.amamanualofstyle.com/view/10.1093/jama/9780190246556.001.0001/med-9780190246556-chapter-11.
3. NIH Style Guide: Race and National Origin, NIH.gov: National Institutes of Health, 2024.
4. A. Donsanjh, Childhood asthma and anthropogenic CO_2 emissions, *J. Asthma Allergy*, 2011, **4**, 103–105.
5. G. Brundtland, *Report of the World Commission on Environment and Development: Our Common Future*, Oxford University Press: United Nations General Assembly, 1987.
6. J. Elkington, Towards the Sustainable Corporation: Win-Win-Win Business Strategies for Sustainable Development, *Calif. Manage. Rev.*, 1994, **36**(2), 90–100.
7. The 17 Goals, https://sdgs.un.org/goals (accessed June 2024).
8. The SDGs wedding cake, https://www.stockholmresilience.org/research/research-news/2016-06-14-the-sdgs-wedding-cake.html (accessed May 2024).
9. Causes and Effects of Climate Change, https://www.un.org/en/climatechange/science/causes-effects-climate-change#:~:text=As%20temperatures%20rise%2C%20more%20moisture,waters%20at%20the%20ocean%20surface (accessed April 2024).
10. 2023: A historic year of U.S. billion-dollar weather and climate disasters, https://www.climate.gov/news-features/blogs/beyond-data/2023-historic-year-us-billion-dollar-weather-and-climate-disasters#:~:text=Adding%20the%202023%20events%20to,376%20events%20exceeds%20%242.660%20trillion (accessed April 2024).
11. J. M. Nordbotten, S. A. Levin, E. Szathmáry and N. C. Stenseth, Ecological and evolutionary dynamics of interconnectedness and modularity, *Proc. Natl. Acad. Sci. U. S. A.*, 2018, **115**(4), 750–755.
12. S. L. Gates, Ecological Interconnectedness: Entwined Selves, Transcendent and Immanent, in *Roots, Routes and a New Awakening: Beyond One and Many and Alternative Planetary Futures*, ed. A. K. Giri, Springer Singapore, Singapore, 2021, pp. 75–97.
13. Religions and environmental protection, https://www.unep.org/about-un-environment-programme/faith-earth-initiative/religions-and-environmental-protection#:~:text=Baha'i%20Faith%3A%20The%20Baha,life%20and%20the%20natural%20world (accessed April 2024).
14. B. Morton and K. Fasching-Varner, "Equity", in *Encyclopedia of Diversity and Social Justice*, ed. S. Thompson, Rowman & Littlefield, Lanham, MD, 2015, vol. 1, pp. 303–304.
15. L. A. Bell, Theoretical foundations for social justice education, in *Teaching for diversity and social justice*, ed. M. Adams, L. A. Bell and P. Griffin, Routledge/Taylor & Francis Group, 2nd edn, 2007, pp. 1–14.

16. M. Liboiron, R. Liu, E. Earles and I. Walker-Franklin, Models of justice evoked in published scientific studies of plastic pollution, *FACETS*, 2023, **8**, 1–34.
17. CDP Technical Note: Relevance of Scope 3 Categories by Sector, CDP Worldwide, 2022.
18. ISO 20400: Sustainable Procurement Guidance, 1 edn, Iso.org: International Organization for Standardization, 2017, pp. 1–52.
19. ISO20400.org, https://www.iso20400.org/ (accessed May 2024).
20. Sustainable Procurement Guidance for Laboratory Supplies and Services, Sustainablepurchasing.org: Sustainable Purchasing Leadership Council, 2022.
21. USING GOVERNMENT PROCUREMENT TO ADVANCE RACIAL EQUITY, Harvard Kennedy School: Harvard Kennedy School Government Performance Lab, 2022 April.
22. International Rights Advocates, https://www.internationalrightsadvocates.org/ (accessed April 2025).
23. I. Marín-Beltrán, F. Demaria, C. Ofelio, L. M. Serra, A. Turiel and W. J. Ripple, *et al.*, Scientists' warning against the society of waste, *Sci. Total Environ.*, 2022, **811**, 151359.
24. Global Resources Outlook: Bend the Trend – Pathways to a liveable planet as resource use spikes, United Nations Environment Programme, Nairobi, 2024.
25. M. A. Urbina, A. J. R. Watts and E. E. Reardon, Labs should cut plastic waste too, *Nature*, 2015, **528**(7583), 479.
26. Global Plastics Outlook: Economic Drivers, Environmental Impacts and Policy Options, Paris, France, OECD, 2022.
27. A. Ragusa, V. Notarstefano, A. Svelato, A. Belloni, G. Gioacchini and C. Blondeel, *et al.*, Raman Microspectroscopy Detection and Characterisation of Microplastics in Human Breastmilk, *Polymers*, 2022, **14**(13), DOI: 10.3390/polym14132700.
28. C. J. Hu, M. A. Garcia, A. Nihart, R. Liu, L. Yin and N. Adolphi, *et al.*, Microplastic presence in dog and human testis and its potential association with sperm count and weights of testis and epididymis, *Toxicol. Sci*, 2024, **200**(2), 235–240.
29. A. J. Jamieson, L. S. R. Brooks, W. D. K. Reid, S. B. Piertney, B. E. Narayanaswamy and T. D. Linley, Microplastics and synthetic particles ingested by deep-sea amphipods in six of the deepest marine ecosystems on Earth, *R. Soc. Open Sci.*, 2019, **6**(2), 180667.
30. S. Allen, D. Allen, V. R. Phoenix, G. Le Roux, P. Durántez Jiménez and A. Simonneau, *et al.*, Atmospheric transport and deposition of microplastics in a remote mountain catchment, *Nat. Geosci.*, 2019, **12**(5), 339–344.
31. M. Campen, A. Nihart, M. Garcia, R. Liu, M. Olewine and E. Castillo, *et al.*, Bioaccumulation of Microplastics in Decedent Human Brains Assessed by Pyrolysis Gas Chromatography-Mass Spectrometry, *Res. Square*, 2024, DOI: 10.21203/rs.3.rs-4345687/v1.

32. D. Posen, P. Jaramillo, A. E. Landis and W. M. Griffin, Greenhouse gas mitigation for U.S. plastics production: energy first, feedstocks later, *Environ. Res. Lett.*, 2017, **12**, DOI: 10.1088/1748-9326/aa60a7.
33. L. M. McKenzie, R. Z. Witter, L. S. Newman and J. L. Adgate, Human health risk assessment of air emissions from development of unconventional natural gas resources, *Sci. Total Environ.*, 2012, **424**, 79–87.
34. Now We're Fracked: Fracking's Public Health Crisis, FoodandWater-Watch.org: Food & Water Watch, 2021.
35. E. N. Mayfield, J. L. Cohon, N. Z. Muller, I. M. L. Azevedo and A. L. Robinson, Cumulative environmental and employment impacts of the shale gas boom, *Nat. Sustainability*, 2019, **2**, 1122–1131.
36. T. L. Tyree, Environmental injustice: Fracking, *Public Health Nurs.*, 2021, **38**(6), 1015–1018.
37. L. Vogel, Fracking tied to cancer-causing chemicals, *Can. Med. Assoc. J.*, 2017, **189**(2), E94–E95.
38. Y. Ogneva-Himmelberger and L. Huang, Spatial distribution of unconventional gas wells and human populations in the Marcellus Shale in the United States: Vulnerability analysis, *Appl. Geogr.*, 2015, **60**, 165–174.
39. K. Zwickl, The demographics of fracking: A spatial analysis for four U.S. states, *Ecol. Econ.*, 2019, **161**, 202–215.
40. E. Livezey, Hazardous Waste, The Christian Science Monitor, 1980.
41. Ionfield Cleaning Systems Sterility Specifications, https://ionfieldsystems.com/products/puretip-sas/ (accessed June 2024).
42. Human Rights Due Diligence in Malaysia's Manufacturing Sector, Coldbath Square, London EC1R 5HL, Ethical Trading Initiative, 2019, vol. 8.
43. I. Ben-David, Y. Jang, S. Kleimeier and M. Viehs, Exporting pollution: where do multinational firms emit CO_2?, *Econ. Policy*, 2021, **36**(107), 377–437.
44. International Energy Output 2016, U.S. Energy Information Administration: U.S. Energy Information Administration; 2016 May.
45. Z. Vaskovich, J. McCreesh and E. Farbstein, How manufacturers can reduce carbon emissions, Normative, 2024. Available from: https://normative.io/insight/reduce-manufacturer-carbon-emissions/.
46. Commission E, Environment D-Gf, J. Fry, R. Sheane, W. Schreiber, S. Royston, *et al.* Study on the environmental impact of palm oil consumption and on existing sustainability standards – Final report and appendices, Publications Office, 2018.
47. B. W. Brook, N. S. Sodhi and P. K. Ng, Catastrophic extinctions follow deforestation in Singapore, *Nature*, 2003, **424**(6947), 420–426.
48. A. E. Anak Erison, Y. H. Tan, N. M. Mubarak, J. Kansedo, M. Khalid and M. O. Abdullah, *et al.*, Life cycle assessment of biodiesel production by using impregnated magnetic biochar derived from waste palm kernel shell, *Environ. Res.*, 2022, **214**(Pt 4), 114149.

49. S. H. Pranolo, J. Waluyo, F. A. Putro, M. A. Adnan and M. G. Kibria, Gasification process of palm kernel shell to fuel gas: Pilot-scale experiment and life cycle analysis, *Int. J. Hydrogen Energy*, 2023, **48**(7), 2835–2848.
50. UNICEF, Palm Oil and Children in Indonesia, United Nations Children's Fund, 2016.
51. S. Z. Al-Mahmood, Palm-Oil Migrant Workers Tell of Abuses on Malaysian Plantations: Global palm-oil industry contributes to human trafficking, rights advocates say, *Wall St. J.*, 2015, **26**, https://www.wsj.com/articles/palm-oil-migrant-workers-tell-of-abuses-on-malaysian-plantations-1437933321.
52. U. R. Gracida-Alvarez, P. T. Benavides, U. Lee and M. Wang, Life-cycle analysis of recycling of post-use plastic to plastic *via* pyrolysis, *J. Cleaner Prod.*, 2023, **425**, 138867.
53. H. H. Jensen, For manufacturers, the circular economy strengthens supply chains. Here's how. World Economic Forum, 2024. Available from: https://www.weforum.org/stories/2024/02/how-manufacturers-could-lead-the-way-in-building-the-circular-economy/.
54. Big Corporations Contribute to Water Shortages — How Can They Fix It?, https://eponline.com/articles/2020/06/03/big-corporations-contribute-to-water-shortages-how-can-they-fix-it.aspx (accessed May 2024).
55. Lawmakers Open Groundwater Fight Against Bottled Water Companies, https://stateline.org/2020/02/12/lawmakers-open-groundwater-fight-against-bottled-water-companies/ (accessed April 2024).
56. O. Rosane, Major Corporations Making the World Water Crisis Worse, Common Dreams, 2024. Available from: https://www.commondreams.org/news/corporations-water-crisis.
57. J. P. Negroni, The Answer to the No. 1 Problem of the Chile Mining Industry: Water Use in Mining. IDE Technologies, 2021. Available from: https://ide-tech.com/en/blog/the-answer-to-the-no-1-problem-of-the-chile-mining-industry-water-use-in-mining/.
58. Drought in Numbers 2022, restoration for readiness and resilience, United Nations: United Nations Convention to Combat Desertification, 2022.
59. W. Leal Filho, U. Saari, M. Fedoruk, A. Iital, H. Moora and M. Klöga, *et al.*, An overview of the problems posed by plastic products and the role of extended producer responsibility in Europe, *J. Cleaner Prod.*, 2019, **214**, 550–558.
60. C. K. Mayers, Strategic, Financial, and Design Implications of Extended Producer Responsibility in Europe: A Producer Case Study, *J. Ind. Ecol.*, 2007, **11**(3), 113–131.
61. Combating Trafficking in Counterfeit and Pirated Goods, U.S. Department of Homeland Security Office of Strategy, Policy and Plans, Washington, D.C., 2020.
62. The Global Slavery Index 2023, WalkFree.org, Walk Free, 2023.

63. Global Estimates of Modern Slavery, Forced Labour and Forced Marriage, International Labour Organization, Walk Free, International Organization for Migration, Geneva, Switzerland, 2022.
64. International Labour Standards, https://www.ilo.org/international-labour-standards (accessed June 2024).
65. The ETI Base Code, Ethical Trading Initiative, London, 2018.
66. Guiding Principles on Business and Human Rights, United Nations: United Nations Human Rights Office of the High Commissioner, 2011.
67. R. Andaya, Symposium on Sanctioning Human Rights Violations: Magnitsky Laws in the World, Human Rights Watch and the University of Tokyo International Law Training and Research Hub, 2021.
68. Glove Supply Chain Traceability, https://eagleprotect.com/blogs/articles/glove-supply-chain-traceability (accessed June 2024).
69. Nitrile Gloves Global Market Report 2024, thebusinessresearchcompany.com: Grand View Research, 2024.
70. J. C. Prata, A. L. P. Silva, T. R. Walker, A. C. Duarte and T. Rocha-Santos, COVID-19 Pandemic Repercussions on the Use and Management of Plastics, *Environ. Sci. Technol.*, 2020, **54**(13), 7760–7765.
71. J. Zhu, M. Saberian, S. T. A. M. Perera, R. Roychand, J. Li and G. Wang, Reusing COVID-19 disposable nitrile gloves to improve the mechanical properties of expansive clay subgrade: An innovative medical waste solution, *J. Cleaner Prod.*, 2022, **375**, 134086.
72. MARGMA CONFIDENT GLOBAL DEMAND FOR RUBBER GLOVES WILL REBOUND IN, 2024, https://www.bernama.com/en/news.php?id=2289382#:~:text=KUALA%20LUMPUR%2C%20April%2019%20(Bernama,307.2%20billion%20pieces%20in%202023 (accessed June 2024).
73. Decent work challenges and opportunities in Malaysia's rubber glove supply chain, International Labour Office: International Labour Organization, Malaysia, 2023.
74. Glove Story: Global Glove Production Amidst the Covid-19 Pandemic [press release], U.S. International Trade Commission, 2021.
75. Trafficking In Persons Report, State.gov: United States of America Department of State Office to Monitor and Combat Trafficking in Persons, 2022.
76. Withhold Release Orders and Findings, https://www.cbp.gov/trade/forced-labor/withhold-release-orders-and-findings (accessed January 2024).
77. Withhold Release Orders and Finding List, https://www.cbp.gov/trade/forced-labor/withhold-release-orders-and-findings (accessed June 2024).
78. Collingsworth T. Mia *et al.* v. Kimberly Cark Corp., *et al.*,. United States Federal District Court for the District of Columbia; Active.
79. SA8000 Standard, https://sa-intl.org/programs/sa8000/ (accessed October 2024).
80. About B Corp Certification, https://www.bcorporation.net/en-us/certification/ (accessed October 2024).
81. L. Nargi, Can a Business Be a "Force For Good" If It Uses Child Labor?, Food Print, 2022. Available from: https://foodprint.org/blog/b-corps-child-labor/.

82. Situation and gap analysis on Malaysian legislation, policies, and programmes and the ILO Forced Labour Convention and Protocol, ILO.org: International Labour Organization, 2019.
83. What is a circular economy, https://www.ellenmacarthurfoundation.org/topics/circular-economy-introduction/overview (accessed July 2024).
84. Extended Producer Responsibility, Washington, D.C., National Conference of State Legislators, 2023.
85. S. Mintz, Decolonizing the Academy. Inside Higher Ed, 2021. Available from: https://www.insidehighered.com/blogs/higher-ed-gamma/decolonizing-academy.
86. Keele Manifesto for decolonizing the curriculum, https://www.keele.ac.uk/equalitydiversity/equalityframeworksandactivities/equalityawardsandreports/equalityawards/raceequalitycharter/keeledecolonisingthecurriculumnetwork/#keele-manifesto-for-decolonising-the-curriculum (accessed July 2024).
87. Decolonize Your Syllabus, https://college.up.edu/gwsm/decolonize-your-syllabus.html (accessed July 2024).
88. B. Hofstra, V. V. Kulkarni, S. Munoz-Najar Galvez, B. He, D. Jurafsky and D. A. McFarland, The Diversity–Innovation Paradox in Science, *Proc. Natl. Acad. Sci. U. S. A.*, 2020, **117**(17), 9284–9291.
89. A. R. Bejerano and T. M. Bartosh, Learning masculinity: Unmasking the hidden curriculum in science, technology, engineering, and mathematics courses, *J. Women Minorities Sci. Eng.*, 2015, **21**, 107–124.
90. 9 Scientists Who Didn't Get the Credit They Deserved, Oxford Royale, 2024. Available from: https://www.oxford-royale.com/articles/9-scientists-didnt-get-credit-deserved/.
91. B. Spellberg, J. Bartlett, R. Wunderink and D. N. Gilbert, Novel approaches are needed to develop tomorrow's antibacterial therapies, *Am. J. Respir. Crit. Care Med.*, 2015, **191**(2), 135–140.
92. R. Veugelers and J. Wang, Scientific novelty and technological impact, *Res. Policy*, 2019, **48**(6), 1362–1372.
93. The Cost of Loneliness: Women, Work and The Invisible Force That's Undermining Them As They Rise, TheLi.st, Berlin Cameron, Benenson Strategy Group, New York, New York, 2023.
94. Decolonizing the AHR, *Am. Hist. Rev.*, 2018, **123**(1), xiv–xvii, DOI: 10.1093/ahr/123.1.xiv.
95. V. Hunt, D. Layton and S. Prince, Diversity Matters, McKinsey.com, 2015.
96. K. Powell, These labs are remarkably diverse – here's why they're winning at science, *Nature*, 2018, **558**(7708), 19–22.
97. S. C. L. Kamerlin, When we increase diversity in academia, we all win, *EMBO Rep.*, 2020, **21**(12), e51994.
98. D. Medlin and C. Lee, Diversity Makes Science Better, Association for Psychological Science, 2012.
99. L. Hong and S. E. Page, Groups of diverse problem solvers can outperform groups of high-ability problem solvers, *Proc. Natl. Acad. Sci. U. S. A.*, 2004, **101**(46), 16385–16389.

100. Diversity and STEM: Women, Minorities, and Persons with Disabilities, Alexandria, VA, National Science Foundation National Center For Science and Engineering Statistics, 2023.
101. A. K. Daniels, Invisible Work, *Soc. Probl.*, 1987, **34**(5), 403–415.
102. S. J. Baserga, Underpaid: Women scientists in the academy, American Society For Biochemistry and Molecular Biology, 2022. Available from: https://www.asbmb.org/asbmb-today/opinions/031022/underpaid-women-scientists-in-the-academy.
103. M. Katherine and S. Sharon, Explaining the Gender Wage Gap in STEM: Does Field Sex Composition Matter?, *RSF: Russell Sage Found. Soc. Sci.*, 2016, **2**(4), 194–215.
104. M. B. Ross, B. M. Glennon, R. Murciano-Goroff, E. G. Berkes, B. A. Weinberg and J. I. Lane, Women are credited less in science than men, *Nature*, 2022, **608**(7921), 135–145.
105. F. Liu, P. Holme, M. Chiesa, B. AlShebli and T. Rahwan, Gender inequality and self-publication are common among academic editors, *Nat. Hum. Behav.*, 2023, **7**(3), 353–364.
106. M. Nguyen, S. I. Chaudhry, M. M. Desai, K. Dzirasa, J. E. Cavazos and D. Boatright, Gender, Racial, and Ethnic Inequities in Receipt of Multiple National Institutes of Health Research Project Grants, *JAMA Network Open*, 2023, **6**(2), e230855.
107. C. Y. Chen, S. S. Kahanamoku, A. Tripati, R. A. Alegado, V. R. Morris and K. Andrade, *et al.*, Systemic racial disparities in funding rates at the National Science Foundation, *eLife*, 2022, **11**, e83071.
108. O. H. Petersen, Inequality of Research Funding between Different Countries and Regions is a Serious Problem for Global Science, *Function*, 2021, **2**(6), zqab060.
109. 2020 Gender Social Norms Index (GSNI): Tackling Social Norms: A game changer for gender inequalities, United Nations Development Programme, New York, 2020.
110. K. B. H. Clancy, K. M. N. Lee, E. M. Rodgers and C. Richey, Double jeopardy in astronomy and planetary science: Women of color face greater risks of gendered and racial harassment, *J. Geophys. Res.:Planets*, 2017, **122**(7), 1610–1623.
111. R. E. Bernard and E. H. G. Cooperdock, No progress on diversity in 40 years, *Nat. Geosci.*, 2018, **11**(5), 292–295.
112. D. J. Nelson and H. N. Cheng, *Diversity in the Scientific Community Volume 1: Quantifying Diversity and Formulating Success*, American Chemical Society, Washington DC, 2017.
113. J. Trejo, The burden of service for faculty of color to achieve diversity and inclusion: the minority tax, *Mol. Biol. Cell*, 2020, **31**(25), 2752–2754.
114. S. M. Kalita, How to End the Unfairness of Invisible Work, Time, 2023. Available from: https://time.com/charter/6317237/how-to-end-the-unfairness-of-invisible-work/#:~:text=So%2Dcalled%20invisible%20work%E2%80%94a,stop%20seeing%20such%20labor%20everywhere.

115. J. E. Rodríguez, K. M. Campbell and L. H. Pololi, Addressing disparities in academic medicine: what of the minority tax?, *BMC Med. Educ.*, 2015, **15**(1), 6.
116. M. W. Rossiter, The Matthew Matilda Effect in Science, *Soc. Stud. Sci.*, 1993, **23**(2), 325–341.
117. D. Chakraverty, Faculty Experiences of the Impostor Phenomenon in STEM Fields, *CBE Life Sci. Educ.*, 2022, **21**(4), ar84.
118. C. Woolston, Pay gap widens between female and male scientists in North America, *Nature*, 2021. Available from: https://www.nature.com/articles/d41586-021-00387-3.
119. W. W. Ding, A. Ohyama and R. Agarwal, Trends in gender pay gaps of scientists and engineers in academia and industry, *Nat. Biotechnol.*, 2021, **39**(8), 1019–1024.
120. The Global Gender Gap Report 2018, World Economic Forum, Geneva, Switzerland, 2018.
121. D. K. Ginther and K. J. Hayes, Gender differences in salary and promotion for faculty in the humanities 1977–95, *J. Hum. Resour.*, 2003, **38**(1), 34–73.
122. D. K. Ginther and S. Kahn, Does Science Promote Women? Evidence from Academia 1973–2001, in Science and Engineering Careers in the United States, University of Chicago Press for NBER Science Engineering Workforce Project, Chicago, IL, 2009.
123. N. Bhalla, Strategies to improve equity in faculty hiring, *Mol. Biol. Cell*, 2019, **30**(22), 2744–2749.
124. D. K. Ginther, Reflections on race, ethnicity, and NIH research awards, *Mol. Biol. Cell*, 2022, **33**(1), ae1.
125. D. K. Ginther, W. T. Schaffer, J. Schnell, B. Masimore, F. Liu and L. L. Haak, *et al.*, Race, ethnicity, and NIH research awards, *Science*, 2011, **333**(6045), 1015–1019.
126. F. T. Odedina and M. C. Stern, Role of funders in addressing the continued lack of diversity in science and medicine, *Nat. Med.*, 2021, **27**(11), 1859–1861.
127. M. S. Lauer and D. Roychowdhury, Inequalities in the distribution of National Institutes of Health research project grant funding, *eLife*, 2021, **10**, e71712.
128. N. Düzgünes, Science by consensus: Why the NIH grant review system must be changed, *Scientist*, 1999, **13**, 13.
129. N. Düzgüneş, 'Science by consensus' impedes scientific creativity and progress: A simple alternative to funding biomedical research, *F1000Res.*, 2022, **11**, 961.
130. T. M. Evans, L. Bira, J. B. Gastelum, L. T. Weiss and N. L. Vanderford, Evidence for a mental health crisis in graduate education, *Nat. Biotechnol.*, 2018, **36**(3), 282–284.
131. F. T. Maestre, Ten simple rules towards healthier research labs, *PLoS Comput. Biol.*, 2019, **15**(4), e1006914.

132. V. B. Chaudhary and A. A. Berhe, Ten simple rules for building an antiracist lab, *PLoS Comput. Biol.*, 2020, **16**(10), e1008210.
133. I. Wojnowska-Baryła, K. Bernat and M. Zaborowska, Plastic Waste Degradation in Landfill Conditions: The Problem with Microplastics, and Their Direct and Indirect Environmental Effects, *Int. J. Environ. Res. Public Health*, 2022, **19**(20), 13223.
134. M. Shams, I. Alam and M. S. Mahbub, Plastic pollution during COVID-19: Plastic waste directives and its long-term impact on the environment, *Environ. Adv.*, 2021, **5**, 100119.
135. University of Georgia Green Labs, https://greenlabs.uga.edu/initiatives/waste-diversion/, (accessed August 2024).
136. CU Green Labs, https://www.colorado.edu/ecenter/programs/cu-green-labs-program/lab-efforts-waste (accessed August 2024).
137. UCL Laboratory Waste, https://www.ucl.ac.uk/estates/our-services/waste-and-recycling/laboratory-waste (accessed August 2024).
138. C. Seydel, Greening the lab, *Nat. Methods*, 2023, **20**(10), 1449–1453.
139. Living near a landfill could damage your health, *Sci. Dly.*, 2016. Available from: https://www.sciencedaily.com/releases/2016/05/160524211817.htm.
140. J. M. Norton, S. Wing, H. J. Lipscomb, J. S. Kaufman, S. W. Marshall and A. J. Cravey, Race, wealth, and solid waste facilities in North Carolina, *Environ. Health Perspect.*, 2007, **115**(9), 1344–1350.
141. School TN, Tishman Environment and Design Center's Work Reveals Most Incinerators Pollute Low-Income and Communities of Color and Worsen Climate Change. The New School, 2022. Available from: https://blogs.newschool.edu/news/2022/05/tishman-environment-and-design-centers-work-reveals-most-incinerators-pollute-low-income-and-communities-of-color-and-worsen-climate-change/.
142. U.S. Municipal Solid Waste Incinerators: An Industry in Decline, Tishman Environment and Design Center at The New School, 2019.
143. N. Tangri, Waste incinerators undermine clean energy goals, *PLOS Clim.*, 2023, **2**(6), e0000100.
144. L. Traven, Busting the myth: waste-to-energy plants and public health, *Arh. Hig. Rada Toksikol.*, 2023, **74**(2), 142–143.
145. P. W. Tait, J. Brew, A. Che, A. Costanzo, A. Danyluk and M. Davis, *et al.*, The health impacts of waste incineration: a systematic review, *Aust. N. Z. J. Public Health*, 2020, **44**(1), 40–48.
146. Landfill Methane Outreach Program (LMOP), https://www.epa.gov/lmop/basic-information-about-landfill-gas (accessed August 2024).
147. P. O. Njoku, J. N. Edokpayi and J. O. Odiyo, Health and Environmental Risks of Residents Living Close to a Landfill: A Case Study of Thohoyandou Landfill, Limpopo Province, South Africa, *Int. J. Environ. Res. Public Health*, 2019, **16**(12), 2125.
148. F. Mataloni, C. Badaloni, M. N. Golini, A. Bolignano, S. Bucci and R. Sozzi, *et al.*, Morbidity and mortality of people who live close to municipal waste landfills: a multisite cohort study, *Int. J. Epidemiol.*, 2016, **45**(3), 806–815.

149. M. Danthurebandara, S. Passel, D. Nelen, Y. Tielemans and K. Van Acker, Environ. Socioeconomic, Impacts Landfills, 2013.
150. P. Stoett, V. M. Scrich, C. I. Elliff, M. M. Andrade, N. de M. Grilli and A. Turra, Global plastic pollution, sustainable development, and plastic justice, *World Dev.*, 2024, **184**, 106756.
151. K. M. Bharath, U. Natesan, R. Vaikunth, R. P. Kumar, R. Ruthra and S. Srinivasalu, Spatial distribution of microplastic concentration around landfill sites and its potential risk on groundwater, *Chemosphere*, 2021, **277**, 130263.
152. M. Kazour, S. Terki, K. Rabhi, S. Jemaa, G. Khalaf and R. Amara, Sources of microplastics pollution in the marine environment: Importance of wastewater treatment plant and coastal landfill, *Mar. Pollut. Bull.*, 2019, **146**, 608–618.
153. S. Ullah, S. Ahmad, X. Guo, S. Ullah, S. Ullah and G. Nabi, *et al.*, A review of the endocrine disrupting effects of micro and nano plastic and their associated chemicals in mammals, *Front. Endocrinol.*, 2022, **13**, 1084236.
154. Advancing Ocean Justice in the Global Plastics Treaty, Earth Law Center: Earth Law Center, 2024.
155. M. Imran, K. R. Das and M. M. Naik, Co-selection of multi-antibiotic resistance in bacterial pathogens in metal and microplastic contaminated environments: An emerging health threat, *Chemosphere*, 2019, **215**, 846–857.
156. J. R. Jambeck, R. Geyer, C. Wilcox, T. R. Siegler, M. Perryman and A. Andrady, *et al.*, Plastic waste inputs from land into the ocean, *Science*, 2015, **347**(6223), 768–771.
157. C. M. Pirkle, G. Muckle and M. Lemire, Managing mercury exposure in northern Canadian communities, *Can. Med. Assoc. J.*, 2016, **188**(14), 1015–1023.
158. M. Nabuurs, S. Schep and M. Pessoa de Miranda, The socio-economic impacts of mercury pollution on fisheries and livelihoods, Secretariat of the Minamata Convention on Mercury, Switzerland, 2023.
159. Toxicological Effects of Methylmercury, National Academies Press (US), Washington, D.C., 2000. Available from: https://pubmed.ncbi.nlm.nih.gov/25077280/.
160. R. Geyer, J. R. Jambeck and K. L. Law, Production, use, and fate of all plastics ever made, *Sci. Adv.*, 2017, **3**(7), e1700782.
161. D. Hoornweg and P. Bhada-Tata, What a Waste: A Global Review of Solid Waste Mangement, World Bank, Washington, DC, 2012.
162. A. L. Brooks, S. Wang and J. R. Jambeck, The Chinese import ban and its impact on global plastic waste trade, *Sci. Adv.*, 2018, **4**(6), eaat0131.
163. A NEOCOLONIAL PLASTICS SCANDAL, Plastic Soup Foundation, Amsterdam, Netherlands, 2022.
164. M. G. Kibria, N. I. Masuk, R. Safayet, H. Q. Nguyen and M. Mourshed, Plastic Waste: Challenges and Opportunities to Mitigate Pollution and Effective Management, *Int. J. Environ. Res.*, 2023, **17**(1), 20.

165. The U.S. Recycling System, https://www.epa.gov/circulareconomy/us-recycling-system (accessed September 2024).
166. Resource Conservation and Recovery Act (RCRA) Regulations, https://www.epa.gov/rcra/resource-conservation-and-recovery-act-rcra-regulations#nonhaz (accessed July 2024).
167. H. Jeswani, C. Krüger, M. Russ, M. Horlacher, F. Antony and S. Hann, *et al.*, Life cycle environmental impacts of chemical recycling *via* pyrolysis of mixed plastic waste in comparison with mechanical recycling and energy recovery, *Sci. Total Environ.*, 2021, **769**, 144483.
168. T. Uekert, A. Singh, J. S. DesVeaux, T. Ghosh, A. Bhatt and G. Yadav, *et al.*, Technical, Economic, and Environmental Comparison of Closed-Loop Recycling Technologies for Common Plastics, *ACS Sustainable Chem. Eng.*, 2023, **11**(3), 965–978.
169. O. Eriksson and G. Finnveden, Plastic waste as a fuel - CO_2-neutral or not?, *Energy Environ. Sci.*, 2009, **2**(9), 907–914.
170. A. Demetrious and E. Crossin, Life cycle assessment of paper and plastic packaging waste in landfill, incineration, and gasification-pyrolysis, *J. Mater. Cycles Waste Manage.*, 2019, **21**(4), 850–860.
171. The Science of Plastic Recycling, Georgia: Science for Georgia, Atlanta, 2023.
172. A. N. Rollinson and J. Oladejo, Chemical Recycling: Status, Sustainability, and Environmental Impacts, Global Alliance for Incinerator Alternatives, Berkeley, Californnia, 2020.
173. A. Somoza-Tornos, A. Gonzalez-Garay, C. Pozo and M. Graells, Espuña A, Guillén-Gosálbez G. Realizing the Potential High Benefits of Circular Economy in the Chemical Industry: Ethylene Monomer Recovery *via* Polyethylene Pyrolysis, *ACS Sustainable Chem. Eng.*, 2020, **8**(9), 3561–3572.
174. M. Gear, J. Sadhukhan, R. Thorpe, R. Clift, J. Seville and M. Keast, A life cycle assessment data analysis toolkit for the design of novel processes – A case study for a thermal cracking process for mixed plastic waste, *J. Cleaner Prod.*, 2018, **180**, 735–747.
175. A. Samuelson, The plastic industry's $30 million lie, HEATED, 2024. Available from: https://heated.world/p/the-plastic-industrys-30-million?utm_source=post-email-title&publication_id=2473&post_id=146898183&utm_campaign=email-post-title&isFreemail=true&r=43m99&triedRedirect=true.
176. V. Singla, Recycling Lies: Chemical Recycling of Plastic is Just Greenwashing Incineration, NRDC.org: National Resources Defense Council, 2022.
177. FACT SHEET: Trash Incinerator Closures 2000–2024, EnergyJustice.net: Energy Justice, 2024.
178. G. Elliott-Smith, Waste Incineration across Europe and the UK: profit at the expense of Climate, EURACTIV, 2020, vol. 2020.
179. M. Anshassi, T. Smallwood and T. G. Townsend, Life cycle GHG emissions of MSW landfilling *versus* Incineration: Expected outcomes

based on US landfill gas collection regulations, *Waste Manage.*, 2022, **142**, 44–54.
180. J. Vahk, The impact of Waste-to-Energy incineration on climate, Zero Waste Europe, 2019.
181. Waste-to-Energy is incompatible with climate mitigation: the EU takes a stand [press release], Zerowasteeurope.eu: Zero Waste Europe, 2020.
182. R. Sharma, M. Sharma, R. Sharma and V. Sharma, The impact of incinerators on human health and environment, *Rev. Environ. Health*, 2013, **28**(1), 67–72.
183. Laboratory Standard 29 CFR 1910.1450. U.S. Occupational Safety and Health Administration: U.S. Occupational Safety and Health Administration; 1990, p. 9.
184. S. Scott and C. Greever, Equity of Hazardous Waste Management in U.S. Higher Education, Making the Research Enterprise More Sustainable and More Equitable, [Methodology for Equity Determinations in Hazardous Waste Management], In press.
185. P. Mohai, P. M. Lantz, J. Morenoff, J. S. House and R. P. Mero, Racial and socioeconomic disparities in residential proximity to polluting industrial facilities: evidence from the Americans' Changing Lives Study, *Am. J. Public Health*, 2009, **99**(Suppl 3), S649–S656.
186. Toxic Wastes and Race at Twenty1987–2007, United Church of Christ Justice and Witness Ministries, Cleveland, OH, 2007.
187. B. Covert, Race Best Predicts Whether You Live Near Pollution, The Nation, 2016. Available from: https://www.thenation.com/article/archive/race-best-predicts-whether-you-live-near-pollution/#:~:text=It's%20not%20just%20water%3A%20A,person%20living%20near%20hazardous%20waste.
188. Ellen Macarthur Foundation: Circular Economy Introduction, https://www.ellenmacarthurfoundation.org/ (accessed June 2024).
189. Circular Economy for Labs Community of Practice, https://www.circulareconomy.i2sl.org/ (accessed May 2024).
190. C. Silvestri, L. Silvestri, A. Forcina, G. Di Bona and D. Falcone, Green chemistry contribution towards more equitable global sustainability and greater circular economy: A systematic literature review, *J. Cleaner Prod.*, 2021, **294**, 126137.
191. UGA Laboratory Waste Diversion Lab Glass Rehoming Program, https://greenlabs.uga.edu/initiatives/waste-diversion/ (accessed August 2024).
192. Seeding Labs, https://seedinglabs.org/instrumental-access/ (accessed May 2024).
193. EPA, Sustainability Primer available from https://www.epa.gov/sites/default/files/2015-05/documents/sustainability_primer_v9.pdf (accessed August 2025).

CHAPTER 8

Sustainable Digital Science†

DEIRDRE BLACK

Royal Society of Chemistry, Thomas Graham House, Cambridge Science Park, Milton Road, Cambridge CB4 0WF, England
Email: blackd@rsc.org

8.1 The Growth of Digital Science

People in the worlds of business and policy sometimes talk about a 'twin transition' to a green and digital future. The idea is that, because the green and digital transitions are happening at the same time, there is a unique opportunity to develop digital societies and economies with reduced greenhouse gas emissions, pollution and biodiversity loss.

Taking a broad view of sustainability – spanning planet, people and profit – digital technologies can potentially also improve everything from safety and security to health outcomes.

Digital transformation, or digitisation, has been happening in science for decades. By developing and using digital technologies scientists make new discoveries and applications that will contribute to a sustainable future in myriad ways – from improving human health through new diagnostics and therapies to developing better batteries and solar energy technologies.

The good news is that, because the use of digital tools in science is increasing, there is a golden opportunity to integrate sustainability thinking and training into new or changing fields. Another positive is that digital technologies can offer ways of reducing the environmental impacts of science compared with the baseline of science today.

The Sustainability of Science: How Science Impacts the Environment, and What Can Be Done
Edited by Martin Farley

Published by the Royal Society of Chemistry, www.rsc.org

However, digitisation brings with it environmental impacts because of the energy, water and materials demands of everything from data storage and artificial intelligence (AI) to microchip manufacturing and electronics waste.

So, the reality is that – as in society and the economy more broadly – digitisation and sustainability are both so complex, dynamic and wide-ranging that it is challenging to advance both in perfect lock step.

I hope that this chapter will have something for everyone who is interested in science and sustainability, including people who may not currently consider themselves to be very involved in 'digital science'. It is a mix of case studies and practical things to try, along with some food for thought and debate about the connections between digitisation and sustainability in science.

I have put 'digital science' in quotes as it is a term that people use in different ways. Here I am using it as a broad catch-all for the many ways in which the roughly 9 million[1] researchers around the world use digital technologies. For concreteness, I have included lots of examples that I have collected and find interesting. If readers have different or better examples I would love to hear about them.

8.1.1 From Everyday Digital Tools to Scientific Computing

Before considering the environmental impacts of digital science and how to minimise them, let's look at some of the ways in which researchers use digital tools.

At a basic level, scientists use digital tools for everyday activities including sending emails, reading online publications and participating in virtual conferences. Some researchers use commercial word processing and presentation products to share their findings, or spreadsheets to record and analyse data.

A recent addition is generative AI tools like *ChatGPT*. *Nature* found that around one in three postdocs who responded to a September 2024 survey[2] use AI chatbots to some degree for tasks like drafting text, writing code and finding or synthesising information. There are new research assistant tools like *Potato AI*[3] that help researchers analyse the scientific literature and build computational experiments.

In certain fields researchers use specialised tools like *LaTeX* and *ChemDraw*® to record and report scientific work involving equations or drawings. Scientists and engineers use mathematical software like *Maple* or *Mathematica* and numerical computing platforms like MATLAB® for everything from modelling the attenuation of wi-fi signals to calculating interactions between elementary particles.

When using these off-the-shelf or 'commoditised' tech tools, scientists are running code, generating data and using semiconductors even if they do not actually develop software, wrangle data or design or build hardware themselves. This is true for the multitude of bits of software and hardware integral to everything from scientific measurement instruments and lab equipment to data analysis and modelling platforms.

There are also large communities of researchers in computational science or scientific computing, whether that is computational biology, chemistry, materials or physics, or of course computer science and engineering, mathematics and statistics. People in these fields are at the vanguard of research developing computing for, and applying computing to, different problem or 'domain' areas. They are tackling challenges from molecular simulations and bioinformatics to modelling the Earth's climate or the collision of galaxies.

There are also important research initiatives with wide applicability within and across scientific domains, including building data repositories, developing electronic lab notebook software and using robotics to automate lab or process workflows.

So pretty much all researchers are doing digital science to some degree, whether that is using digital platforms to share their science, developing their own computer architectures and AI agents, or something in between.

It is beyond the scope of this chapter to discuss research in computer science itself and allied fields such as deep learning and quantum computing or robotics and vision, but important to note that there are fertile interfaces at which advances in 'compute' and 'science' drive forward research in both. This energy has been present for a long time in multidisciplinary areas of computational science. It is accelerating and more prominent now with interest in '*AI for science*' from technology companies like Google,[4] IBM[5] and Microsoft.[6]

8.1.2 The Promise of Digital Science

In 2019 the Royal Society of Chemistry asked hundreds of scientists around the world what they were most excited about when it came to progress and promise in their research fields. There was huge excitement about data and digital technologies – from computational modelling and advanced measurement to AI and automation.

The pace of adoption of digital tools has increased since then, but the vision for how data and digital technologies enable new scientific discoveries – and solutions to environmental, social and economic sustainability challenges – has been clear for some time in many communities.

> *"There are remarkable tools that are changing the face of science .... advances in measurement are critical – and we are getting better at using multiple measurement techniques to solve formidable problems by collating and merging different types of data"*
>
> Sarbajit Banerjee, Professor of Chemistry, Texas A&M University[7]

> *If you were a student in crystallography 20 years ago, you solved the structure for a single interesting protein in your PhD. Now computers and robots routinely solve 10s or 100s a month. This means you can ask totally different questions. In my field that allows you to think in terms of multi-component*

systems – you move from designing something that binds or interacts at one site in a particular way to designing something that impacts in exactly the way I would like it to with several different components of a system.

Charlotte Deane, Professor of Structural Bioinformatics, University of Oxford

"With automation, we can do some materials synthesis experiments 1000 times faster than an individual person can. This kind of step-change in the number of experiments we can do means we can afford to tackle bigger questions. We can be more ambitious in the challenges we take on rather than incrementally exploring around what we already know."

Andy Cooper, Professor of Chemistry, University of Liverpool

My students work in a field that has been revolutionised in the last 20 years. From computing one property of one material in a year, they can now perform hundreds or thousands of computations in a week. This allows you to answer different types of questions – higher-level questions. Not just how an ion moves in one crystal structure, but what kinds of structures enable fast ionic movement. This level of abstraction was not available 20 years ago in the field.

Kristin Persson, Professor in Materials Science and Engineering, UC Berkeley[8]

There has been huge growth in the number of researchers developing and using digital tools, as well as the ways in which they are using them. Starting with the number of researchers working in the field of Artificial Intelligence (AI) itself, an example is the *Annual Conference on Neural Information Processing Systems* (NeurIPS). Founded in 1987, the 2010 edition of the conference was held in Vancouver, Canada and had 1354 delegates. The 2022 edition had 9825 in-person delegates in New Orleans, USA and a further 5555 online.[9]

One place to see trends in the adoption of AI by natural scientists is scientific publications. The most recent large-scale bibliographic analysis I could find was published in May 2024.[10] By analysing 80 million research publications across 20 diverse scientific fields including biology, chemistry and physics, the authors found that scholarly engagement with AI had increased thirteen-fold between 1985 and 2022. They classified nearly 9% of all 2022 scientific publications in their dataset as 'AI-engaged'.

8.1.3 Proteins, Black Holes and Robots

To get a flavour for the depth and diversity of ways in which digital technologies are integrated into science research, let us look at some examples. Even though tiny proteins and massive black holes are wildly different in terms of their size, structure and origins, the ways in which researchers study them have a lot in common when viewed through a data-driven or digital science lens:

- **Advanced measurement** techniques developed over many decades to detect signals, reveal structures and visualise data. Today's instruments use – and even involved inventing – multiple digital technologies in areas including optoelectronics, sensing, precision control systems and image processing.
- **Computing** in combination with mathematics, statistics and theory to predict structures and interactions, and to compare with and interpret experimental observations.
- **Data** from measurements, modelling and simulation is generated in huge volumes and harnessed to gain insights.
- **'Multi-mode' measurement** brings together data from different measurement techniques applied to the same object or system. We will see that by combining measurements from X-ray crystallography, nuclear magnetic resonance (NMR) spectroscopy and cryogenic electron microscopy (cryo-EM) scientists have revolutionised our understanding of protein structure and interactions. Using 'multi-messenger astronomy' scientists are combining measurements from telescopes, gravitational wave and neutrino detectors to study everything from supernovae to coalescing neutron stars.

Multidisciplinary and global collaborations from scientists around the world are key to all of this, with contributions over decades from thousands of computational scientists, theorists, engineers and experimentalists across domains including biology, chemistry and physics.

8.1.3.1 Measuring and Predicting Protein Structure

Understanding proteins, and how they interact with one another and with other structures such as the small molecules in medicines, is important for many reasons – from breathtaking discoveries about the building blocks of life and the mechanisms of ageing to preventing and treating human health conditions.

Since 1971 the Protein Data Bank (PDB) has served as a repository of information about the structure of proteins and other biomolecules.[11] It contains all the publicly available structure measurements, made using techniques such as X-ray crystallography, NMR spectroscopy and cryo-EM.

The PDB is free to access and currently has over 215 000 entries deposited by more than 65 000 research groups. The expertise, facilitation and infrastructure to maintain it is provided by universities and research centres in Europe, Japan and the United States.

A protein is made from a chain of tens to thousands of amino acids, which twist and fold in three dimensions. Predicting, as opposed to measuring, the shape or 3D structure of proteins involves computation combined with experimental and theoretical expertise in biology and chemistry. An important initiative in improving our ability to predict the structure of proteins, and other biomolecules including RNA, is the *Critical Assessment of Structure Prediction (CASP)* which started in 1994 as '*a community wide experiment to determine and advance the state of the art in computational structural biology*'.[12]

CASP participants start with a chain of amino acids and use the computational techniques they have developed to predict the 3D shape of the protein. This 3D structure has been determined experimentally and is known to the *CASP* organisers, but not to the participants. Eventually there is a big reveal and participants compare how close they got to predicting the experimentally determined structure. All of the researchers in the field then go on to hone their techniques based on their collective learning. In 1994, 35 groups took part, submitting over 100 predictions. By 2022 and the fifteenth iteration of *CASP* nearly 100 groups submitted more than 53 000 models on 127 modelling targets.[13]

In 2018 a team of researchers that used *AlphaFold 1*, an AI system that uses deep machine learning to predict protein structure, placed first in the overall *CASP* rankings. By 2020 *AlphaFold* was the top-ranked protein structure prediction method by a large margin. The *AlphaFold Protein Structure Database*, a partnership between Google DeepMind and EMBL's European Bioinformatics Institute, provides access to the now more than 200 million structures predicted by *AlphaFold*.[14]

In its citation for the 2024 Nobel Prize in Chemistry, the Nobel Foundation states that[15] '*... AlphaFold 2 has been used by more than two million people from 190 countries. Among a myriad of scientific applications, researchers can now better understand antibiotic resistance and create images of enzymes that can decompose plastic*'.

In November 2024 Google DeepMind and Isomorphic Labs released the *AlphaFold 3* model code and weights for academic use.[16]

Researchers used the known protein structures from the PDB, with their corresponding amino acid sequences, to train the *AlphaFold 2* neural network.[17] So, in addition to reflecting advances in computation, biology and chemistry, the recent breakthroughs in protein structure prediction were made possible by the long-term deposition, curation and machine-readability of data in the PDB.

The PDB data, in turn, come from measurements made using continually evolving instruments and techniques, themselves involving advanced software and hardware. An example is cryo-EM, for which the 2017 Nobel Prize in Chemistry citation reads '*for developing cryo-electron microscopy for the high-resolution structure determination of biomolecules in solution*'.[18]

In 1975 Richard Henderson published the first, fuzzy, picture of a protein using an electron microscope. By 2013 researchers had overcome numerous hurdles and could produce high resolution images of the 3D structures of the Zika virus and the protein complexes that govern the circadian rhythm. This almost forty-year journey involved finding ways to preserve delicate biological structures in the harsh conditions inside an electron microscope for long enough to measure the structure. It also depended on the development of computational image processing methods in which hundreds or thousands of the microscope's fuzzy 2D images are analysed and merged to reveal a sharp 3D structure.

There have been many further advances since 2017, including applying AI-based methods to data processing and analysis of cryo-EM images[19] and automation of cryo-EM sample preparation.

X-ray crystallography and NMR spectroscopy are also multi-Nobel Prize winning techniques. Joined now by cryo-EM these techniques together provide unprecedented amounts of data about the structure and interactions of a huge range of molecules and materials.

8.1.3.2 Predicting and Measuring Gravitational Waves from Black Hole Collisions

At 5:51 am EDT on 14 September 2015, the Laser Interferometer Gravitational-wave Observatory[20] (LIGO) made the first direct detection of gravitational waves arriving on Earth.[21] In doing so they confirmed a century-old prediction from Einstein's theory of general relativity that accelerating massive objects cause ripples or vibrations in space-time itself. The waves detected by observatories in Louisiana and Washington, USA, were generated by the dramatic merger of two black holes.

Astronomers had previously identified multiple isolated candidates for black holes, but the LIGO measurement was the first observation of two black holes spiralling around each other and then coalescing to create a larger one. The observation opened a new field in astronomy based on detecting gravitational waves, which cannot be seen in telescopes that detect electromagnetic signals such as radio waves, visible light or X-rays.

Detecting the tiny space-time ripple that reached the Earth, having travelled 1.3 billion light years from the point where the black holes coalesced, required what the 2017 Nobel Prize for physics citation calls *the most sensitive instrument ever made*.[22] It involved developing new laser technology, inventing new materials, constructing 4 km vacuum tube interferometer arms, achieving immaculate synchronisation between laser beams travelling in each arm and a level of seismic isolation beyond anything accomplished previously. Each LIGO observatory is studded with thousands of sensors – accelerometers, seismometers, microphones, magnetometers and many others – which enable scientists to isolate the signal from an incoming gravitational wave from tiny local disturbances.

Numerical general relativity is vital to gravitational wave detection and astronomy. In simple terms, scientists build models of astrophysical objects and use advanced mathematical analysis and computational techniques to solve Einstein's equations. In this way they can predict the form of the gravitational waves that will be produced when, say, two black holes or neutron stars merge.

These predictions are crucial for gravitational wave detection as they are a guide for what to look for when designing the detector. Later, these predictions are essential to having confidence that a signal is indeed a gravitational wave. For gravitational wave astronomy, the ability to compare a measured waveform with a calculated one is the key to figuring out the properties of the system that emitted the waves, for example figuring out the mass and spin of a black hole resulting from a merger.

Another aspect of LIGO, from a digital science perspective, is that the detectors produce huge amounts of data, with hundreds of terabytes generated and analysed per year.[23] As a result, researchers use numerous data processing and analysis techniques, and even invented new ones like the 'fast chirp transform'.

The collaborative nature of the gravitational wave community also involves distributed computing, with scientists contributing to and accessing both data sets and computing resources from multiple locations around the world. LIGO's initial funding was approved in 1992 and since then more than 1000 researchers from 18 countries have been involved.

8.1.3.3 Molecular and Materials Discovery: Data, AI and Automation

Expanding out from the example of measuring and predicting protein structure, there are several major avenues of inter-connected research converging across multiple domains.

- '*In Silico*' Design of Molecules and Materials

  There is huge effort, with new scientific papers appearing daily, in using computing to discover and design molecules, materials and processes. As well as advancing fundamental scientific understanding, this effort promises lower-cost and higher speed routes to discovering everything from new medicines to biodegradable plastics and solar fuels.

  For example, if scientists know that a material has properties of interest for a new type of lithium-free battery, they can use computational techniques to explore the space of materials with related structures and similar properties. This can provide promising candidates for new battery materials, as well as indications of materials that will not work, all before lengthy or costly experiments to make and study lots of new materials in a lab.

  Another example connects protein science with small molecule drug discovery. Some diseases are caused by proteins changing or interacting with other proteins. An active area of medicines discovery involves developing small molecules that bind to a protein in a specific way to prevent or reduce these undesirable changes or interactions. If scientists know of a small molecule that binds to a protein in a particular way, they can run simulations to see if, by making changes to the molecule, they will get stronger or weaker binding. Once they identify the most promising molecules, they then discuss with medicinal chemists to decide which compounds to try to make in the lab.

Academic researchers have been active in these fields and using machine learning and AI for decades. Today there is increasing corporate R&D investment from the tech sector as well as sectors such as pharmaceuticals, chemicals and consumer products.

- Measuring and Predicting Small Molecule Crystal Structures

  Predicting the possible structures of a crystal made of small molecules is important for deepening our understanding of matter. It is also key for sectors such as pharmaceuticals and agrochemicals that make and seek ingredients for medicines, pesticides and fertilisers because the packing of molecules in the crystal controls their solubility and bio-availability.

  The Cambridge Crystallographic Data Centre (CCDC) is a global repository where X-ray crystallography researchers deposit their experimental determinations of the crystal structures of small organic and metal-organic molecules.

  The CCDC organise the Crystal Structure Prediction (CSP) Blind Test.[24] Starting with the 2D representation of a molecule and information about experimental conditions, participants use the methods they have developed to predict crystal structures. At the end they compare their results with structures which have been determined experimentally but are not publicly available. The first workshop was held in 1999 with a total of 14 authors listed on the associated paper.

  The seventh CSP Blind Test, with results published in October 2024, had 150 researchers from 28 groups spanning 14 countries. A nice illustration of the combination of expertise in deep computing and deep chemistry is the conclusion by the authors that the success of some of the predictions '*... can be attributed to the use of highly reliable quantum chemical calculations, cloud computing, machine learning techniques, tailor-made force fields, careful accounting of thermal effects, and efficient conformational sampling algorithms ...*.'[25]

  Initiatives such as the CCDC are, like the Protein Data Bank, crucial in acting as a repository for data in a format that can be accessed by researchers globally for many purposes including training AI models.

- Lab Automation

  Many companies sell automated platforms for experimental biology, chemistry and materials science, and there is also active research in developing next generation automated lab technologies. There are automated systems to manage different types of samples, solvents, catalysts and starting materials, to carry out chemical reactions or formulate mixtures, and to purify and analyse the products. Automated platforms also enable researchers to run experiments and to retrieve the data from them remotely.

Labs use automation at different scales, starting with automation of a single instrument or workflow. Researchers have also developed mobile robots which can move around a lab performing tasks at different stations.[26]

A large-scale example is the Carnegie Mellon University Cloud Lab,[27] a \$40 M facility[28] which has 200+ interconnected scientific instruments running 24/7. The lab is staffed by specialised technical professionals who maintain instruments and ensure supplies of consumables. Researchers send experiments to the lab and retrieve data from them *via* an online platform.

Instruments are connected *via* tracks along which samples or reagents are moved automatically. These instruments include general laboratory automation (liquid handling robotics), chemistry analytical instrumentation (high performance liquid chromatography, mass spectrometers), structural characterisation (X-ray crystallography, NMR spectrometers) and biological instrumentation (flow cytometers, high-throughput sequencing).

- Iteration Within and Between Experimental Cycles

  Researchers are increasingly combining different approaches, for example combining automation with AI to carry out wet lab experiments in an iterative way, using data gathered at one stage to decide the next experimental step.

  They are also using automation and AI to connect different experiments. Data from automated labs can feed into building a data repository, developing scientific hypotheses, making predictions in combination with modelling and simulation, or deciding to progress to a new phase of analysis or testing.

'The key thing that digital will enhance in science is the connectivity between steps, making best use of all existing knowledge so that you go maximally informed into each stage in the discovery, development and innovation process, storing and sharing knowledge each time you do new experiments'.

Dr Horst Weiss, Vice President, Knowledge Innovation, BASF SE[29]

8.1.3.4 Three Final Examples

Before moving on to discuss the environmental impacts of digital science and how to reduce them, here are some final vignettes.

Mapping the Brain of a Fruit Fly. In October 2024 the FlyWire consortium[30] published an amazing map showing the 'wiring' between neurons in a brain. This 'connectome' captures the more than 54.5 million synapses between the roughly 140 000 neurons in a fruit fly brain.[31] It took neuroscientists more than four years, with the help of AI tools, to stitch together data from electron microscopy images of slices of the fly's brain, to form a full map of the brain.[32]

Higgs Boson Detection at CERN.[33] ATLAS is one of the two collaborations that confirmed the existence of the Higgs boson in 2012. The ATLAS detector, which is approximately the size of a cathedral, includes a Silicon Semiconductor Tracker with over 6 million implanted readout strips. Hardware such as field-programmable gate arrays (FPGAs), along with specialised software, sequentially reduce raw data by a total factor of 200 000 to record the remaining 320 megabytes of ATLAS data per second.

Great Internet Mersenne Prime Search (GIMPS).[34] In October 2024 GIMPS announced that volunteers had found a prime number 41 024 320 digits long, which is more than 16 million digits longer than the previous largest known prime number. By using powerful Graphics Processing Units (GPUs), available in the cloud at discounted prices during periods of low use, the volunteers created a kind of "cloud supercomputer" spanning 17 countries to search for the new prime. It is only the 52nd prime number of the from 2^n-1, or Mersenne prime, ever discovered.

8.2 Environmental Impacts of Digital Science

We have seen the variety and growth in the use of digital technologies in science as well as some of the wonderful discoveries and applications that flow from this research.

The environmental impacts of this digital science are perhaps less immediately obvious than those of activities where you can see material resources such as paper, plastic or chemicals, and hear energy-intensive ventilation, heating and cooling systems at work.

Understanding the environmental impacts of the digital sector is challenging in general because of the complexity and variety in how and where digital technologies are made, used and disposed of. There are also limitations on what data is available. To quote a 2023 case study from the MIT Schwarzman College of Computing '*In the age of machine learning, cryptocurrency mining, and seemingly infinite data storage capacity enabled by cloud computing, the environmental costs of ubiquitous computing in modern life are obscured by the sheer complexity of infrastructures and supply chains involved in even the simplest of digital transactions*'.[35]

There are also wider social impacts, including displacement of indigenous people, labour rights and impacts on human health and well-being. These important questions are beyond the scope of this chapter which focusses primarily on environmental impacts. Let us turn to some examples of environmental impacts.

8.2.1 Semiconductor Manufacturing

Microchips, printed on semiconductor wafers, are ubiquitous. In science they are in everything from laptops to supercomputers and from small sensors and instruments to robots and large-scale facilities.

Semiconductor manufacturing has significant environmental impacts including emission of greenhouse gases (GHGs) and particulate matter into the air and atmosphere. It involves withdrawal of fresh water, which has particularly serious implications for areas where water is scarce. Semiconductor fabrication plants also produce general waste including manufacturing waste and sludge from waste-water treatment, as well as hazardous waste such as certain chemicals and heavy metals requiring careful management.

Quantifying these environmental impacts is highly technical, and necessarily relies on assumptions, averages and limited primary data. In one study researchers at the University of Salento estimated the impacts associated with the fabrication plants or 'foundries' of four companies with a combined 70% share of world foundry revenue.[36] Based on analysis of corporate CSR reports and other sources, the researchers found that combined scope 1 and 2 GHG emissions increased from about 8 Mt CO_2e in 2012 to 15 Mt CO_2e in 2021. Water withdrawal or intake increased from about 90 Mt to 220 Mt over the same decade. General waste generated increased almost seven-fold from about 60 kt to 415 kt, and hazardous waste more than tripled from 140 kt to 460 kt. The volume of wafers produced almost tripled over the same period, and the increases in environmental impacts reflect both this increase and the greater complexity of the wafers being manufactured.

8.2.2 Data Centres and Data Transmission Networks

According to the International Energy Agency (IEA), in July 2023 data centres and data transmission networks each accounted for 1–1.5% of global electricity use. Together they accounted for 1% of global energy-related greenhouse gas emissions.[37] This electricity use is increasing as services and technologies like streaming, gaming, blockchain, AI and virtual reality are all boosting demand for data services. The IEA forecasts that datacentres' total electricity consumption could double from 2022 levels to 1000 terawatt hours by 2026.[38]

This plays out in different ways in different countries. For example, in a 2024 study[39] researchers at University College Cork found that, as of 2023, data centres accounted for 21% of Ireland's electricity consumption. Between 2017 and 2023, all new wind energy generation installed in Ireland was absorbed by data centres. Electricity demand from data centres had grown at an annual rate of 22.6% since 2015, compared to 0.4% for other sectors.

In the United States, in September 2024 Microsoft announced[40] a 20-year power purchase agreement (PPA) that will enable the restart of the 835 megawatt 'Three Mile Island' nuclear facility in Pennsylvania that was retired in 2019. In October 2024 Google announced a corporate agreement to purchase nuclear energy from small modular nuclear reactors to be developed by Kairos Power, headquartered in California.[41]

There are also concerns about water – used in data centre cooling – especially with the increase in data centre capacity associated with the growth of AI.

8.2.3 AI

The environmental impacts of AI specifically are a live subject of concern and discussion, with a focus on energy, climate and water. The trend is that, without intervention, energy and water use associated with AI will increase. This is in part simply because more people and applications are using AI. It is also because as AI technologies develop, for example becoming larger or more accurate, they have typically used more energy.[42]

This is a rapidly evolving area in many countries. For example, in early 2024 a US senator introduced a bill requiring the US Environmental Protection Agency to carry out a study on the environmental impacts of AI, and the National Institute of Standards and Technology to convene a consortium to develop voluntary reporting standards for those impacts.[43]

There are some detailed studies available. In 2019 researchers from the University of Massachusetts Amherst presented their calculations[44] of the carbon footprint of 'training AI' at the *57th Annual Meeting of the Association for Computational Linguistics (ACL) in Florence, Italy.* As for other estimates of carbon emissions this study is highly technical and involves assumptions about the electricity mix, type of processors used, cloud service providers as well as the specifics of the computational methods. Looking at two commonly used Natural Language Processing models, the researchers found that the carbon footprint of training one model was about 5 times that of the entire lifetime of a car, including fuel.

Readers interested in learning in more detail may like to look at the more recent *Environmental Impacts of AI* primer.[45]

There are also concerns about water use. According to one study, global AI demand is projected to account for 4.2–6.6 billion cubic meters of water withdrawal in 2027. This is more than the total annual water withdrawal of 4–6 Denmark or half of the United Kingdom.[46]

8.2.4 Scientific Computing

The environmental impacts of research that solely involves computing and instrumentation, often called 'dry lab' research, are different to those of 'wet labs' where research might involve making new molecules or analysing the properties of cells.

A couple of examples of the scale of scientific data generation are:

- The *Electron Microscopy Public Image Archive* (EMPIAR) is a public resource for raw images underpinning 3D cryo-EM maps and tomograms, which themselves are archived in the Electron Microscopy Databank.[47] Its purpose is to provide easy access to state-of-the-art data to facilitate methods development, validation and re-use. As of late March 2025 EMPIAR contained 2077 entries, taking up more than five petabytes (PB), that is five million gigabytes (GB), of storage.[48]
- In 2022 the *Event Horizon Telescope* published the first image of Sagittarius A*, the supermassive black hole at the centre of our galaxy.[49] Depending on instrument setup and other factors a five-day observing campaign can produce about a petabyte of raw data per observatory.[50]

Estimating the climate and wider environmental impacts of scientific computation is highly technical. A scientific computing example is the Geant4[51] toolkit, based on Monte-Carlo methods, which is used in wide-ranging applications from simulating what happens in the ATLAS detectors at CERN to understanding how radiation therapy impacts patients.

In 2021 the *Green Algorithms* team applied their carbon calculator to a piece of computational science research that was studying the damage caused to human DNA by radiation. They estimated that a single experiment using Geant4 emits about 50 kg of CO_2e.[52] The study involved running the simulation for 11 different energy levels and so the total carbon footprint of the study is about 500 kg CO_2e. This is equivalent to driving about 3000 km or flying economy from New York to San Francisco.

8.2.5 Electronic Waste

Electronic waste or e-waste is one of the fastest growing solid waste streams in the world. It covers a range of products, including many that are used in scientific research, which are discarded after use – from computers, printers and phones to medical and scientific instruments.

In the EU the amount of new electrical and electronic equipment put on the market rose from 7.6 million tonnes in 2012 to 13.5 million tonnes in 2021.[53] Global shipments of personal computing devices were forecast to reach over 398 million in 2024.[54] The total collected electrical and electronic equipment in the EU increased from 3.0 million tonnes in 2012 to 4.9 million tonnes in 2021.

Globally, according to the World Health Organization, in 2022, an estimated 62 million tonnes of e-waste was produced.[55] Only 22.3% was documented as formally collected and recycled.

Lead is commonly released into the environment when e-waste is recycled, stored or dumped using informal activities, including open burning. It is estimated that millions of women and child labourers involved in the informal recycling sector globally may be at risk of hazardous e-waste exposure.

8.3 Reducing Environmental Impacts of Digital Science

In 2020 the Royal Society of Chemistry surveyed practising researchers to ask for their experiences and attitudes on the environmental impacts of research.[56] There was an overwhelming desire to reduce these impacts as well as a wide variety of experiences – from people who said that they simply don't know where to start to those who are very active in sustainable research.

In an ideal world, as in other areas of sustainability, we would know where 'hotspots' of environmental impact[57] are and make interventions there. In reality, many people have neither the knowledge and data nor the time, money or power to identify and take action to reduce the environmental impacts of digital research.

In the end I think it is important to start somewhere, so in that spirit I am including a variety of examples that will be relevant to different people depending on their research area, proficiency with computing and the resources available to them in their organisation.

8.3.1 Quick Tips

There are many excellent resources and initiatives aiming to support more environmentally sustainable digital science. For those who are new to thinking about this, here is an initial list, with more detail and links in the following sections.

- Be aware that digital hardware and software, including computing and instruments, have an environmental footprint
- Consider what emails, data and documents to generate, download, share or save, and for how long
- Connect with the Green/Sustainability team or IT services at your institution
- Join in-person or online communities of practice related to different aspects of sustainable digital science
- Access online resources, presentations and training webinars about lab sustainability and green computing
- Try tools like carbon calculators to help understand the emissions associated with your computing
- Where possible and relevant use Electronic Lab Notebook (ELN) software
- Use research equipment for as long as possible, taking into consideration its energy efficiency and research performance

- Share research instruments
- Repair, sell, donate or buy used computers and instruments
- Ask suppliers about the lifetime and environmental performance of their computing and instrumentation products and services, and factor this into decision-making when purchasing new equipment
- Ask suppliers about their Waste Electrical and Electronic (WEEE) take-back schemes
- Recycle or dispose of e-waste appropriately

8.3.2 Green Computing

There is an active community of researchers in computational science who are developing tools and creating communities of practice and incentives to support researchers in reducing the environmental impacts of computing.

An important example is the *Green Algorithms* Initiative,[58] started by researchers at the University of Cambridge, UK in 2021.

Green Algorithms has open access calculators that enable researchers to calculate the carbon footprint of computations, including those run on High Performance Computing platforms. It also includes many tips and resources for green computing. One example is a set of '*Ten simple rules to make your computing more environmentally sustainable*'[59] developed by researchers from the University of Cambridge and the European Molecular Biology Laboratory. Another is a set of *GREENER principles for environmentally sustainable computational science.*[60] Both publications are packed with detailed evidence and tools to support researchers in applying the rules.

There are also resources such as *GES 1point5* to assess the carbon footprint of research,[61] *Code Carbon* to estimate the carbon footprint of executing Python code[62] and the section on computational labs in the guidebook for sustainability in laboratories[63] from the University of Groningen. In addition, there are studies carried out by scientists in specific fields such as astrophysics,[64] bioinformatics[65] and neuroscience.[66]

The Software Sustainability Institute, funded by UK Research and Innovation, was founded in 2010 and its mission is to help people build better and more sustainable research software to enable world-class research.

In 2024 the institute launched the *GreenDiSC*[67] certification scheme which provides a roadmap for research groups and institutions who want to tackle the environmental impacts of their computing activities. The scheme is collaborative, open access and relevant across research domains.

Reflecting the need to support both 'bottom-up' initiatives by research groups and 'top-down' or centralised efforts by research institutions, GreenDiSC offers certification for both. The criteria reflect the many opportunities and actions that people can take, including those who are new to considering environmental sustainability. The certification provides both a framework for action and an incentive to support action (Table 8.1).

Table 8.1 Current criteria categories for bronze level certification for Green DiSC certification for research groups [for complete descriptors, and for criteria categories for central teams, see the Software Sustainability Institute[68]].

General	• Nomination of a Green DiSC representative • Computing sustainability is part of the induction procedure • Information and resources are shared with other groups engaging with sustainability. • Dissemination of this program in the institution
Offices	• Inventory of office computing hardware • Identification of electronic waste processing streams • Identification of unused computing equipment
Data storage	• Inventory of main data resources • Implement regular cleaning of users own data directories
Compute	• Inventory of computing infrastructures used • Inventory of the most demanding computing pipelines • Training on (green) computing best practices is provided

8.3.3 Making Data FAIR

Initiatives that enable the useful sharing of scientific data are an essential way of getting the maximum scientific impact from the financial, human, material and energy resources invested in obtaining the data.

Sometimes there are ethical, privacy or commercial reasons why data cannot be shared, but where possible the ideal is that data is Findable, Accessible, Interoperable and Reproducible (FAIR) so that researchers in different places can access and use the data in a straightforward way.

Repositories such as the PDB and the CCDC, discussed above, are examples of highly curated sets of data in standardised formats that can be used by researchers in many fields. The development of these repositories is a tribute to early community leadership and commitment to the benefits of sharing data in a research field. The repositories also depend on different models for direct and indirect financial support from research institutions, funding agencies and companies in several countries.

Larger collaborations typically build data sharing and management into their planning and governance, and this is now a default requirement of most research funders. For example, the LIGO gravitational wave collaboration discussed above has a data management plan[69] that sets out transparently its approach to assuring the quality, preservation and public release of data.

In some fields the submission of a Data Availability Statement in association with publications is either mandatory or a norm. Even if not mandatory, publishers are increasingly encouraging and supporting authors to make the data associated with a journal article available.[70]

There are barriers to sharing data and more again to making it FAIR. Developing and maintaining repositories takes time and money, and researchers need to have the time and incentives to submit current and historical data to repositories in a standardised format. In some fields there is more than one repository or there is no standardised format for data.

One initiative that has been working to bring coherence to data in the physical sciences since 2022 is the Physical Sciences Data Infrastructure,[71] funded by UK Research & Innovation. It aims to support researchers to find, access and combine data from different sources, share their own data and software, use AI to explore data and learn how to make the results of their research open and FAIR.

8.3.4 Digital Scientific Publishing

In 2020 nine partners, including the BBC and Sky, came together with computer scientists at the University of Bristol to form the DIMPACT coalition which works to help the media industry understand and manage the significant carbon impacts of digital content.[72]

The members of DIMPACT now include media and streaming companies as well as academic publishers such as Cambridge University Press and the Institute of Physics. Its activities including convening communities and expert working groups, and the development of resources and tools such as the DIMPACT tool[73] to assess the emissions of streaming and digital publishing.

As for calculations of carbon footprints in other sectors, the development of this kind of tool is quite technical and based on a methodology that standardises the scope and boundaries between companies in the digital publishing supply chain. The tool also supports companies to do company-specific data collection and analysis.

In addition to supporting scientific publishers to make changes that reduce their emissions, the development and availability of tools like DIMPACT can support wider community and sector-level initiatives. Using the tool can also enable the publishing industry or individual publishers to provide data to help the editorial boards, authors and readers of scientific journals to change their own use patterns and to request changes from publishers.

8.3.5 Hardware and Instruments: Sharing, Selling and Donating

It makes sense to use computing hardware and scientific instruments as much as possible, assuming that their energy and scientific performance is acceptable. This can be done by using equipment for as long as possible and by sharing equipment so that it is not idle.

Sharing scientific resources is integral to large collaborations and measurement or computing facilities, but it can be more complicated to share smaller pieces of research equipment as this involves having an inventory of available items, an efficient system for booking them and mechanisms to ensure that they are used and maintained appropriately.

In the UK, the Equipment Data Service[74] enables searching across all published UK research equipment databases in one place. To improve the visibility and use of research equipment, the terms and conditions for UK Research & Innovation grants now require that all new equipment purchased over £138 000 be registered on the Equipment Data national database. The database currently logs almost 21 000 pieces of equipment from 65 institutions.

There are also regional initiatives such as the Southeast Equipment Sharing Consortium[75] which gives researchers from the universities of Cambridge, Oxford, Imperial, UCL and Southampton access to a combined 4600 items.

When a lab is looking to acquire a new piece of equipment, or once a lab or institution no longer has use for a piece of equipment, there are options to buy, sell and donate used and refurbished equipment.

The UniGreen Scheme[76] is an asset resale service for university sellers and buyers, offering both parties a report on carbon savings. Warp It[77] is another option used by universities and government to find, sell or give away furniture as well as lab equipment and consumables. A quick Google search of 'used scientific equipment marketplace' yields many results and the University of Cambridge has a useful set of options for equipment donations, disposal and recycling on its Equipment Sharing[78] pages.

There are also direct sales. A large-scale example is the 11 500-square-foot *Lilly Life Sciences Studio* laboratory designed and installed by Eli Lilly as part of a $90 million R&D investment in 2017 in San Diego. In 2024 Arctoris, a drug discovery services company headquartered in Oxford, UK and Boston, USA, announced that it had acquired the majority of the lab, including multiple automated biochemistry and cell biology workflows as well as automated storage capacity for up to 4 million compounds.

Other opportunity areas, from an equipment supplier perspective, are extended warranties and software upgrades to ensure that hardware and instruments can be used for as long as possible. Many suppliers also offer electronic equipment 'take-back' schemes voluntarily or to comply with local producer and distributor responsibility regulations.

8.3.6 Reducing the Environmental Impacts of Data Centres and Hardware

There are many efforts to reduce the environmental impacts of data centres.

One option to reduce the use of electricity generated from fossil fuels is to use energy from renewable sources. However, we have seen for Ireland that if data centres buy renewable energy on a large scale this can slow down broader decarbonisation efforts.

People and organisations are working to reduce the environmental impacts of data centres in many ways. An example is the Wellcome Genome

Campus (WGC), home to leading research centres in genomics and computational biology including Wellcome Sanger Institute and EMBL's European Bioinformatics Institute.

The WGC data centre has about 50 PB of storage and receives 3–4 million 'hits' or requests for data daily. The Campus has installed on-site energy generation and currently 10% of the total energy consumed is self-generated from solar photovoltaic arrays, air source heat pumps and a combined cooling, heating and power (CCHP) plant. The waste heat from the CCHP unit is captured and used within an absorption chiller to cool the data centre.[79]

Power Usage Effectiveness (PUE) is a measure of how efficiently a data centre directs energy into computing as opposed to other energy draws, especially cooling. Examples of technologies that can help to increase PUE are cooling fans powered by more energy efficient *electronically commutated (EC) motors*, and *hybrid cooling* technologies which combine dry and evaporative cooling to reduce both water and energy consumption.

Another option is to use heat from data centres to providing heating and hot water for buildings or facilities like swimming pools and greenhouses, although this relies on being within close proximity of sites which require heating. Some data centres are located in colder regions where the lower ambient temperatures give more efficient cooling. This can however introduce additional environmental impacts associated with data transmission networks connecting to users in more densely populated regions.

There is also huge investment in new and faster chips to bring down the computational, environmental and financial cost of computing. Examples are combinations of Central Processing Units (CPUs), Tensor Processing Units (TPUs), GPUs and FPGAs optimised for specific types of computation.[80]

It is important to note however that historically access to faster or more efficient computing, or increased data storage, has resulted in people doing more computing as well as generating and storing more data. Without active intervention in the use of computing and data this 'rebound effect' (Jevons Paradox) is set to continue, even with improvements in the environmental performance of algorithms,[81] microchips or data centres.[82]

8.3.7 Using Digital Tools to Make Science More Environmentally Sustainable

Digital tools can also help make science more environmentally sustainable. Here are a few examples:

Emailing and reading digital documents reduces the consumption of paper and print materials, and associated impacts including deforestation. However, as with tablets used for note-taking,[83] it is important where possible to minimise frequent procurement of multiple devices used for a limited range of purposes.

Virtual meeting tools can reduce business travel and the carbon footprint of air travel.

Sensors, timers and control systems can reduce energy consumption by optimising everything from the level of lighting or heat needed in a lab to the fan and sash settings of a fume hood. Building Management Systems can also enable significant utility savings if set up efficiently.

Electronic inventories can enable efficient use of consumables and better management of samples within labs and centralised sharing across labs.

Microfluidics and 'lab on a chip' technologies can reduce the scale of experiments and therefore the consumption of material resources and energy.

Automated ultra-low temperature (ULT) freezers can save energy and help improve sample tracking and storage within the freezer so that samples are not lost or compromised. Once the freezer is brought down to the required temperature, labelled samples are loaded *via* a small chamber, minimising energy loss. The chamber is then closed and purged of air to reduce frost. It rotates so that the sample can be stored inside the freezer by a robotic arm, which retrieves it when needed. Automated ULT freezers are available at different scales, from those suitable for a university or industry lab to high-capacity systems such as the Hamilton BiOS storage system procured last year by the UK Biobank[84] to store its millions of samples.

Digital tools integrating 'green-ness' into the development of experimental programmes. An example is AI4Green,[85] an Electronic Lab Notebook (ELN) software including collaboration tools, data archival and management, and green and sustainability metrics for organic chemistry. This tool has interactive features, for example enabling a researcher to get recommendations, impose constraints and draw on their own experience when it comes to choosing solvents that present varying degrees of hazard.

Computational modelling and simulations can reduce the need for physical experiments that use energy and material resources, as discussed above for molecular and materials discovery. In areas like environmental and climate science, modelling and simulation is the only option, as it is not possible to do the equivalent of controlled experiments to study large-scale impact.

Data reduction strategies increasingly involving AI mean that instruments only store data that is needed and for as long as it is useful. This reduces the environmental and financial costs of data storage and is particularly important in fields like particle physics, neuroscience and cryo-EM with large datasets from measurements and imaging. We saw previously that, using a combination of hardware and software, the ATLAS particle physics collaboration records 200 000 times less data for analysis than the raw data from the detector.

8.3.8 Using Digital Tools to Get the Greatest Value from Research

Digitisation can also help make science more sustainable by maximising the value obtained from whatever research is done in several inter-connected ways.

8.3.8.1 Research Design and Quality

Online experimental design tools can help researchers ensure that only the necessary number of experiments or steps are carried out, which means that no material or energy resources are wasted in unnecessary steps. These tools can also help ensure that the findings are statistically significant and so ensure that the experimental results are useable. An example in the area of animal research is the Experimental Design Assistant.[86]

8.3.8.2 Reproducibility

Detailed documentation and reporting of results, including data, helps ensure that other researchers can access and reproduce experiments. In addition to confirming results found by others, they can build on them in their own research programmes. Systematic recording and reporting can be facilitated by Electronic Lab Notebook (ELN) software, used to capture data and detailed information about experiments in a standard format that can be shared and used by other people in a research group or more broadly. Automated labs also enable reproducibility because the inputs, experimental steps, local conditions and data read from instruments are captured in detail.

8.3.8.3 Negative Results

Digital tools make it easier to record and share information about what has not worked, sometimes called 'failed experiments' or 'negative or nil results', as well as results that would typically be included in a scientific research publication. Sharing negative results avoids waste of effort and resources because others do not repeat experiments that have already been done but for which the findings were not published. Recording and sharing negative results also provides more complete datasets for training AI models, helping to make them more accurate and useful.

8.3.8.4 Recording and Sharing Data

ELN software, physical automation and data repositories facilitate the generation, storage and sharing of data in standard formats that can easily be found and used by as many researchers as possible. These technologies can all support efforts to make data FAIR, discussed above.

8.3.8.5 Making the Most of Historical Data

AI tools can pull out key information from large volumes of historical scientific articles, patents and reports, saving time and ensuring that future research builds on all previous knowledge in a field. There are also tools that can reliably pull out detailed technical data which historically has not been structured in a standardised way. An example is ChemDataExtractor[87] which can pull out numerical quantities along with their identifiers, units and errors, even from tables nested inside scientific articles in pdf format.

8.3.9 Deciding Whether Digital Options Reduce Environmental Impact

It may seem obvious that the examples of digital technologies in the previous section will improve the 'environmental performance' of science. However, we know from earlier sections that the environmental impacts of digital technologies themselves are significant. So, in order to decide whether digitisation actually brings a net benefit from an environmental perspective, we need to compare quantified environmental impacts of one thing with another.

For some situations this kind of comparison is possible, at least in principle. You can do a life cycle assessment (LCA) to estimate, say, the carbon footprint of a paper copy of a scientific article. You can also estimate the carbon footprint of producing and reading that article on an online platform or of downloading, storing and reading it on a computer. Then you can numerically compare the carbon footprint of the paper and digital options.

The same is true for ultra-low-temperature freezers. With the right expertise, information and data, you can estimate the embedded, operational and end-of-life environmental footprint of a conventional and of an automated freezer. Then you can decide how best to reduce the footprint when considering all facets of a freezer's lifetime in balance with each other.

In practice, most people do not have the expertise, information or time to do an LCA for each of their activities, and it makes sense to have where possible general guidance to avoid duplication of effort.

However, sometimes it is complicated even in principle to compare the environmental impacts or gains of a digital option with the 'non-digital' equivalent because there is no real equivalent.

For example, we have seen that computational modelling and simulation can reduce the need for wet-lab experiments. And also that physical automation can bring benefits from the perspectives of reproducibility and maximising the value derived from scientific data.

It is possible in principle to do an LCA of computational experiments, as we saw above for the carbon footprint of running Monte Carlo simulations. It is also possible in principle to do an LCA of experiments run in automated wet labs, including those that integrate AI.

However, what is not clear is what to compare these LCAs to, because generally computational or automated experiments do not replace a single wet-lab experiment or set of experiments like-for-like.

There are situations such as high-throughput screening (HTS) where such a comparison is possible in theory. Used commonly in scientific areas like drug discovery, HTS involves testing millions of compounds in automated labs to identify those with the most potential for medicines development. It is possible in principle to compare the environmental footprint of a high-throughput experiment with the equivalent set of experiments done manually. It is also possible to try to quantify improvements in reproducibility across the set of experiments. However, this comparison is purely theoretical because, without high-throughput approaches, those millions of experiments would not have been done! HTS has enabled a new approach to scientific discovery, so there is no previous benchmark to compare it with.

Turning to modelling and simulation, computational experiments can predict molecules or materials with interesting properties. The work of David Baker, 2024 Nobel Laureate, on computational protein design is a wonderful example. It is impossible however to know how many wet-lab experiments it might have taken to discover a given molecule or material, or indeed if it ever would have been discovered without computational approaches.

As we heard in the quotes about the promise of digital science earlier, computation and automation enable scientists to design research programmes based on higher level thinking and abstraction. Along with advanced measurement, these techniques and their scientific outputs open exciting new avenues to achieving research impacts.

In the end I think that what is important, noting the discussion about rebound effects earlier, is to avoid 'moral licensing' and a sense that computation or automation has a lower environmental impact than wet lab experiments. Science is always evolving as researchers develop new techniques and, irrespective of what those techniques are, the principle of maximising research sustainability applies.

8.4 Conclusions, Opportunities and Open Questions

We have seen some of the amazing discoveries and applications made possible by integration of digital technologies in science. We have also seen that, notwithstanding their potential to reduce environmental impacts in some ways, digital technologies have significant environmental impacts of their own.

The amount of data being generated from scientific measurements, modelling and simulations is increasing. So is the computing power needed for computational science and digital research tools used across many scientific disciplines.

This means that, without active intervention or transformational improvements in the environmental performance of everything from data centres to AI algorithms and processors, the environmental footprint of digital science will continue to increase.

There is a need for investment in systems and resources to support more environmentally sustainable research. This will minimise the duplication of effort and cost involved in multiple researchers or organisations trying to figure out where to focus. It will also make it easier for people and organisations with limited computing or sustainability expertise to make changes.

Alongside this there is an exciting opportunity to integrate environmental considerations into the training of scientists that develop or use computing, design experiments and analyse data. In addition to benefiting individuals who acquire these valuable skills, this will increase overall awareness of and competence in environmental sustainability in the businesses and organisations they go on to work in.

Tools like ELN software and approaches like systems thinking[88] can enable researchers to incorporate environmental sustainability into their research design at different scales. Building in constraints associated with reducing computational, financial and environmental costs can spur innovation, for example the development of new machine-learning methods for genetics[89] and 'frugal modelling' of molecular spin dynamics.[90]

Thinking about sustainable science in a broader sense, there are a few things to watch. First is if and how the increased involvement of corporate R&D in 'AI for science' influences norms and solutions related to the publication and sharing of data, code and results. There are also issues related to equity and the 'digital divide' between researchers that do and do not have access to advanced instrumentation, computation and automation, although in principle automated labs can support the democratisation of research by giving remote access. A third is how AI can potentially undermine research integrity and therefore lead to wasted investment of resources in research. For example, there are concerns that AI makes it easier to generate 'fake' results and papers, leading to time wasted trying to reproduce or build on those results as well as compromising the quality of systematic reviews. Fake results also compromise the training data for AI tools designed to accelerate literature review and analysis.

There is huge diversity in the kinds of science people do, the discoveries and applications they are pursuing and the resources they have available to them. There is also tremendous diversity in where they work, whether that is in universities, companies and hospitals or out at field stations, on boats or aircraft.

This means that there are numerous avenues for reducing the environmental impacts of research, and digital science is no exception. I have tried to give some concrete suggestions as well as links to explore the resources and communities developed by experts in different fields. It can be daunting to get started, especially if you are trying to figure out the 'best' place to start, in which case my suggestion is to start where you can.

Acknowledgements

While any errors or misunderstandings are mine, I would like to thank several people variously for helpful conversations and comments on drafts of this chapter: Mark Bastable (Hamilton Storage GmbH), Martin Farley

(UK Research & Innovation), Jeremy Frey (University of Southampton), Andy Gray (Wellcome Genome Campus), Loïc Lannelongue (University of Cambridge), Brien Nolan (Dublin City University), Jeremy Sanders (University of Cambridge), Nick Souter (University of Sussex) and my colleagues Gemma Humby-Smith, Richard Kidd, Carolina Ojeda Porras and Kate Tustain at the Royal Society of Chemistry.

References

1. *UNESCO Science Report*, 2021. https://unesdoc.unesco.org/ark:/48223/pf0000377433?posInSet=6&queryId=N-b7b632d1-da27-4e03-b295-962333343229 (accessed 23 March 2025).
2. *HowChatGPT is transforming the postdoc experience*, Nature Career Feature, 2024. https://www.nature.com/articles/d41586-023-03235-8 (accessed 23 March 2025).
3. *Potato AI – AI research assistant for biology*. https://readysetpotato.com/ (accessed 23 March 2025).
4. *A new golden age of discovery: Seizing the AI for Science opportunity Google DeepMind Essay*, 2024. https://deepmind.google/public-policy/ai-for-science/?utm_source=linkedin&utm_medium=social&utm_campaign=&utm_content= (Accessed 23 March 2025).
5. *IBM's multi-modal foundation model for materials.* https://huggingface.co/collections/ibm/materials-673465deacbdf38d9c0c6303 (accessed 23 March 2025).
6. Microsoft Research AI for Science. https://www.microsoft.com/en-us/research/lab/microsoft-research-ai-for-science/ (accessed 23 March 2025).
7. *Science Horizons*, Royal Society of Chemistry report, Science Horizons, 2019 (accessed 23 March 2025).
8. *Digital Futures*, Royal Society of Chemistry report and videos, Digital Futures, 2020 (accessed 23 March 2025).
9. 37th Annual Conference of Neural Information Processing Systems (NeurIPS), 2023, Fact Sheet https://media.neurips.cc/Conferences/NeurIPS2023/NeurIPS2023-Fact_Sheet.pdf (accessed 23 March 2025).
10. *Oil & Water? Diffusion of AI Within and Across Scientific Fields* E. Duede, W. Dolan, A. Bauer, I. Foster and K. Lakhani, arXiv:2405.15828v1 [2405.15828] Oil & Water? Diffusion of AI Within and Across Scientific Fields.
11. *Worldwide Protein Data Bank* (PDB). https://www.wwpdb.org/ (accessed 23 March 2025).
12. *15th Community Wide Experiment on the Critical Assessment of Techniques for Protein Structure Prediction*, Home - CASP15 (accessed 23 March 2025).
13. *16th Community Wide Experiment on the Critical Assessment of Techniques for Protein Structure Prediction*. https://pinot3.genomecenter.ucdavis.edu/casp16/index.cgi (accessed 23 March 2025).
14. *AlphaFold Protein Structure Database*. https://alphafold.ebi.ac.uk/ (accessed 23 March 2025).

15. Nobel Prize in Chemistry Announcement 2024, Royal Swedish Academy of Sciences https://www.nobelprize.org/prizes/chemistry/2024/press-release/ (accessed 23 March 2025).
16. *AlphaFold 3 predicts the structure and interactions of all of life's molecules*, Isomorphic Labs, 2024. https://www.isomorphiclabs.com/articles/alphafold-3-predicts-the-structure-and-interactions-of-all-of-lifes-molecules// and AI protein-prediction tool AlphaFold 3 is now more open, Nature News, 2024. https://www.nature.com/articles/d41586-024-03708-4 (Accessed 23 March 2025).
17. *AlphaFold: A practical guide*, EMBL-EBI Training. https://www.ebi.ac.uk/training/online/courses/alphafold/an-introductory-guide-to-its-strengths-and-limitations/what-is-alphafold/ (accessed 23 March 2025).
18. Nobel Prize in Chemistry Announcement, Royal Swedish Academy of Sciences, 2017. https://www.nobelprize.org/prizes/chemistry/2017/press-release/ (accessed 23 March 2025).
19. A. Dhakal, R. Gyawali, L. Wang and J. Cheng, Artificial intelligence in cryo-EM protein particle picking: recent advances and remaining challenges, *Briefings Bioinf.*, 2025, **26**(1), bbaf011.
20. Laser Interferometer Gravitational-Wave Observatory. https://www.ligo.caltech.edu/page/about (accessed 23 March 2025).
21. B. P. Abbott, *et al.*, Observation of Gravitational Waves from a Binary Black Hole Merger, *Phys. Rev. Lett.*, 2016, **116**, 061102.
22. *Scientific Background on the Nobel Prize in Physics*, Royal Swedish Academy of Sciences, 2017. https://www.nobelprize.org/uploads/2018/06/advanced-physicsprize2017-1.pdf.
23. Shawhan P S for the LIGO Scientific Collaboration, LIGO data analysis, *Nucl. Instrum. Methods Phys. Res., Sect. A*, 2003, **502**(2–3), 396–401.
24. Cambridge Crystallographic Data Centre, Crystal Structure Prediction Blind Tests. https://www.ccdc.cam.ac.uk/community/ccdc-for-the-community/partnerships-and-initiatives/csp-blind-test/ (accessed 23 March 2025).
25. L. M. Hunnisett, *et al.*, The Seventh Blind Test of Crystal Structure Prediction: Structure Generation Methods, *J. Acta. Cryst. B*, 2024, **80**, 517–547.
26. An example is B. Burger, P. M. Maffettone and V. V. Gusev, *et al.*, A mobile robotic chemist, *Nature*, 2020, **583**, 237–241.
27. Carnegie Mellon University, AI at CMU. https://ai.cmu.edu/research-and-policy-impact/ai-for-science (accessed 23 March 2025).
28. Tradeline Facilities Planning and Space Strategies Conference, 2023. https://www.tradelineinc.com/reports/2023-1/carnegie-mellon-open-first-ever-academic-cloud-lab (accessed 23 March 2025).
29. Digital Futures, Royal Society of Chemistry report, 2020. rsc.org/globalassets/22-new-perspectives/discovery/digital-futures/rsc-digital-futures-report—digital.pdf (accessed 23 March 2025).
30. FlyWire Brain. https://flywire.ai/ (accessed 23 March 2025).
31. S. Dorkenwald, *et al.*, Neuronal wiring diagram of an adult brain, *Nature*, 2024, **634**, 124–138.

32. Largest brain map ever reveals the fruit fly's neurons in exquisite detail, Nature News,2024. https://www.nature.com/articles/d41586-024-03190-y (accessed 23 March 2025).
33. ATLAS Fact Sheet https://cds.cern.ch/record/1457044/files/ATLAS%20fact%20sheet.pdf (accessed 23 March 2025).
34. The Great Internet Mersenne Prime Search. https://www.mersenne.org/ (accessed 23 March 2025).
35. S. G. Monserrate, The Cloud Is Material: On the Environmental Impacts of Computation and Data Storage, *MIT Case Stud. Soc. Ethical Responsib. Comput.*, 2022, DOI: 10.21428/2c646de5.031d4553.
36. M. Ruberti, Environmental performance & trends of the world's semiconductor foundry industry, *J. Ind. Ecol.*, 2024, **28**, 1183–1197.
37. International Energy Agency. https://www.iea.org/energy-system/buildings/data-centres-and-data-transmission-networks (Accessed 23 March 2025).
38. *Electricity 2024: Analysis and forecast to 2026*, International Energy Agency. https://iea.blob.core.windows.net/assets/18f3ed24-4b26-4c83-a3d2-8a1be51c8cc8/Electricity2024-Analysisandforecastto2026.pdf (accessed 23 March 2025).
39. H. Daly, University College Cork, Energy Policy & Modelling Group, Data Centres in the Context of Ireland's Carbon Budgets. https://www.friendsoftheearth.ie/assets/files/pdf/data_centrres_and_the_carbon_budgets_-_prof_hannah_daly_dec_2024.pdf (accessed 23 March 2025).
40. Microsoft Sustainability News, 2024. https://www.microsoft.com/en-us/microsoft-cloud/blog/2024/09/20/accelerating-the-addition-of-carbon-free-energy-an-update-on-progress/ (accessed 23 March 2025).
41. *Will AI's huge energy demands spur a nuclear renaissance?* Nature News, 2024. https://www.nature.com/articles/d41586-024-03490-3?utm_source=Live+Audience&utm_campaign=4941c770ea-nature-briefing-daily-20241029&utm_medium=email&utm_term=0_b27a691814-4941c770ea-50122980 (accessed 23 March 2025).
42. G. Tamburrini, The AI Carbon Footprint and Responsibilities of AI Scientists, *Philosophies*, 2022, 7, 4.
43. S.3732 – Artificial Intelligence Environmental Impacts Act of 2024, United States Congress. https://www.congress.gov/bill/118th-congress/senate-bill/3732/text (accessed 23 March 2025).
44. E. Strubell, A. Ganesh and A. McCallum, Energy and Policy Considerations for Deep Learning in NLP, https://arxiv.org/abs/1906.02243.
45. S. Luccioni, B. Trevelin and M. Mitchell, The Environmental Impacts of AI – Policy Primer, 2024. https://github.com/sashaluccioni/sashaluccioni.github.io/blob/main/assets/AI%20%2B%20Environment%20Primer%20(Hugging%20Face).pdf (accessed 23 March 2025).
46. P. Li, J. Yang, M. A. Islam and S. Ren, Making AI Less "Thirsty": Uncovering and Addressing the Secret Water Footprint of AI Models, https://arxiv.org/abs/2304.03271.
47. Electron Microscopy Databank EMDB < Home (accessed 23 March 2025).

48. Electron Microscopy Public Image Archive (EMPIAR) EMPIAR - Electron Microscopy Public Image Archive (accessed 23 March 2025).
49. *Event Horizon Telescope, News Release*. https://eventhorizontelescope.org/blog/astronomers-unveil-strong-magnetic-fields-spiraling-edge-milky-way%E2%80%99s-central-black-hole (accessed 23 March 2025).
50. *Event Horizon Telescope, FAQs*. https://eventhorizontelescope.org/faq/how-much-data-recorded-during-observation-and-how-it-transferred-central-processing (accessed 23 March 2025).
51. S. Agostinelli, *et al.*, GEANT4 - a simulation toolkit, *Nucl. Instrum. Methods Phys. Res., Sect. A*, 2003, **506**(3), 250.
52. L. Lannelongue, J. Grealey and M. Inouye, Green Algorithms: Quantifying the Carbon Footprint of Computation, *Adv. Sci.*, 2021, **8**(12), 2100707.
53. *E-waste in the EU: facts and figures*, European Parliament, 2024 update. https://www.europarl.europa.eu/topics/en/article/20201208STO93325/e-waste-in-the-eu-facts-and-figures-infographic (accessed March 2025).
54. *Personal computing devices shipments worldwide from 2024 to 2028*, Statistica. https://www.statista.com/statistics/272595/global-shipments-forecast-for-tablets-laptops-and-desktop-pcs/ (accessed march 2025).
55. *Electronic Waste, Key Facts*, World Health Organisation, 2024. https://www.who.int/news-room/fact-sheets/detail/electronic-waste-(e-waste)#:~:text=Key%20facts,collected%20and%20recycled%20(2).
56. *Sustainable laboratories: A community-wide movement towards more sustainable laboratories*, Royal Society of Chemistry report, 2023. Sustainable laboratories - practices to reduce the environmental effects of research (accessed March 2025).
57. P. G. Jessop and A. R. MacDonald, The need for hotspot-driven research, *Green Chem.*, 2023, **25**, 9457–9462.
58. *Green Algorithms: Towards environmentally sustainable computational science*. https://www.green-algorithms.org/ (accessed 23 March 2025).
59. L. Lannelongue, J. Grealey, A. Bateman and M. Inouye, Ten simple rules to make your computing more environmentally sustainable, *PLOS Comput. Biol.*, 2021, DOI: 10.1371/journal.pcbi.1009324.
60. L. Lannelongue, H. E. G. Aronson and A. Bateman, *et al.*, GREENER principles for environmentally sustainable computational science, *Nat Comput Sci*, 2023, **3**, 514–521.
61. J. Mariette, *et al.*, An open-source tool to assess the carbon footprint of research, *Environ. Res.: Infrastruct. Sustainability*, 2022, **2**, 035008.
62. Code Carbon. https://codecarbon.io/ (accessed March 2025).
63. T. Freese, R. Kat, S. D. Lanooij, T. C. Böllersen, C. M. De Roo, N. Elzinga, *et al.*, A guidebook for sustainability in laboratories, *ChemRxiv*, 2024, DOI: 10.26434/chemrxiv-2023-g3lmq-v4.
64. S. Portegies Zwart, The ecological impact of high-performance computing in astrophysics, *Nat. Astron.*, 2020, **4**, 819–822.
65. J. Grealey, L. Lannelongue and W.-Y. Saw, *et al.*, The Carbon Footprint of Bioinformatics, *Mol. Biol. Evol.*, 2022, **39**(3), msac034.

66. N. E. Souter, N. Bhagwat and C. Racey, *et al.*, Measuring and reducing the carbon footprint of fMRI preprocessing in fMRIPrep, *Hum. Brain Mapp.*, 2024, **45**(12), e70003.
67. Green DiSC: a Digital Sustainability Certification. https://www.software.ac.uk/GreenDiSC (accessed 23 March 2025).
68. Green DiSC certification criteria. https://www.software.ac.uk/GDC090724 (accessed 23 March 2025).
69. LIGO Data management Plan Data_Management_Plan-v29.pdf (accessed 23 March 2025).
70. *An example is the Data sharing guidance and policy for preparing your journal article*, Royal Society of Chemistry. https://www.rsc.org/journals-books-databases/author-and-reviewer-hub/authors-information/prepare-and-format/data-sharing/#choosingarepository (accessed 23 March 2025).
71. *Physical Sciences Data Infrastructure*, UK Research & Innovation. https://www.psdi.ac.uk/ (accessed 23 March 2025).
72. University of Bristol, News and features, *Pioneering tool to manage media industry's digital carbon footprint.* https://www.bristol.ac.uk/news/2020/january/dimpact.html (accessed 23 March 2025).
73. DIMPACT tool. https://dimpact.org/tool (accessed 23 March 2025).
74. Jisc UK Equipment Data service. https://equipment.data.ac.uk/about (accessed 23 March 2025).
75. UK Science and Engineering South Equipment Sharing. https://www.ses.ac.uk/get-involved/equipment-sharing/ (accessed 23 March 2025).
76. UK UniGreen Scheme: Asset resale service for universities. https://www.unigreenscheme.co.uk/ (accessed 23 March 2025).
77. UK Warp It reuse network. https://www.warp-it.co.uk/ (accessed 23 March 2025).
78. Equipment Donations, Disposal and Recycling, University of Cambridge. https://www.equipment-sharing.cam.ac.uk/equipment-donations-disposal-and-recycling (accessed 23 March 2025).
79. *Energy Management Policy*, Wellcome Genome Campus. https://www.wellcomegenomecampus.org/aboutus/sustainability/energy-saving-schemes/ (accessed 23 March 2025).
80. *How cutting-edge computer chips are speeding up the AI revolution*, Nature News, 2024. https://www.nature.com/articles/d41586-024-01544-0 (accessed 23 March 2025).
81. MIT Technology Review, 2025. DeepSeek might not be such good news for energy after all | MIT Technology Review (accessed 23 March 2025).
82. C. Freitag, M. Berners-Lee, K. Widdicks, B. Knowles, G. S. Blair and A. Friday, The real climate and transformative impact of ICT: A critique of estimates, trends, and regulations, *Patterns*, 2021, **2**(9), 100340.
83. A. Suksuwan, A. Matossian and Y. Zhou, *et al.*, Environmental LCA on three note-taking devices, *Procedia CIRP*, 2020, **90**, 310–315.
84. Hamilton Storage Press Release, 2024, UK Biobank Selects Hamilton BiOS (accessed 23 March 2025).

85. S. Boobier, J. C. Davies and I. N. Derbenev, *et al.*, AI4Green: An Open-Source ELN for Green and Sustainable Chemistry, *J. Chem. Inf. Model.*, 2022, **63**(10), 2895.
86. The Experiment Design Assistant – EDA, UK National Centre for the Replacement, Refinement & Reduction of Animals in Research. https://nc3rs.org.uk/our-portfolio/experimental-design-assistant-eda (accessed 23 March 2025).
87. ChemDataExtractor v2. http://www.chemdataextractor2.org/ (accessed 23 March 2025).
88. K. Widdicks, *et al.*, Systems thinking and efficiency under emissions constraints: Addressing rebound effects in digital innovation and policy, *Patterns*, 2023, **4**(2), 10067.
89. J. Mbatchou, L. Barnard and J. Backman, *et al.*, Computationally efficient whole-genome regression for quantitative and binary traits, *Nat Genet*, 2021, **53**, 1097–1103.
90. G. M. Gutiérrez-Finol, A. Ullah, M. Gonzáles-Béjar and A. Gaita-Arino, A call for frugal modelling: two case studies involving molecular spin dynamics, *Green Chem.*, 2025, **27**, 3167–3177.

CHAPTER 9

Certifications and Standards†

M. RODRÍGUEZ-MARTÍNEZ AND B. A. ROUSE*

European Molecular Biology Laboratory (EMBL), Meyerhofstraße 1, 69117 Heidelberg, Germany
*Email: brendan.rouse@embl.de

9.1 Introduction – Redefining Scientific Excellence

In the last few years, the need to carry out research in an environmentally sustainable way has emerged as a focus area for the research community. The need to achieve a research model compatible with the preservation of our environment should be included in the definition of scientific excellence. Environmental sustainability in research is arising as a topic that not only covers the way in which research is conducted, but that implies the redefinition of the whole scientific system.

We are currently observing a paradigm shift in the research model, with open access science and environmental sustainability being two of the significant changes that are happening. Although both concepts go hand in hand, establishing parameters for the accessibility of published research is achieved through a wide variety of initiatives for *e.g.* FAIR principles – scientific sustainability needs its own set of parameters. It is in this context that it could be argued that the need for sustainability standards and certifications emerges as a mechanism to drive environmental sustainability into scientific operations.

The road to establishing and implementing standards for science sustainability is far from straight but it is fortunately paved by examples we can draw from other industries and other areas within scientific research. Several pre-established factors and limitations need to be taken into account,

The Sustainability of Science: How Science Impacts the Environment, and What Can Be Done
Edited by Martin Farley

Published by the Royal Society of Chemistry, www.rsc.org

such as scientific quality, existing supply chains and health and safety regulations. It is therefore essential to have a clear view of the 7 "*W Questions*" concerning the research being evaluated. What is being researched, which considerations need to be made, who is responsible for the research conducted, and where, when, why and how is it conducted. Several parties should be involved in considering these questions and all should be involved in the development of sustainability standards and relevant certifications. These include, but are not limited to, scientists, research organisations, technical staff, administrators, funding agencies, policy makers and suppliers. In this chapter we will delve into the current state of sustainability standards and certification, the involvement of the aforementioned players, and the potential avenues to achieve a sustainable model for scientific research.

9.2 Learning from Open Access in Academic Science Research

Open access science in academic research has been a topic of extensive discussion in the past decade. As a clearly interconnected subject, we can use it as inspiration to find ways to bring environmental sustainability to the forefront of relevant subjects in the current research model. Open access science is in itself a pillar of research sustainability, as it promotes collaboration and accessibility to scientific knowledge, reducing the redundancy of ongoing science and therefore the resources consumed.

Worldwide, several funding agencies and governments are endorsing open access science by making funding dependent on the compliance with their open access policies. In 2022, the White House launched a policy to have all its federally funded research open access by 2025. The EU's initiatives like the European Open Science Cloud (EOSC) and the Open Science Monitor are clear concerted efforts to embrace open access science. We can apply several principles that are used to implement open science worldwide. Transparency, being one of the key tenets of Open Science, can be followed almost to the letter in research sustainability by prioritising, for example, transparent reporting of research project carbon emissions, as well as methodologies utilised. Moreover, encouraging collaborative approaches, including data sharing to reduce redundancy of scientific projects is key to research sustainability. Open Science encourages inclusive and participatory decision-making processes. Drawing from this, environmental sustainability initiatives can incorporate community engagement, including scientists, institutions, funding bodies, and other stakeholders. Furthermore, global engagement and standardisation of open science requirements, which are deeply embedded in Open Science initiatives are key to environmental sustainability. Cross-border collaboration, adherence to global standards, and the sharing of best practices are crucial for addressing planetary challenges. When we evaluate Open Science, we can see the impact of a global standard as the foundation of a universal, consistent and relevant model for research. It is therefore clear that sustainable science should strive for its own standard, ensuring consistency across the field.

Nevertheless, whereas for Open Access publishing there is some clear criterion to comply with, to *e.g.*, publish in a free and accessible form to all, there is no clear set of criteria research needs to fulfil to be considered environmentally sustainable. It is here where the sustainable laboratory programmes or frameworks are playing an important role in helping to define, from a bottom-up perspective, the aspects that should be considered when assessing research sustainability.

9.3 Bottom-up Certifications for Research Laboratories

In the past few years, there has been an increase in the number of "green" *grassroots* initiatives[1] aiming to drive sustainability in research laboratories – particularly in life sciences – following a bottom-up approach. These initiatives have gained momentum and do not only include laboratory members trying to push sustainability within their laboratory setting, but have grown to institutional, regional and even Pan-European networks.[2] The creation of sustainable lab certifications and programmes derives from the need of these bottom-up initiatives to have a framework where individuals can find specific actions to follow to make their research more environmentally friendly, as well as a mechanism of recognition for having taken action. Because of the need to define the steps required to become a more sustainable laboratory, many research facilities and universities have established their own green certifications (*e.g.* the National Institute of Health, the University of Washington, and The University of Edinburgh). Alternatively, some programmes have been adapted to be relevant to laboratories and other settings (*e.g.* Green Impact). However, in the past few years, two sustainable lab certifications have gained relevance in the international landscape, the Laboratory Efficiency Assessment Framework (LEAF), from University College London (UCL)[3,4] and My Green Lab certification[5] from the non-profit organisation My Green Lab (MGL). In 2024, in recognition of the increasing importance of computing-based research, a new certification has been initiated called Green DiSC, which focusses on computational science ("dry laboratories") and digital impacts. Furthermore, in 2025 European funders and other stakeholders began a project to develop open-access standards and certification for a broad range of research activities through a UKRI initiated effort titled SPARKHub (short for Sustainable Practice and Research Knowledge).

Both wet-lab certifications (MGL & LEAF) cover similar areas; laboratory operations, laboratory management, communications and collaboration. Within laboratory operations, the certifications look at energy consumption, carbon footprint, water usage, and waste management amongst other areas. Within laboratory management, both certifications underlie the importance of consumable and sample management, as well as experimental planning.

Recognised as one of the main emitters of scientific research, emissions from business travel are also included in both programmes. These, among other important areas, are widely explored in both certifications in different but comparable ways.

In the next sections we will describe these certifications, the differences between them (summarised in Table 9.1) and give an overview of the relevance and impact of these programmes in the current landscape of laboratory sustainability. We have included Green-DISC also for comparison of its mechanism, though note it focusses on a different research activity (computing).

9.3.1 My Green Lab Certification

Founded in 2013 by neuroscientist Allison Paradise, MGL is a non-profit organisation dedicated to promoting sustainability in laboratories. Over the years, it has evolved into a recognised authority on laboratory sustainability, offering certification programmes tailored to the unique challenges and requirements of laboratory settings. MGL certification is a programme aimed at promoting sustainability within laboratory operations. It has been adopted by over 2000 laboratories in different sectors and is recognised by the United Nations Race to Zero campaign as a measure towards a zero-carbon future. It provides individual laboratories with a framework for assessing and improving their environmental performance, promoting collaboration and building a strong sustainability culture. In 2024, the price to access the programme ranged from 500 USD for individual academic labs and 4000 USD for individual industry labs, with this reducing as more labs join the programme.

The certification process entails a thorough assessment of laboratory practices in relation to sustainability standards through requiring 50% of lab members to complete an extensive survey. It comprises multiple stages, each aimed at evaluating and enhancing environmental performance.

1. Application and assessment: laboratories interested in certification begin by submitting an application to MGL. MGL conducts an evaluation of the laboratory's practices across various sustainability criteria. These criteria encompass a wide range of topics, including energy use, water consumption, waste management, chemical handling, procurement practices, and more. The assessment process involves data collection, analysis, and evaluation to determine a baseline of the laboratory's current environmental impact. Laboratories can select which criteria are relevant to them or they want to address.
2. Feedback, action plan and implementation: following the assessment, laboratories receive feedback reports from MGL. These reports provide insights into the laboratory's performance, highlighting areas of strength and prioritising opportunities for improvement. Based on the feedback received, laboratories develop an action plan tailored to their

Table 9.1 Comparative table of LEAF, GreenDiSC, and MGL certifications.

Programme	Fees	Management	Criteria selection	Methodology	Who can run it	Institutional engagement	Auditing	Open access
MGL	Yes	MGL	User selected	Survey	50% of the lab minimum	Not required	MGL Subsidiary	No
LEAF	Yes	Internal	Preset	Criteria responses	No minimum user engagement stipulated	Required for registration	Internal (management or local peer-to-peer)	No
Green DiSC	None	Systemic	Preset	Criteria responses	No minimum user engagement stipulated	Certification is possible for institutions, research groups or both, with stand-alone criteria for institutions and research groups	External peer-to-peer	Yes

specific needs and goals. Laboratories put their action plan into action, implementing the identified sustainability measures. This may involve making operational changes, adopting new practices and procedures, investing in equipment upgrades, and engaging staff in sustainability initiatives.

3. Verification and certification: once the sustainability measures have been implemented, MGL verifies that the laboratory has successfully aligned with certification standards. Auditing takes place remotely and verification is done *via* virtual meeting and document submission assessment which is reviewed by MGL's subsidiary "Impact Laboratories". This verification process promotes that the laboratory has made tangible improvements in its environmental performance and is eligible for certification. Upon successful completion of the verification process, MGL offers different award levels based on the extent of sustainability achievements, ranging from Bronze, Silver, Gold, Platinum to Green, which is their highest level. An independent study of this process highlighted some perceived risk of greenwashing, partly due to a "low threshold to overstate green behaviours, particularly in a multiple-choice questionnaire".[22]

Overall, the MGL certification process empowers laboratories to improve their environmental performance, reduce resource consumption, and minimise their environmental impact. Importantly, MGL certification has reached the pharmaceutical and biotech companies which are significant carbon emitters in life sciences.

As part of writing this chapter users were asked for their perspective. MGL certification is in general found to be a straightforward and user-friendly tool and helps build a strong sustainability community in research laboratories. While some labs find the programme easy to use and appreciate its feedback mechanisms, others mentioned that more targeted assessment questions could improve the programme. As with any certification scheme there is recognition of the need for robust auditing processes to ensure transparency and accountability. MGL certification is a valuable tool for laboratories seeking to demonstrate their commitment to sustainability and contribute to global environmental efforts. Highlighted as one of the strongest values of the programme is the requirement of creating a sustainability task force to run the programme with 50% engagement of the laboratory. This implies the involvement of several laboratory members in the running of the programme and automatically renders the actions widely accepted and implemented, creating at the same time a common sustainable mindset in the team.

9.3.2 Laboratory Efficiency Assessment Framework (LEAF)

Established by Martin Farley and Joanna Marshall-Cook at University College London (UCL), The Laboratory Efficiency Assessment Framework

(LEAF) is a programme designed to enhance the efficiency and sustainability in laboratory operations globally. LEAF was piloted in the UK and Ireland between 2018 and 2020 with a reported 235 lab groups across 23 institutes taking part and assessing cost savings of £641 K and 648 tCO_2e of avoided emissions.[3] At the time of writing, there are over 100 institutions signed up to LEAF, including the European Molecular Biology Laboratory (EMBL) and many from the Green Labs Netherland network.[4]

Participation in the LEAF programme not only reduces carbon emissions but also fosters an environment conducive to high-quality research by specifically aligning some criteria on research quality with sustainability. Laboratories are recognised with bronze, silver, or gold levels based on the number of sustainability actions which are included in the LEAF criteria the labs have undertaken.

The LEAF programme comprises of the framework of criteria, and includes equipment and waste calculators to measure financial and carbon impact and track improvements. These are made accessible through the online platform which is managed by UCL. The framework contains links to encompassing guides regarding sustainable lab equipment, consumables, and procedures. The programme has an institutional fee depending on the organisation's size.

The certification process involves the evaluation of laboratory practices against specific sustainability criteria and consists of several steps, which start with engagement and then lead to individual laboratory action. Notably, only non-commercial (or publicly funded) institutions can participate.

1. Application and registration: institutions interested in certification begin by engaging UCL, where they are required to pay the registration fee, and agree to UCL's contract of usage, as the programme is operated through the university itself. Institutions are supported to define the hierarchical structure of the organisation, which allows individual laboratories to then be mapped within the institution.
2. Distribution and training: the framework and criteria are made available to all labs, should they choose to register, within the participating institution, and resources are made available to all members to explain what the participation entitles and how the framework and all tools work. The platform contains guides and easy-to-follow steps for the users.
3. Implementing measures and completing LEAF criteria: participating labs are able to implement measures to comply with the LEAF criteria in their own time. The criteria are fixed with an award level, and so to accomplish Bronze, all relevant criteria must be addressed. They are expected to describe how they comply with each criteria using an online platform. They can optionally use inbuilt calculators to estimate the impact of their measures in terms of carbon and cost saving. They can also optionally describe other measures they are taking which are not part of the LEAF criteria. Once all the criteria for a specific level

(bronze, silver or gold) are completed by the lab they can submit their responses for audit.

4. Audit, feedback and certification: this step relies on the administrators of the framework within the participating institution. The organisation can choose to carry out laboratory audits using a central resource, such as a sustainability officer, or they can implement a local peer-to-peer system whereby participating labs audit each other, providing feedback and confirming compliance or rejecting application. The chosen auditors review the responses provided and provide initial feedback, including recommended changes during the audit. The auditors (peers or administrators) can then visit the lab in-person or conduct the audit remotely and either grant the award or go through a second round of feedback until all criteria are fulfilled.

Users told us that they find LEAF particularly useful as a behavioural change campaign and consider the institutional engagement essential to push sustainability measures forward. Moreover, taking part in the programme has in many cases resulted in the establishment of other sustainable initiatives that are not included in the criteria but scientists have developed on their own. Although the process and content is straight-forward and requires minimal administrative tasks, users fed back that the platform could be improved to improve the overall experience. Here, the lack of external auditing arises as a concerning point that enables the use of this tool as a greenwashing strategy by the institutions using it,[22] in part due to the local peer-to-peer auditing system. This is particularly relevant as some UK funders have started to request “green lab certification or equivalent”, and so the assurance of positive impact. However, despite the concerns, LEAF is the most widely used scheme across Europe for academic organisations, particularly in the UK, and overall has received positive feedback from a variety of institutions.

9.3.3 EMBL Case Study – Use of Certification Programmes for Research Organisations

In 2022 the European Molecular Biology Laboratory (EMBL) piloted LEAF as a behaviour change tool and proceeded with a full rollout of the framework across five sites in 2023. Throughout the year, all 81 of EMBL’s wet research and service labs signed up to the framework with over 70% achieving a certification.

A successful behaviour change campaign needs to recognise what motivates the target audience and what barriers there are stopping individuals from behaving in the desired way. Like many life science organisations there are many scientists at EMBL who are environmentally conscious and want to live and work in a sustainable way. However, there is sometimes a disconnect between the desire to be more sustainable and the actual behaviour of scientists who continue to work in a resource intensive way. The Sustainability Office identified a number of barriers that were stopping the scientists at EMBL from carrying out their research in a more sustainable manner. These were primarily an unwillingness to change practices which had been established as the new normal, a lack of clear guidance and a lack of some key facilities which enable more sustainable practices. LEAF proved to be a successful tool in breaking down these barriers.

One of the key criteria for a successful behaviour change campaign is that the desired behaviour is broken down into manageable steps providing a clear roadmap for individuals to follow. The LEAF criteria of ∼47 actions provide this roadmap and is easy for individuals and research groups to follow, as they set out to achieve bronze, silver and then gold certification. Adding to the general resources provided by the LEAF platform, the Sustainability Office produced a Green Labs guide for EMBL's laboratories which provided EMBL specific guidance, ensuring the users were better informed on how to quickly implement actions. Producing such local guidance would be a recommended step for any institution seeking to streamline processes. Another key aspect to successful behaviour change is to provide solutions and support to help the individuals overcome the challenges they face. At EMBL, one key obstacle was the lack of a suitable recycling stream for packaging waste being available to all labs. In response, a new recyclable laboratory waste stream was set up and made available to labs which were not previously able to recycle their packaging waste. Furthermore, other teams across EMBL also provided new services to the LEAF labs, including the EMBL lab kitchens providing additional support in cleaning and sterilising glassware as well as refilling tip boxes and EMBL's scientific instrumentation maintenance team which provided energy audits to those labs who had signed up to LEAF. This also showed the importance of having the institute administration involved in the implementation of certification programmes as well as a solid support infrastructure of scientific support facilities, which might not be available to all laboratories.

A third key criteria for behaviour change campaigns is using positive reinforcement to highlight and celebrate successes. The action of handing out LEAF certificates as well as the recognition of EMBL's director general for labs who achieved gold certification provided a positive reinforcement for green lab behaviours across the organisation. This creates a positive reinforcement loop that encourages continued behaviour change.

Finally, the LEAF auditing process used at EMBL facilitated 81 conversations about green lab practices taking place between research groups and the Sustainability Office in 2023. These conversations took place as each submitting laboratory participated in a local audit by the Sustainability Office, providing an opportunity for research and sustainability staff to engage directly. Each one of these conversations was an opportunity for the Sustainability Office to share best practice, to praise the labs taking place and to reinforce the need for sustainable laboratory behaviours. This further normalises the desired behaviour.

There are tangible outcomes from the use of LEAF as a behaviour change tool at EMBL, where we can see normalised behaviour has been changed and very quickly a 'new normal' is created. This is probably most evident in the use of ULT freezers, where over 70% are now set to −70 °C rather than the previous normal of −80 °C. The success of the LEAF rollout has been a significant contributor to EMBL reducing its energy consumption, reducing its total carbon footprint (Scope 1, 2 & 3) and seeing an increase in its recycling rate.

9.3.4 General Considerations

Upon analysis of the current landscape of laboratory certifications and standards several considerations seem of the highest relevance. In the following sections we delve into the following questions that will be relevant in the future model for research sustainability. The accessibility of sustainability criteria to allow for the open access of sustainable science, the need for standardised auditing in the certification process to avoid the use of these certifications as a greenwashing strategy for research organisations, the importance of conducting social and empirical science on laboratory sustainability to feed into the certification criteria to follow and the need to cover all scientific disciplines and stakeholders in the certification process.

9.3.5 Open Access in Lab Sustainability

As previously mentioned, the parallel with Open Access Science may help in understanding the concept of environmental sustainability in laboratories and the path to follow for its implementation. In particular, one of the aspects that is missing from all existing certification is precisely the public availability of the criteria. Restraining laboratory sustainability only to those who can either afford it or have the support (from administration as well as from a sustainability office) of their institution is limiting the main drive that has brought laboratory sustainability this far: the bottom-up approach. Until standards are established and promoted by higher regulatory bodies, lab sustainability relies on social responsibility and reputational risk for the institutions, and is often driven by the volunteer care of research staff themselves. Opening the access to the criteria and resources required to achieve a sustainable laboratory will improve transparency as well as increase access to good-practice, as while certification is a mechanism to recognise and standardise, some may have such programmes currently. Beyond open access of the criteria themselves, there should be a transparent governance process in place to ensure the creation and development of these criteria reflect best-practice. This is crucial to ensure that good-practice evolves in time as new innovation is developed. Furthermore, appropriate governance can provide stewardship to ensure potential greenwashing is filtered out, as some resources/content may be commercially driven as opposed to environmentally.

9.3.6 Research on Laboratory Sustainability – an Insider's Perspective

Research on sustainable laboratory practices is essential for identifying effective strategies to minimise environmental impact while maintaining research quality. By investigating energy-efficient technologies and practices, waste reduction methods, green chemistry principles and other laboratory practices, laboratories can enhance sustainability without compromising scientific integrity. Ultimately, investing in research on sustainable laboratory practices not only benefits the environment but also promotes cost savings, current and future regulatory compliance, and a culture of environmental stewardship within the scientific community. Having this research accessible is essential to establish standardised criteria for sustainable laboratory certifications that would be backed up by scientific data and will therefore be easier to apply globally.

Fortunately, this matter is gaining importance and recognition within the scientific community, and research on the topic has been published by renowned scientific journals in the past few years. This topic is broadly discussed in other chapters of this book and will not be examined here. However, of relevance for the understanding of the acceptance of sustainability practices in the scientific community is the fact that PhD studies are being conducted on laboratory practices to identify the best criteria and standards to be implemented. The scientist Bianca Schell is conducting her doctorate research on sustainability practices in laboratories under the supervision of Professor Nico Bruns at the Technical University of Darmstadt and the University of Konstanz. This research is funded by the Evonik Stiftung, a foundation that funds, amongst other activities, PhD fellowships in the chemical sciences. The foundation was established by one of the world's leading specialty chemistry companies, indicating the relevance of the study in the current landscape. Moreover, a specific part of the doctorate research includes a project to compare and analyse the current and future directions of lab sustainability certifications which was recently published.[22] This project is funded by the German Federal Environmental Foundation (DBU) under the acronym NachLabs.

Reflecting on the research she has conducted about sustainable laboratory certifications, Bianca Schell recognises that, fortunately, scientists are increasingly understanding the substantial carbon footprint associated with scientific research. However, addressing sustainability concerns in laboratory practices necessitates more than individual efforts; it requires a collective approach guided by established guidelines and oversight mechanisms, akin to those addressing health and safety or diversity. Achieving climate objectives demands a transformative shift within the scientific community, facilitated by specialised programmes such as MGL certification or LEAF. These initiatives offer essential expertise and frameworks to guide laboratories towards more sustainable practices, contributing to broader environmental conservation efforts. According to Ms. Schell, understanding the strengths and weaknesses of programmes like MGL certification or LEAF is key to improving their effectiveness and gaining wider acceptance among researchers. Moreover, Ms. Schell points out that since scientists rely on data, it is crucial to provide scientific evidence on the impact of conducting research in a sustainable way. She also underscores the importance of everyone's – including funding bodies and policymakers – understanding and recognition of the importance of sustainable practices in scientific research.

Ms. Schell emphasises the importance of ensuring universal accessibility to knowledge on sustainable laboratory practices, considering that not all laboratories have the financial means or willingness to invest in sustainability programmes. In an era characterised by open access, withholding valuable information behind paywalls is unnecessary. Much of this knowledge already exists within the public domain, making it inefficient for individual researchers to independently seek out the most sustainable options. There is a pressing need for an open-access repository of best practices, sustainable alternatives, and practical guides. For instance, optimising ventilation systems—an area outside the core expertise of many wet lab scientists—requires readily available guidance. Efforts are underway to develop tools that assist laboratories in prioritising actions, striking a balance between easily implementable initiatives and more time-intensive yet impactful measures. However, Ms. Schell recognises that a simple catalogue cannot entirely replace the certification process. Engaging individuals, motivating them, and providing external recognition and acknowledgment for their efforts, such as through certification are crucial aspects.

In closing, Ms. Schell highlights the necessity for funding agencies to play a pivotal role in advancing sustainability within laboratories. By mandating minimum standards and focusing on impactful measures like optimising ventilation,[6] sustainable purchase[7] and promoting recycling, funding agencies can drive significant progress towards a more sustainable future. Certification programmes offer a practical avenue for labs to meet these standards, but external validation is crucial to ensure accountability and effectiveness. Ultimately, by integrating sustainability requirements into funding criteria, agencies can catalyse widespread adoption of sustainable practices, fostering a culture of environmental responsibility within the scientific community.

9.3.7 The Importance of Including All Research Based Scientific Disciplines

Laboratories across various scientific fields contribute significantly to resource consumption, waste generation, and environmental impact. Therefore, adopting sustainable practices in all disciplines is essential for mitigating these effects and promoting environmentally responsible research within the scientific community. By encompassing diverse scientific disciplines in certifications, such as chemistry, biology, computational science, and others, certification programmes would ensure that sustainability standards are universally applicable and relevant. Laboratories often collaborate on research projects spanning multiple disciplines, and it is therefore imperative that all research areas in one project are covered and adhere to sustainability standards. This cross-disciplinary collaboration strengthens the effectiveness of sustainability initiatives and maximises their impact. Currently, sustainable laboratory certifications apply only to wet-lab science (*i.e.* bench-based research) carried out in a typical laboratory. However, there are other research settings that are equally resource intensive and contribute greatly to the environmental impact of research. Animal experimentation for example, requires animal housing, which functions in a different way to typical laboratories but can require even greater amounts of energy per m^2 as well as significant resources. Likewise, computational science is gaining importance in the current research landscape and is often the biggest contributor to the carbon footprint of a research project. Fortunately, in the past years, efforts have been made to develop sustainability criteria for these scientific settings. Moreover, in 2021 and following the development of a sustainability programme within the European Molecular Biology Laboratory (EMBL) and partnership with the Sustainable European Laboratories Network (SELs), sustainable computational science has been gaining attention. In 2023, Dr. Loïc Lannelongue, together with scientists from EMBL and other research institutions published a collection of what should be the principles of sustainable computational science.[8] Since then, a set of criteria and standards has been piloted in several laboratories within EMBL and other scientific institutions and has been launched as a public resource called Green DiSC in 2024. Importantly, the model used by the Green DiSC Certification is open-access, free, and functions at multiple levels (research group, as well as institutional).

More recently, UKRI has funded the development of the SPARKHub, which aims to provide a single open-access resource for certification of all research activities. Launched recently, it will be governed by research funders (and not owned by UKRI), with content developed and curated by the academic community. This is to ensure there is quality control on content, and allowing it to evolve as new solutions are developed. The model proposed is peer-to-peer, but with a percentage that is randomly validated as a check. The development of this platform highlights that addressing standards in sustainable research are still changing, and we look forward to where the topic will go.

In conclusion, the inclusion of all scientific disciplines in sustainable laboratory certifications is essential for promoting environmental responsibility and ensuring research projects are following sustainable practices in all fields involved. By embracing a holistic approach to sustainability, certification programmes can empower laboratories across diverse fields to contribute to a more sustainable future for all.

9.4 Top-down Engagement – Players Needed for the Sustainable Transformation

Until now, scientists, students, and technical staff have primarily spearheaded grassroots efforts to foster sustainability in scientific practices. However, broader engagement from stakeholders such as policymakers, funding bodies, research institutions, and suppliers is imperative, and has begun. These entities are increasingly recognising the urgency of sustainability in science and the need to actively contribute to its advancement. Policymakers can enact regulations to incentivise sustainable practices, while funding bodies can allocate resources to support research on sustainable practices and an incentivise action by integrating sustainability standards into their funding requirements. Research institutions play a vital role in integrating sustainability into their policies and operations as well as supporting their research staff and scientists in their sustainability journey. Additionally, suppliers must bring eco-friendly products and services to the market at competitive prices, avoiding the 'eco premium' often applied to sustainable products. Collaborative efforts across all sectors are essential for driving meaningful change toward a more sustainable scientific landscape (see Figure 9.1).

9.4.1 Scientific Institutions

Scientific institutions, such as universities, research industry and research institutes play a major role in science sustainability. On the one hand, they are responsible for the research carried out within their organisation and the

Figure 9.1 Important players in laboratory sustainability. Scientists, research performing organisations, governments, funders and suppliers all play an important role in advancing sustainability in scientific research. They all must work together and share the responsibility of making research more sustainable. Credits: Creative Team/EMBL.

way this is conducted, and on the other hand, they have the responsibility to provide the required infrastructure to facilitate and support sustainability efforts of the scientists and to be aligned with the current international landscape. Moreover, they must lead by example, striving to operate in the most sustainable way possible.

There are many benefits to operating as a sustainable organisation, as well as the moral obligation to respond to the climate crisis, environmental degradation and to ensure organisations operate in a way that does not cause social damage. One of the positive impacts that can come from sustainability is the reduction of waste and therefore a reduction in operating costs. This can come from operating in a more energy efficient way, reducing energy waste, a reduction in water costs or a reduction in waste management costs. These three areas of expense are covered in the aforementioned certifications, LEAF and MGL, as well as new ones being developed.

A less direct impact, though no less important, is the improved reputation of the organisation, especially with the current and future workforce. Studies have shown that employees have a positive perception towards organisations[9] that have developed sustainability actions, employees have a preference for eco-friendly organisations[10] and sustainable organisations are highly regarded by potential employees.[11] However, it should be noted that this effect is not always significant if the organisation lacks legitimacy. It is therefore important for organisations to find ways to achieve internal and external legitimacy of their sustainability actions (summarised in ref. 12). We argue that including sustainability certification and standards is one very effective way of achieving this, as they provide a common means for recognition externally.

Current certifications and standards that have been discussed in this chapter are targeted at the laboratory level and do not apply themselves well at an institutional level. However, there are many other certification schemes and standards which are available to research organisations to legitimise their sustainability credentials.

The International Organization for Standardization has a number of standards which are relevant to sustainability,[13] including very common ones on Environmental Management Systems (ISO 14001) and Energy Management Systems (ISO 50001). International Standards on sustainability offer a clear and practical way to achieve operational improvements, comply with legal requirements, and meet stakeholder expectations in a flexible way which can accommodate a variety of organisation types and sizes.

For organisations who have decided to be transparent about their ecological and social footprint by publishing reports, there are many standards that ensure the data being reported is accurate and comparable with others. For carbon footprint reporting, a commonly used standard is the Greenhouse Gas Protocol (GHG) which sets out methods and expectations for reporting on carbon emissions.[14] For broader sustainability topics, such as diversity and equal opportunities, health and safety and responsible sourcing, the Global Reporting Initiative's standards[15] or the International

Sustainability Standards Board are commonly used.[16] There are also new compulsory reporting requirements for organisations operating in the European Union called the Corporate Sustainability Reporting Directive (CSRD) and in the UK there is the now long running Streamlined Energy and Carbon Reporting (SECR) policy.[17]

Another interesting standard associated with sustainability has been created to ensure that carbon reduction targets are transparent and suitably ambitious. The standard has been created by the Science Based Target initiative (SBTi) who provide a framework to organisations to define an emission reduction target which is in line with the Paris Agreement.[18] The SBTi also validates the targets that companies set with over 4000 companies having worked with the SBTi to date. Unfortunately, SBTi does not assess the targets for public sector, educational or non-profit organisations.

The commitment of research organisations to sustainable development, not only mitigates reputational risk but also fulfils a broader social responsibility as promoters or scientific progress. Embracing sustainability aligns with the ethical need of scientists to minimise their environmental footprint and advance knowledge in a way that respects our planet's finite resources.

9.4.2 Funders and Their Role in Advancing Sustainable Science

Funders play a crucial role in advancing sustainability in scientific research through two main avenues. First, as mentioned before, funding can support research and innovation enabling sustainable practices is necessary to define standards applicable to specific research settings. Second, funders can significantly influence scientific behaviour by mandating or promoting sustainability impacts as a prerequisite for grant applications or core funding for institutions. This requirement would not only encourage research organisations to prioritise sustainability in their operations but also foster a culture of accountability and responsibility towards environmental responsibility shared by the research project, the scientists and the research organisation. In this model, funders serve as key drivers in promoting sustainable practices across the scientific community. Funders face challenges though to achieve this, as their primary focus to enable world-class research must be balanced with the impacts of doing this research. This is not always easy to balance, as can be seen for example with funding the sending of resource-intensive satellites into space which then enable us to better understand climate change. Funders must also consider that the administrative burden beset on the research community is already significant, and so actions must be streamlined but impactful. It is in this sense that certifications may be of particular use, as they allow funders to know that environmental impacts have been considered prior to application of fund stages, as opposed to during. Certifications also provide a standard understanding, helping grant assessment panels who may not be environmental experts. Furthermore, certifications and systems ideally should be aligned and harmonised between funders.

Thankfully, over recent years, we have witnessed a growing involvement of funding bodies in environmental sustainability efforts. More and more funders are now requesting sustainability statements or seeking explicit commitments to sustainable practices from organisations and laboratories and research on sustainable practices is being increasingly supported.

Below is a list with some of the most relevant funders adhering that are including sustainability into their funding calls:

Horizon Europe: the European Union's flagship research and innovation programme, successor to Horizon 2020, emphasises sustainability across its various funding streams.

European Research Council (ERC): the ERC, under the European Commission, supports researchers in various fields and encourages consideration of societal and environmental impacts.

Marie Skłodowska-Curie Actions and Green Charter: encourages applicants to detail the sustainability impacts of their research project as well as the host organisations.

UKRI: some councils within UKRI, *e.g.* MRC, NERC, and ESRC, have included questions requiring applicants for some grants to consider how their work can mitigate environmental impacts. Furthermore, UKRI is supporting the development of SPARKHub.

Swiss National Science Foundation (SNSF): SNSF places importance on societal impact and sustainability in some research proposals, considering broader implications beyond scientific advancements.

Wellcome Trust: while primarily a biomedical research funder, the Wellcome Trust encourages researchers to consider the societal and ethical implications of their work. They now require UK organisations applying for their funding to be signatories to the UK's Concordat on Environmental Sustainability in Research, as well as requiring laboratories to be "green lab certified or equivalent".

United Nations Development Programme (UNDP): while not a traditional research funding body, UNDP supports research initiatives that contribute to sustainable development goals.

Deutsche Forschungsgemeinschaft (DFG): has introduced mandatory considerations on environmental sustainability at the proposal level for applicants to answer, but of which may not be assessed as part of the overall research quality.

National Science Foundation (NSF): in the United States, the NSF encourages researchers to address broader impacts, including considerations some of sustainability, in their grant proposals.

National Institute of Environmental Health Sciences (NIEHS): in the United States, NIEHS funds research with a focus on environmental health, including projects that address sustainability challenges.

Australian Research Council (ARC): ARC promotes responsible research practices, including sustainability considerations, in its funding programmes.

Research Ireland (RI): currently are assessing the outcomes of a pilot sustainable laboratory certification programme, in which they utilised the MGL certification.

On the other side of the funders' responsibility lies the need to fund enabling research about how research can further decarbonise, to help understand the standards and criteria that should be implemented to increase research sustainability without affecting science quality. Fortunately, from 2020 on, there has been an increase in the engagement of research societies and related organisations in funding research about sustainability practices in the lab, through grants or project awards. Among others, The Royal Society of Chemistry, the European Molecular Biology Organisation (EMBO), the NIHR, the MRC, the Evonik foundation and the German Federal Environmental Foundation are recognising and funding ongoing efforts to research environmental responsible research.

Moreover, in September 2023 and with the aim of engaging with funding agencies and promoting the importance of the transition to sustainable research, Green Labs Netherlands hosted a "funders meeting". In this event, representatives of international funding agencies got together to deep dive into the topic "Sparking Green Movement in the Dutch funding landscape".[19] The event was funded by the Knowledge Centre Global Health and well attended by Dutch and international funding organisations and Dutch Ministry representatives. Importantly, in May 2024, EMBO followed up this initiative and hosted a closed policy workshop bringing together a wide range of European funders and representatives from laboratories, grassroot initiatives, institutes and other organisations working on sustainable issues to discuss how funders can promote sustainability in scientific research. As a first result of these efforts the Heidelberg Agreement was published in October 2024.[23] Endorsed and supported by several funding bodies and relevant stakeholders, the agreement establishes the points of agreement and recommendations to move forward in the transition to a more sustainable research model. Among these are the need for establishing sustainability standards and well-regulated certifications. These and future related events are highlighting the importance of working together to achieve a more sustainable model for scientific research.

9.4.3 The Role of Governments as Policy Makers and Funders in Sustainable Science

We think of governments as having a key role to play in the move towards a more sustainable research environment. As science relies so heavily on funding from public bodies, they have the ability to influence the community in a positive way. The majority of countries have signed up to the Paris Agreement and have carbon reduction or Net Zero targets in place. It is therefore logical that they would target their spend on research and development on most sustainable research projects, particularly as scientific research grows in investment.

Governments are also providers of signals to communities on the societal expectation on them. Their support of sustainable research standards and certifications would give them credibility and surely lead to an increase in their uptake. They can do this by requiring auditing of net zero targets of national funders which facilitate research funding or of the institutes themselves which fall within their ownership.

They also have a part to play in ensuring the sustainable science movement is fair and transparent. An example of this is the upcoming legislation proposed by the European Parliament on green marketing by companies which will look to ban greenwashing practices. This legislation will protect consumers of goods from being mis-sold products due to spurious green claims. This will surely be an issue for the many life science suppliers who already market some of the products as green.

9.4.4 Laboratory Suppliers – Key Players in Lab Sustainability

Improving the sustainability of laboratory practices goes hand in hand with the choice of consumables, reagents and equipment that scientists have available to conduct their research. Therefore, suppliers play a crucial role in advancing sustainability within the scientific community and directly influence the environmental impact of research activities. Their choices in materials, production methods, packaging and shipping can significantly affect the overall sustainability of laboratory operations in a direct and indirect way. Suppliers can contribute to sustainability by offering eco-friendly alternatives to traditional products, such as energy-efficient equipment, recyclable materials or products with less single-use plastic components, and biodegradable consumables among other options. Additionally, they can implement sustainable manufacturing practices, reduce waste generation, and optimise transportation logistics to minimise carbon emissions. Suppliers also have a role to play in providing information and resources about sustainable procurement practices. This includes transparent reporting on the environmental footprint of their products and guidance on best practices for reducing environmental impact. In line with this, the EU had a proposition in 2023 for establishing a horizontal European climate label for products.[20] At the moment there are several eco-labels to attest to a product's sustainability level based on their production process and other related considerations.

Collaboration between suppliers and scientific institutions is essential for driving innovation and promoting the adoption of sustainable solutions.

By working together, they can identify opportunities for improvement, develop new environmentally friendly products and technologies, and establish partnerships to advance sustainability throughout the supply chain. An example of that can be found in the collaboration between EMBL scientists and the equipment supplier Beckman to achieve a more efficient and sustainable use of their pipetting robot.[21] Overall, suppliers have a significant responsibility and opportunity to contribute to making science more sustainable, and their active involvement is essential for achieving meaningful progress in this area.

Certifications are a means of ensuring purchasers can transparently understand the innovation regarding sustainability that some manufacturers have achieved, and duly allow manufacturers a common means to communicate. Equally though, if misgoverned or utilising the wrong mechanism, they can be a tool for greenwashing to promote commercial goals. Careful consideration and transparency of mechanisms are necessary to ensure certifications achieve their intended goals.

9.5 What Might the Future of Sustainable Laboratory Certifications Look Like

In a model where scientists, funders, governments, and suppliers collaborated effectively on laboratory sustainability, the future could hold remarkable progress in environmental stewardship, resource efficiency, and scientific innovation. Laboratories would operate with minimal environmental impact, adopting sustainable practices in waste management, energy consumption, and material usage. Funders would prioritise grants for research projects that integrate sustainability into their methodologies and outcomes and develop new grants to fund enabling research for more sustainable practices. Governments would enact policies to incentivise sustainable practices in research institutions and provide funding for green initiatives. Suppliers would develop and supply environmentally friendly products and technologies tailored to the needs of sustainable laboratories through active collaboration with scientists and funders. Together, these stakeholders would create a culture of sustainability in scientific research, leading to a healthier planet and more resilient scientific community. In this model, sustainable research certifications could play a central role in providing frameworks for establishing and recognising sustainable practices. Certifications could be added to the benchmarks for scientific excellence, enhancing their reputation and attracting further support from funders. Ultimately, sustainable research certifications would serve as catalysts for driving systemic change across the scientific community, accelerating progress towards a more sustainable future.

However, we must acknowledge that sustainable research certifications have room to evolve. In order to achieve this, several aspects of the existing certifications would have to change and fit a more strict scheme in terms

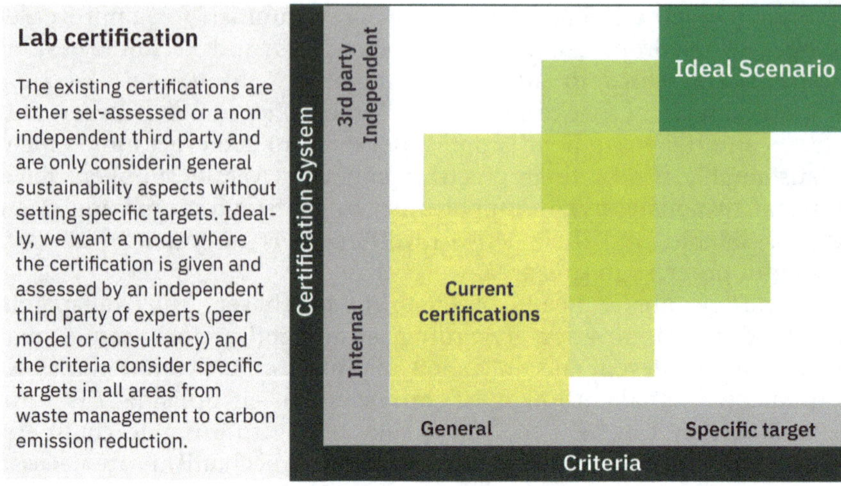

Figure 9.2 Graph explaining the current state of laboratory certifications and the ideal model we should work toward in order to grant a reliable certification model. Type of criteria, from general to specific (*X* axis) and type of auditing process from self-review to certified third party (*Y* axis) are illustrated together with the current most commonly used certifications and the planned ones. Credits: Creative Team/EMBL.

of the type of criteria evaluated and the auditing process (see Figure 9.2). They will also have to be open access with an established governance, sector buy-in and resource availability to ensure they are in accessible platforms.

9.5.1 Criteria for a Solid Sustainable Certification

If we were to characterise the two existing schemes which are most prevalent in the community, LEAF and MGL, we would say that the criteria they use are suggestive and generalistic. In the future we believe it could be beneficial for a certification scheme to be developed which is more precise and prescriptive to complement the existing schemes. To illustrate the difference between generalistic and prescriptive, we could look at criteria for business travel. The current certification schemes ask laboratories to state how they are taking steps to reduce their business travel impact. These open text questions can invite responses which are vague and low impact (they can also provide space for ambitious detailed responses). An alternative, more prescriptive and impactful criteria would set limits, state minimum requirements or demand more detailed responses. In the business travel example, the criteria could set a maximum carbon footprint for business travel, for the sake of this example let us say one tonne of CO^2e per member per year. These targets could get progressively harder as labs move from bronze to gold.

This evolution would provide greater certainty that a certified laboratory was making real reductions in their environmental impact, but it would also allow labs to be compared more easily. However, we must recognise that this evolution would require a level of governance to ensure the criteria are appropriate and are updated to keep them ambitious enough to remain at the forefront of moves to make research more sustainable. In the example above, there would need to be agreement that one tonne of CO_2e per member per year was appropriate. The minimum standards, targets or prescriptive criteria would need to be set through the work of a group of sustainable science experts who would act as guardians of the criteria and who have the required reputation for their criteria to be accepted by the research community.

From the point of view of the research laboratory or institution, the implementation of this criteria would also require a greater level of administrative support to help calculate the environmental impact of their actions and make sure the documents and reports necessary are produced for the auditing process to take place. In the travel example above, this administrative work would require recording the flights and trips of all staff and establishing a method to detect when they go over the selected limit.

Moreover, these requirements could not be restricted to the laboratory user level, as in many cases their responsibility and freedom of action would be limited. The certification and its evaluation would have to be extended to an institutional level. This would either require the adherence of the research organisation to a separate certification scheme or the consideration of this prerequisite into the auditing and evaluation of the laboratory criteria fulfilment.

9.5.2 Models for Auditing of Certifications

As mentioned in previous sections, in the current schemes the assessments are carried out by the organisations or through an internal peer review (as is the case with LEAF) or by a for-profit subsidiary (as is the case with MGL). To increase transparency, increase reliability and reduce the chance of abuse and greenwashing a system of third-party assessments for applying laboratories with an independent host of the certification scheme would help. Third-party audits provide an independent and objective evaluation of a laboratory's compliance with sustainability standards, offering assurance to stakeholders that the certification is based on real action and results which are evidenced and which have been observed by the third party. The development of sustainable laboratory schemes in this way would enable these schemes to be used by policy makers and funders to allocate their funds towards more sustainable laboratories and projects.

However, this development would be costly both in terms of money and time. It would require the creation of a new role for assessors and auditors for the goal of having a well-defined and managed certification scheme and imply a greater cost for the laboratories or research institutions applying. In their current state, having certifications schemes which do not have independent third-party auditors are low-cost options, and they have therefore grown a lot in a short period of time, have engaged a lot of scientists in sustainable laboratory practices, and have facilitated a reduction in research's environmental impact. There is space for both types of certification schemes in the future to collaborate into creating an engaging platform with the right level of stringency and governance.

To achieve this goal and reduce the cost of developing the necessary tools and scaffolding necessary for the above model to succeed one could make use of the peer review scientific model. In the field of sustainability, peer review could serve as a valuable means of certification, ensuring the credibility and validity of sustainability practices and the compliance with standardised criteria to achieve a certain level of certification. Similar to its role in scientific research, peer review involves the evaluation of sustainability practices by scientists who would have themselves gone through the certification process successfully. This model would still require some level of a third-party verification system but the first peer review step could help reduce the cost and speed up the process of establishing this kind of model by providing a preliminary assessment report by peers. While peer review involves evaluation by other laboratories certified in the same scientific field, a verification party serves as an additional layer of scrutiny, providing independent validation of the assessment process. This verification party, often consisting of accredited certification bodies or independent auditors, would verify that the peer review process was conducted impartially and in accordance with established standards and guidelines, just like the figure of the editor would do in the peer review process of scientific publication. Their involvement helps to mitigate potential conflicts of interest and ensures that sustainability certifications are based on rigorous evaluation and objective criteria.

Although there is certainly a lot of work to be done, the ongoing efforts observed from bottom-up and top-down approaches are paving the way for a sustainable model of scientific research. Scientists are starting to take responsibility for the research they conduct and this is being reflected in the engagement of their institutions with sustainability and transparency. Likewise, funders are seizing the opportunity to engage with more sustainable policies and require certain compliance with sustainable practices to access their grants and fellowships. There is a very long way ahead, but the journey has already begun.

Acknowledgements

We would like to acknowledge Bianca Schell for her work on a supporting publication comparing relevant certifications. We also thank Eva de Benito Garragori, Begoña Ugarte Uribe, Zuzana Koskova, Swerlana Krez, Irena Provaznikova, Moritz Niemann, Jane Kilcoyne and Siobhan Gaughan for their input as LEAF and MGL users.

References

1. J. Dobbelaere, J. B. Heidelberger and N. Borgermann, Achieving sustainable transformation in science – green grassroots groups need nurturing from the top, *J. Cell Sci.*, 2022, **135**, jcs259645.
2. SELs Network - SELs Network. https://sels-network.org/.
3. UCL. LEAF impact. Sustainable UCL. https://www.ucl.ac.uk/sustainable/take-action/staff-action/leaf/leaf-impact, 2022.
4. T. Franssen, L. Giannini and H. Johnson, LEAF in the Netherlands: Adopting Sustainability in Our Laboratories, 2022. https://zenodo.org/records/7457118, DOI: 10.5281/zenodo.7457118.
5. My Green Lab Certification, *My Green Lab*. https://www.mygreenlab.org/green-lab-certification.html.
6. T. Kitzberger, J. Kotik and T. Pröll, Energy savings potential of occupancy-based HVAC control in laboratory buildings, *Energy Build.*, 2022, **263**, 112031.
7. M. D. Paepe, L. Jeanneau, J. Mariette, O. Aumont and A. Estevez-Torres, Purchases dominate the carbon footprint of research laboratories, 2023. 2023.04.04.535626, Preprint at DOI: 10.1101/2023.04.04.535626.
8. L. Lannelongue, *et al.*, GREENER principles for environmentally sustainable computational science, *Nat. Comput. Sci.*, 2023, **3**, 514–521.
9. H. Tsai, N. K. F. Tsang and S. K. Y. Cheng, Hotel employees' perceptions on corporate social responsibility: The case of Hong Kong, *Int. J. Hosp. Manage.*, 2012, **31**, 1143–1154.
10. P. Zientara and A. Zamojska, Green organizational climates and employee pro-environmental behaviour in the hotel industry, *J. Sustainability Tour.*, 2018, **26**, 1142–1159.
11. D. B. Turban and D. W. Greening, Corporate Social Performance and Organizational Attractiveness to Prospective Employees, *Acad. Manage. J.*, 1997, **40**, 658–672.
12. G. Cachón-Rodríguez, A. Blanco-González, C. Prado-Román and F. Diez-Martin, Sustainability actions, employee loyalty, and the awareness: The mediating effect of organization legitimacy, *Manage. Decis. Econ.*, 2021, **42**, 1730–1739.
13. ISO – Environmental sustainability, *ISO* https://www.iso.org/sectors/environment.
14. Homepage|GHG Protocol. https://ghgprotocol.org/.
15. GRI – Home. https://www.globalreporting.org/.
16. IFRS – International Sustainability Standards Board. https://www.ifrs.org/groups/international-sustainability-standards-board/#resources.
17. Streamlined Energy and Carbon Reporting (SECR) for academy trusts, *GOV.UK*. https://www.gov.uk/government/publications/academy-trust-financial-management-good-practice-guides/streamlined-energy-and-carbon-reporting.
18. Ambitious corporate climate action – Science Based Targets. https://sciencebasedtargets.org/.

19. Sparking green movement in the funding landscape – Green Labs – NL. https://www.greenlabs-nl.eu/sparking-green-movement-in-the-funding-landscape/.
20. Establishing a horizontal European climate label for products|Think Tank|European Parliament. https://www.europarl.europa.eu/thinktank/en/document/EPRS_STU(2023)747453.
21. Do we really need that many tip boxes? How a simple question can lead to greener omics – Sustainability. https://www.embl.org/about/info/sustainability/do-we-really-need-that-many-tip-boxes-how-a-simple-question-can-lead-to-greener-omics/.
22. R. Bianca, Schell, Nico Bruns Lab sustainability programs LEAF and My Green Lab®: impact, user experience & suitability, *RSC Sustainability*, 2024, **2**, 3383–3396.
23. P. M. Weber, *et al.*, The Heidelberg Agreement on Environmental Sustainability in Research Funding (1.0), *Zenodo*, 2024, DOI: 10.5281/zenodo.13938809.

CHAPTER 10

Education for Sustainability and Equity in Research[†]

JOPKE JANMAAT,[a] ARTE GROENEWEGEN,[a] JEROEN DOBBELAERE[b] AND FLORIJN DEKKERS*[a]

[a] Education Center, Faculty of medicine, University Medical Center Utrecht, Utrecht University, Utrecht, The Netherlands; [b] Institute of Science and Technology Austria (ISTA) and University of Vienna, Vienna, Austria
*Email: j.f.dekkers-6@umcutrecht.nl

10.1 The Imperative for Sustainability and Equity in Research Education

10.1.1 The Paradox of Science and Health and Wellbeing

Global environmental crises, such as climate change, biodiversity loss, pollution, and related social inequalities, are profoundly affecting the health and well-being of humans and other life on earth (Figure 10.1).[1] Examples include the increased risks of cardiovascular disease due to air pollution, the changing spread of infectious diseases such as malaria, health risks related to climate migrants and the rise of eco-anxiety driven by climate change.[1] On the other hand, nature has profound healing effects. For example, it is estimated that 70 percent of drugs used for cancer treatment originated from natural sources or are synthetic compounds inspired by nature.[2] Additionally, a study in 1984 already suggested that hospital patients with a view of nature recovered faster from surgery and required less pain medication compared to those without.[3] It is now well-established that green spaces

The Sustainability of Science: How Science Impacts the Environment, and What Can Be Done
Edited by Martin Farley

Published by the Royal Society of Chemistry, www.rsc.org

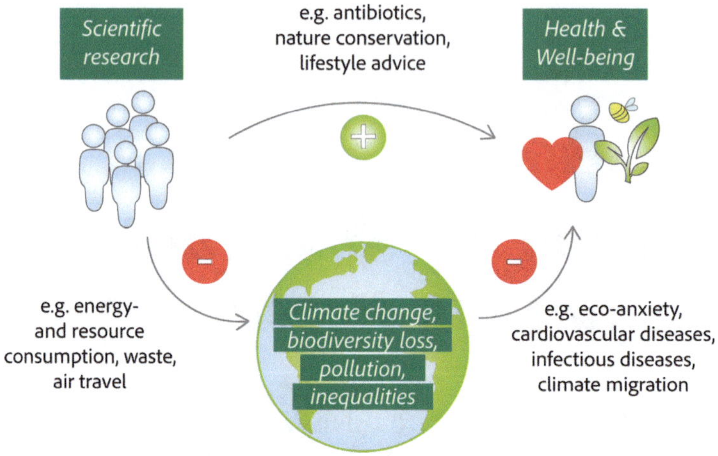

Figure 10.1 The paradox of science and health and well-being. The relationships between scientific research, global crises and health and well-being of humans and their environment are depicted. A transition for scientific research is needed where positive impacts on health and well-being are maximised, while negative environmental and social effects are minimised. This illustration was made by the authors of this chapter.

have beneficial effects on perceived stress levels, overall mental health, quality of life and subjective well-being.[4]

Scientific research plays a paradoxical role in affecting health and well-being (Figure 10.1). Scientific research deepens our understanding of nature and advances health and well-being. However, research also contributes to environmental harm through resource-intensive (laboratory) practices, waste production, data storage and computation, and extensive air travel.[5,6] This impact is further exacerbated by 'wasted research efforts', which are research outcomes that yield little or no societal benefit. Contributing factors include scientists investigating questions that could be adequately addressed with existing knowledge, using unrepresentative samples, or lacking appropriate controls in laboratory experiments.[7,8] Therefore, in addition to reducing the environmental footprint of research practices, minimising 'wasted research efforts' is crucial. This can be achieved, for example, by adopting evidence-based research approaches, more collaboration, or co-developing research questions with relevant stakeholders (*e.g.*, patients or practitioners) to ensure alignment with real-world priorities.[7–10]

Scientific research not only contributes to global crises but also exacerbates global inequalities. For example, 50 percent of the collected European plastic waste is shipped for recycling outside of the EU without accountability, posing social and environmental risks.[11,12] Moreover, scientific practices can lead to epistemic injustice, which occurs when certain groups are unfairly disadvantaged in the production, access, or recognition of knowledge.[13] For instance, the relevance, generalisability and applicability of scientific findings are limited when research studies primarily focus on populations of European and North American descent. Likewise, when scientific data remains inaccessible to large parts of the world's population, existing power imbalances in knowledge producing practices are reinforced. To address these inequalities and enhance academic performance, it is beneficial to promote equity, diversity, and inclusion, recognising personal and research biases at every stage of the research process.[14]

To ensure scientific research meets the needs of the present and the future, a transition is needed to maximise research's positive impact on health and well-being, while minimising its negative environmental and societal effects (Figure 10.1). This transition must prioritise environmental sustainability as well as equity, ensuring that research benefits all populations and averts harm to both people and the planet as much as possible. Especially, when the social landscape shapes the direction of research through funding cuts, political agendas, or by restricting academic freedom, the role of education for just and sustainable scientists becomes even more vital.

10.1.2 The Power of Education

> "Education is the most powerful weapon which you can use to change the world." – Nelson Mandela, 1990[15]

Education plays a crucial role in the transition towards sustainable and equitable research practices. By integrating both sustainability and equity into research training, learners – here comprised of undergraduate students to postdocs – can be empowered to critically assess current practices, make sustainable choices, and challenge the status quo. Such education enables scientists, academics and research institutes to act as societal and environmental stewards, taking responsibility for minimising negative impacts of research while actively contributing to sustainable and equitable solutions. At the same time, training in and for sustainability also fosters agency, empowering researchers to influence policies, advocate for change, and design more responsible research practices.[16]

Present-day scientific research education focuses on a wide range of competencies, such as managing research data, scientific writing, developing networks, applying research ethics and integrity principles, disseminating results, and managing professional development. Such education successfully addresses cognitive abilities such as critical and analytical thinking and creativity.[17] However, research education does not yet adequately prepare (future) researchers to ensure their work is environmentally and societally sustainable. To address this gap, research education requires transformation toward better equipping (future) scientists with competencies that align with sustainability and equity topics in research (see Figure 10.2 and the section 'Competencies for sustainability and equity in research'). A fundamental prerequisite for this

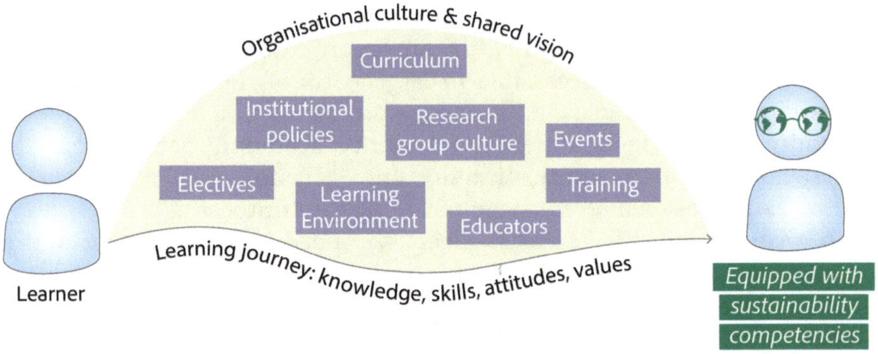

Figure 10.2 **Learning journey of (future) researchers.** Research education requires equipping (future) researchers with competencies (a set of knowledge, skills, attitude and values) that align with sustainability and equity. This equipped researcher is wearing the metaphoric 'planetary health glasses' (Box 10.1). In addition to content of fixed curricula, there are many factors impacting the learning journey, such as availability of elective courses and training, culture within a research group, institutional policies and the learning environment. This illustration was made by the authors of this chapter.

transformation is that education emphasises the responsibility of researchers to apply their acquired skills in service of society.[18] Given the interconnectedness of scientific research, global environmental crises, and related inequalities (Figure 10.1), we emphasise equipping (future) researchers with a

Box 10.1 A planetary health perspective for research.

As stated by the Planetary Health Alliance,[19] planetary health is defined as 'a solutions-oriented, transdisciplinary field and social movement focused on analysing and addressing the impacts of human disruptions to Earth's natural systems on human health and all life on Earth'. In recent years, planetary health has gained growing interest, illustrated by an ever-increasing world-wide community,[20] tailored scientific journals,[21] development of tailored research agenda's[22] and development of vision, frameworks, trainings and learning materials to transform education.[23] As part of the Planetary Health Integration Team[24] – which is fully dedicated to accelerating integration of planetary health principles into the Faculty of Medicine at the University Medical Centre Utrecht since 2023 – the authors of this chapter developed the 'planetary health glasses' (Figure 10.3).[25] The team uses these metaphorical glasses to invite people during lectures and training to envision a healthy and equitable world and encourage them to take action to move towards this future (Figure 10.1).

The team also advocates for (future) scientists to wear these glasses and perform research from the *perspective* of planetary health. In their view,

Figure 10.3 Planetary health glasses. Available *via* Kenniscentrum Global Health.[25]

this would mean, for example, that scientists (1) recognise the interconnectedness of human health and well-being with natural and social ecosystems, (2) embrace the power of diversity and different cultural perspectives, (3) nurture values such as integrity, humility, transparency, care, respect for diversity, collaboration, compassion, and gratitude, (4) engage relevant stakeholders in all phases of the research process, (5) adhere to principles of open science, (5) minimise negative environmental and social impact of research practices, while maximising positive societal impact, and (6) refuse to do research that is not addressing an urgent knowledge gap. The Planetary Health Integration Team co-developed various training tools (see the section 'Tools and resources').

holistic perspective; namely to think and act through a planetary health perspective (Box 10.1).

The educational innovation we advocate for involves two key aspects: sustainability and equity *in* research practices, meaning that researchers are trained to conduct science in the most sustainable and equitable way (*e.g.*, applying evidence-based research practices and minimising resource consumption); and research practices *for* sustainability and equity, meaning that researchers actively engage in projects, internships, and topics that enable sustainability and social justice. While this chapter occasionally touches on the latter, its primary focus is on training for integrating sustainability and equity within research practices. This chapter mainly focuses on scientific bachelor, master's and PhD programmes.

Education for sustainable research practices ultimately aims to be transformative by design, empowering researchers to actively reshape scientific practices. This can be pursued through a combination of incremental changes to existing curricula, so-called 'low-hanging fruits', and more ambitious efforts such as developing new courses, challenges, or even entire academic programmes. While the latter requires greater investment, it may also lead to greater impact. Sustainability literacy should be embedded in mandatory education for all students, while opportunities to deepen knowledge should be made available to those who seek it, for example through elective courses.

10.2 What to Educate?

10.2.1 Competencies for Sustainability and Equity in Research

Sustainability and equity in research can be cultivated by the development of sustainability competencies, as they enable individuals to act sustainably and address relevant complex challenges. Various frameworks outline sustainability competencies,[16,26–28] and while differences exist among them, there is broad agreement within education for sustainable development on key competencies.[16] These commonly recognised sustainability competencies[16] include:

- *Systems thinking competency:* the ability to understand relationships within complex systems, recognise how systems are interconnected across various scales and domains, and cope with uncertainty.
- *Anticipatory competency:* the ability to evaluate possible, probable, and desirable futures, create personal visions for the future, apply the precautionary principle, and assess and deal with risks and consequences.
- *Normative competency:* the ability to reflect on the values and norms guiding one's actions, and negotiate sustainability values, goals and principles, also amidst a context of conflicts of interests, trade-offs, and uncertainty.
- *Strategic competency:* the ability to collaboratively develop and implement innovative actions that promote sustainability locally and globally.
- *Collaboration competency:* the ability to empathise with others, respect differing perspectives, resolve conflicts, and facilitate collaborative and participatory problem-solving.
- *Critical thinking competency:* the ability to question established norms, practices, and opinions, reflect on one's own values and actions, and engage in the sustainability discourse.
- *Self-awareness competency:* the ability to reflect on one's role in society, evaluate and motivate one's actions, and manage personal emotions and desires.
- *Integrated problem-solving competency:* the ability to apply diverse problem-solving approaches to complex sustainability challenges, integrating all the aforementioned competencies to develop inclusive and equitable solutions.

These sustainability competencies can serve to reflect on curricula. Competencies can be defined by an interplay of corresponding knowledge, skills, attitudes, and values, which can further guide education (*e.g.* to tailor learning goals or activities). The following section illustrates what knowledge, skills, attitude and values for sustainability and equity in research can be nurtured through guiding questions.

- *Knowledge:* 'What do learners need to know?'
  For example: How do research practices affect the global ecological crisis, positively and negatively?
- *Skills:* 'What do learners need to know how to do?' or 'What do learners need to be able to do?'
  For example: How to adopt evidence-based research and resource-efficient practices in research? How to balance competing interests when trying to act environmentally sustainable, societally impactful and with scientific validity?
- *Attitude:* 'What are the learners' dispositions?'
  For example: How can a learner's professional and personal identity relate to sustainability and equity? How can a researcher question norms, practices and opinions?
- *Values:*
  For example: What values would the learner like to see expressed in their daily life, working life and research? What does 'do good research' mean? For examples of values to nurture in research education see Box 10.1.

10.2.2 Educational Topics for Sustainability and Equity in Research

A holistic perspective is important when developing sustainability and equity literacy, therefore a diverse array of topics should be addressed. It is often stated that cultivating awareness of our **interconnectedness within nature,** thus the relatedness of health and well-being of humans, other animals, plants and planetary ecosystems, lies at the heart of the sustainability challenge. To cultivate this sense of connection, education should address not only its cognitive but also its affective ('to care') and behavioural ('the commitment to act') dimensions.[23] Additionally, learners should be taught how to minimise resource- and energy-intensive research practices and to reduce waste through circular interventions and energy-saving strategies that promote **environmental sustainability**. As changing established research routines can be challenging, extra attention should be given to demonstrating how green actions can be incentivised.[29] For instance, green science certification tools such as LEAF,[30] My Green Lab,[31] or GreenDiSC[32] offer recognition for sustainability efforts. Moreover, highlighting the co-benefits of green initiatives, such as cost savings, improved research efficiency, and enhanced lab infrastructure, can further motivate researchers to adopt sustainable practices.[29]

Social and environmental sustainability can further be enhanced by promoting transparency, reproducibility, and data accessibility while minimising 'wasted research efforts'.[7] Education plays a key role in this process by fostering knowledge, skills, attitudes and values associated with **evidence-based**

research[8] and **open science** principles[18] including making data Findable, Accessible, Interoperable, and Reusable (FAIR).[33] Equitable participation of community stakeholders in research and its education nurtures awareness of **diversity** and fosters a commitment to and skills, attitudes and values for **inclusion.** Correspondingly, recognising and preventing the broad range of **bias in science** helps to mitigate its influence on research design and interpretation. Bias in science to make learners aware of are for example gender bias (*e.g.* young, white, able-bodied 70 kg male individuals are the norm in research practices)[34] or confirmation bias (*e.g.* scientists favour data that support their *a priori* hypotheses).[10] To consider affective components of learning and reactions to learning, it is important to acknowledge and cope with **eco-emotions** that might arise when facing planetary challenges (addressed in the section 'Activating hope and empowerment').

10.2.3 Learning Goals

The earlier mentioned topics and required competencies can be translated into learning goals that guide educational activities, which in turn are to be assessed by educational bodies (examples are provided in Table 10.1). Applicability of these learning goals might differ between levels, functions and

Table 10.1 Example learning goals for sustainability and equity literacy in research education.

Theme	#	Learning goal
Interconnectedness within nature	1	Can provide examples of how the health of humans, plants, animals, and planetary ecosystems are interconnected.
	2	Can explain the paradoxical role of scientific research in health and well-being (Figure 10.1).
Environmentally sustainable research practices	1	Can apply the R-ladder of circularity to a research project.
	2	Can identify how stakeholders are driving the transition to environmentally friendly laboratory practices.
	3	Can locate and utilise tools and platforms for green data management and sustainable computing.
	4	Can optimise the sustainable use of energy-consuming equipment such as freezers, fume hoods, and benchtop devices.
	5	Can facilitate the sharing of materials, lab spaces, and data.
	6	Can implement ways to minimise the environmental impact of international conferencing.
	7	Can advocate for and adopt animal-free research methods, including alternatives to animal models and animal-derived products.

Table 10.1 (*Continued*)

Theme	#	Learning goal
	8	Can navigate sustainability challenges by aligning personal sustainability efforts within an unsustainable system, for instance by opening dialogues and coping with eco-emotions.
	9	Can navigate in sustainability challenges, for instance by making informed choices while dealing with scientific uncertainties or competing interests of scientific validity and societal and environmental impact.
Evidence-based research	1	Can justify the need for a new research project by conducting a systematic and comprehensive analysis of previous studies.
	2	Can design a study informed by a systematic review.
	3	Can plan and conduct robust scientific experiments.
Open science	1	Can implement open science principles in research, for example by making scientific output and its processes Findable, Accessible, Interoperable and Reusable (FAIR).
	2	Can describe how open science can be applied throughout the entire research process.
	3	Can differentiate between publications and preprints, including their roles and benefits.
	4	Can explain the balance between open science and sustainability, considering trade-offs and synergies.
Diversity and inclusion	1	Recognises the importance of diversity and inclusion in science, including in research teams. For instance, how more inclusive practices influence research questions, methodologies, and interpretation of results.
	2	Can give examples of ways to systemically, transparently and genuinely include stakeholders' perspectives when formulating a research question, designing and conducting research.
	3	Can use bias-free and inclusive written and spoken language, images and symbols.
	4	Can create a research proposal in which relevant voices systematically are included in production, use and circulation of knowledge (principles of epistemic justice).
Bias in science	1	Can explain how biases, such as implicit or confirmation bias, affect both scientific research and society.
	2	Can recognise own biases in science and apply strategies to mitigate them.
	3	Can identify relevant diversity characteristics for baseline variables and categories, and conduct subgroup analyses appropriately.

disciplines of learners. In the process of adjusting and creating learning goals to a learner's level, Bloom's taxonomy may be of help.[35] In this taxonomy learning is categorised into six cognitive levels: remembering, understanding, applying, analysing, evaluating, and creating.[35] In the context of education for sustainability and equity in research, the intention is to move students beyond basic knowledge acquisition (*e.g.*, recognising sustainability challenges) toward higher order thinking skills such as evaluation (*e.g.*, critically questioning the status quo of research practices). In addition, theory of 'constructive alignment' emphasises that learners will learn more effectively when learning goals and (active) teaching methods are purposefully connected to assessment strategies.[36]

10.2.4 Pedagogical Approaches and Educational Activities

Transformative teaching methods that aim to develop learners' understanding of the world can be of guidance, and nurture learners' critical thinking about the world.[37] To empower learners to become active and critical sustainable researchers, key pedagogical approaches include action-oriented learning and a learner-centered approach.[16] In such approaches, the role of an educator often transitions from being an expert on educational content (teaching) towards being a facilitator and coach. Different educational activities can aid these approaches, and are dependent on the context of learning, learner group, and the availability of support and resources.

- **Action-oriented learning** allows students *to work on practical solutions for pressing global challenges*, while nurturing acquisition of knowledge, clarification of values and development of competencies.[16] By orienting towards action, learners move beyond understanding theoretical concepts and acquire the ability to apply theories in practice. Sinakou *et al.*[38] frame five components of sustainability education that build towards taking action (*i.e.* competency of integrated problem-solving), namely (1) action-taking (in projects), (2) students' leadership in their learning and teaching, (3) peer interaction, (4) community involvement and (5) interdisciplinarity. These components can be integrated in challenge- or project-based learning (see the section 'Inspiring practical examples').
- **Learner-centered approaches** are characterised by the *active development of knowledge*, rather than a passive transfer of knowledge, positioning learners as autonomous participants who reflect on their learning processes. *Co-creation*, for example, involves collaborative approaches, in which a curriculum is seen as the joint making of meaning by educators and learners. This can involve learners designing their own educational programme, developing teaching resources prior to teaching, or contributing to curriculum development during a course.[39] Furthermore, active learning methods such as think-pair-share, concept mapping, peer teaching, or group discussions have been shown to actively deepen learners' understanding of the studied subject.[40]

- **Educational activities** for sustainability and equity in research can range from low- to high-effort interventions. Shorter, easily implemented approaches include inviting an *expert guest-speaker* or adding a *single lecture-slide* or *by integrating a sustainability case* while learning another skill (*e.g.* maths can be thought with giving climate examples). *Workshops* or *challenges* on the environmental footprint of scientific research can raise awareness about lab energy use and waste while providing hands-on solutions. A *flipped-classroom approach* can involve an at-home e-module (see 'Tools and Resources' section) followed by a session on planetary health and sustainable research, where participants reflect on ways to reduce their environmental impact. A *stakeholder analysis* and *science-policy communication exercise* can help embed sustainability goals in research agendas. *Peer discussions* on sustainability dilemmas foster collaboration, ethical reflection, and emotion-coping, while dedicated reflection sessions encourage researchers to examine their personal values and the broader societal impact of their work.

10.2.5 Activating Hope and Empowerment

Growing awareness of global crises, such as climate change, negatively impacts mental health worldwide.[41] As such, it is important to consider that eco-emotions may arise when (future) scientists start to better understand the negative impacts of research practices on global crises (Figure 10.1). Education for coping with eco-emotions can help to balance negative and positive emotions of both learners and educators and increase their resilience.[42,43] Coping with eco-emotions can be addressed in education in various ways.

First, getting equipped with actionable sustainability knowledge, skills, attitudes and values might already empower learners to move beyond despair and act more sustainably.[44] This is enhanced when learners engage in internships or challenge-based projects that address real-life problems (see the sections 'action-oriented learning' & 'inspiring examples'). When students can see the direct impact of their work, this induces a sense of agency and responsibility. To further empower learners, it is also of value to create a learning environment that encourages continual reflection in which learners are challenged to question the norm, such as: 'What surprises you in the light of considering sustainability in our research activities?' Additionally, starting or joining a green initiative or green team creates a 'sense of belonging' and allows both learners and educators to make a change with like-minded people. Such actionable steps for stakeholders are provided in Table 10.2, offering practical guidance for incorporating sustainability into educational settings.

Second, various methods have been developed that can be used in education to empower and spark hope. For example, principles of **active hope** developed by Joanna Macy and Chris Johnston have successfully motivated learners worldwide.[45,46] Unlike traditional hope-based frameworks, active hope does not require individuals to feel hopeful; instead, it emphasises

Table 10.2 Actionable steps for various stakeholders to foster sustainability and equity in research education.

Target group	Actions
Learners (*e.g.* undergraduate students, PhD students and postdocs)	• Start or join a green initiative focused on sustainability in research (education). • Ask critical questions to challenge existing research and education practices. • Open dialogue with peers, educators, and decision makers about sustainability in research. • Provide feedback (even when not asked) on curricula and advocate for the integration of sustainability and equity topics, for example through post-course surveys. • Choose impact-driven projects, for example by selecting internships or research projects on sustainability-related topics. • Select sustainability-related (elective) courses or trainings as part of your learning journey. • Attend local or international events on the topic of sustainability and equity in research.
Educators and educational innovators	• Integrate sustainability and equity topics into existing education, for example by adding new content to a lecture, inviting expert guest speakers, adding tailored cases in group discussions, and making use of existing tools. • Foster action-oriented learning, for example through developing a new interdisciplinary, challenge-based course (element). • Foster a learner-centered approach, for example by using active learning methods 'think-pair-share', 'concept mapping' or 'peer-teaching'. • Ensure and attend training for professional development related to sustainable and equitable science. • Start or join a green initiative within or outside the institution. • Advocate for change by discussing about the topic of sustainability with colleagues, for example by structurally adding the topic of sustainability to meeting agendas.
Institution (*e.g.* policy officers and decision makers)	• Dedicate staff to developing and implementing a shared vision, policies, and innovations related to sustainability and equity in research education. • Implement training programmes for the professional development of both educators and researchers, for example, by offering relevant e-learnings or other resources. • Form partnerships with local businesses, non-governmental agencies, or community organisations to bring in relevant expertise and provide opportunities for internships and research projects. • Foster thematic research and education programmes on topics such as planetary health, circular research practices, and sustainable technology development. • Create a learning environment where sustainability and inclusivity are visibly integrated, for example, through signposts, art, campaigns or themed events. • Adopt and implement a Whole Institution Approach.

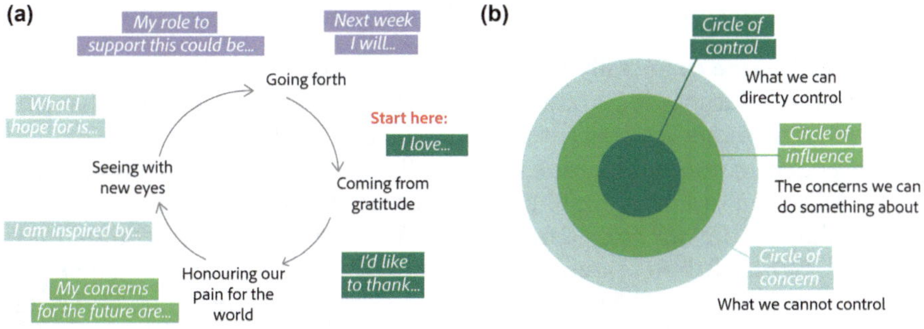

Figure 10.4 Activating hope and empowerment. (a) The spiral of active hope can be used in education as a sentence-starter exercise. The active listening practice invites learners to move through four stages, including (1) gratitude, (2) honouring our pain for the world, (3) seeing with new eyes and (4) going forth. The illustration was adapted from an open-access active hope training.[49] (b) Stephen Covey's circle of influence[48] helps people to shift their focus and energy from concerns that they cannot control (circle of concern), toward things they can directly control (circle of control) or influence (circle of influence). The illustration was adapted from Covey's book.[48] Methods of both (a) and (b) can be used in education to help balance positive and negative emotions.

intention as the driving force for action. Through this approach, learners identify their personal values and intentions, interconnectedness with the natural world, hope, and actions which address their intentions.[45] This can be facilitated by inviting learners to move through four stages of active hope, for instance by answering and discussing a set of 'active hope' questions (Figure 10.4a). Another tool is the **hope wheel** that can assist in engaging with hope-based pedagogies that protect learner well-being while empowering agency through creativity, collective action, and a culture of care.[47] Lastly, Stephen Covey's **circle of influence**[48] helps people to shift focus and energy from concerns that they cannot directly control, toward things they can directly control or influence (Figure 10.4b). It also acknowledges the positive ripple effect that actions of individuals can have on others. All these activating methods have in common that they help people to move into a proactive space, reduce feelings of helplessness and direct energy toward tangible change.

10.2.6 Tools and Resources

There are various tools available to support current and future researchers in developing the knowledge, skills, attitude, and values needed for sustainability and equity in research. Additionally, trainers and educators can use tools to foster transformative education. Below, a selection of relevant tools are highlighted.

10.2.6.1 Sustainability-related Tools

- A **'Guideline to Socially Responsible Research'**[50] has been developed to support (future) researchers in designing, conducting, and disseminating (bio)medical research in a socially responsible manner. This practical guideline includes a checklist and reflective questions by following the research cycle. It also refers to various supportive resources, including the Sex and Gender Equity in Research (SAGER) guidelines,[51] the Genderful Research World resource platform,[52] the James Lind Alliance website for priority setting of research questions,[53] Bias Free language guidelines,[54] the R-ladder for circularity,[55] and the Equator network for standardising reporting guidelines.[56]
- The interactive and playfully illustrated 60-minute e-module **'Just & Sustainable Research Practices'** provides researchers with the knowledge and tools to conduct socially responsible research[57] (open access, login required). Using the research cycle as a framework, the module explores key themes such as evidence-based science, OPEN science, bias in research, and data & computing.
- The interactive and playfully illustrated 60-minute e-module **'Environmentally Friendly Laboratory Research'** equips lab-users with practical strategies to minimise their environmental impact, addressing themes such as energy consumption, resource consumption, waste management and green labs community[57] (open access, login required).

10.2.6.2 Tools to Develop Competencies

- **The Planetary Health Education Framework**[23] is a guiding tool for integrating the planetary health perspective into education. The framework helps institutions, educators, and learners understand the interconnections between environmental, social, and health challenges. The framework emphasises five key domains: interconnection within nature, the Anthropocene and health, systems thinking, equity and justice, and movement building.
- **A Rounder Sense of Purpose**[58] is a framework for educators to teach on sustainable development. The framework consists of 12 key competencies, such as creativity, action and responsibility, each with learning outcomes and practical teaching activities. The framework promotes action-oriented, transformative learning and it encourages critical, creative, and systemic thinking.
- **The Inner Development Goals (IDGs)**[59] provide a framework with the essential 'skills and qualities that help to live purposeful, sustainable and productive lives'. The IDGs can be used in education and in institutions to empower individuals to drive sustainable change across various sectors. By reflecting on being (relationship to self), thinking (cognitive skills), relating (caring for others and the world), collaborating (social skills) and acting (enabling change), the IDG framework

encourages a collective commitment to ethical responsibility, equity, and sustainable development in research and beyond. *A Transition Makers Toolbox*[60] is available, structured around the IDGs including step-by-step learning activities, goals, and outcomes.
- **Co-creation and active learning methods** can be fostered by using the *co-creation navigator* by Waag,[61] and Utrecht University's Teaching and Learning Collection[40] which also comprises activating learning activities.

10.3 A Whole Institution Approach

Ensuring that the content of education aligns with a shared vision of sustainability and equity is essential. However, it is equally important to align this vision with the learning environment to bring education for sustainable research to life. Training future scientists requires creating a learning environment that fosters relevant competencies. To move beyond *ad-hoc* efforts such as adding a single lecture slide on sustainability, toward a more systemic and lasting integration of sustainability and equity within educational and research institutions, a Whole Institution Approach can provide valuable guidance.[62] This approach considers the importance of social and physical contexts in the processes of effective learning. In a systematic review by Holst[63] Whole Institution Approaches are described as 'continuous and participative organisational learning processes aimed at institutional coherence on sustainability, consistently linking the formal and informal (hidden) curricula'. In Whole Institution Approaches, five core principles and seven interconnected organisational areas of action can serve as a guideline to mainstream sustainability as a fundamental principle within all activities.[62] The core principles emphasise (1) *coherence* throughout the institution on what is to be learned and experienced, (2) an environment of *continuous learning*, (3) inclusive *participation*, (4) which comes with individuals' *responsibility* within the organisation and (5) long-term organisational *commitment*. The Approach's seven connected areas of action within an institution[62] consist of:

- **Participatory and proactive governance.**
  E.g. dedicating FTE to integrate sustainability in both curriculum and the institution.
- **Sustainable capacity building.**
  E.g. active capacity building for educators and non-teaching staff, such as a 'Community of practice' or 'Green Teams', have the potential to connect the like-minded, to provide a shared context in which sharing information and experiences (shared learning) can build understanding and insights, and to inspire collaborative actions.[64]
- Education for sustainability and equity in research in **curriculum and learning.**
  E.g. a holistic, action-oriented, situated, just approach to curriculum and learning, with incorporation of both social-emotional and cognitive learning.

- **Sustainable operations and campus management.**
  E.g. providing a future-proof menu in canteens, such as through adopting EAT lancet diet principles,[65] providing recycling bins and infrastructure, and implementing green lab certification schemes.[66]
- **Embedded in community and networks.**
  E.g. partner with local communities and foster inter-organisational networks as collaborative real-world learning grounds.
- **Active communication on sustainability and equity.**
  E.g. clear, consistent and active communication on (un)sustainability both within and outside the organisation ("show and tell").
- **A research agenda that aligns with sustainable development.**
  E.g. stimulate the active involvement of learners into thematic (institutional) research programmes that focus on planetary health, circular research practices, or sustainable technology development.

An underlying holistic organisational culture (*i.e. shared* beliefs, values and meanings that are the fertile soil of the organisational system) connects all principles and areas of action. Literature shows important external factors for succeeding of a Whole Institution Approach, namely prioritisation of sustainability and equity in policies, sufficient and long-term funding, and providence of access to expertise and support (also listed in Table 10.2).[62]

10.4 Inspiring Practical Examples

When aiming to bring theory into practice, learning from others can be helpful and inspiring. In the following section, integration of sustainability is illustrated through practical examples, varying in scope and method.

10.4.1 Transdisciplinary Student Challenge: Using Lab Waste for 3D Printing

During her PhD, May Buchholz explored ways to make 3D printing more sustainable.[67] As the 3D printing process relies on plastic, she wondered: Could we recycle discarded tip-box inlets into filament for our printers? At the Princess Máxima Center, she set up a recycling pipeline, but it was not fully optimised. An interdisciplinary student team with a background in energy and environmental sciences helped by conducting waste audits to identify suitable plastics and provide recommendations to improve the process. The students' insights and recommendations led to new collaborations and shaped her research to develop recycled 3D-printed lab tools further. *Maj Buchholz, PhD Student, Princess Máxima Center, Utrecht, the Netherlands.*

10.4.2 The CO_2-assistant: A Student Initiative Transforming Medical Education

Climate change was named the greatest global health threat of the upcoming century.[1] A group of medical students of the Academical Medical Centre of Utrecht (UMC Utrecht) wondered: why isn't this crisis part of our education? This question sparked the creation of the CO_2-assistant in 2019, a student-led initiative dedicated to integrating planetary health into medical education at UMC Utrecht and promoting planetary health in society. The CO_2-assistant was acknowledged and appreciated by their faculty and university, through winning the university's Student Award Exceptional Extracurricular Achievements. What started in Utrecht has since grown into a national movement, now active in seven other locations across the Netherlands. Through advocacy, workshops, guest lectures, and co-developing educational innovations, the CO_2-assistant collaborates with faculty, other student-advocacy groups such as the International Federation of Medical Students Associations (IFMSA)-NL Act4Health groups, and stakeholders to shape the curriculum of the future. *The CO_2-assistant, Utrecht, the Netherlands, via:* https://co2assistent.nl/

10.4.3 The Carbon Literacy Project: An International Course for Researchers

Courses accredited by the Carbon Literacy Project can engage stakeholders in low carbon action and behaviour in their professional setting.[68] The Carbon Literacy Project is a UK registered charity governed by the Carbon Literacy Trust.[69] Courses are designed for a specific sector or community and are typically short courses (1 to 2 working days) either in-person, online, or a mixture. Courses relevant to the academic and research sector include those offered by The Alliance for the Sustainability Leadership in Education (EAUC) course,[70] the Zeroverse course for professionals and post-doctoral students in research settings,[71] and the Big Zero course for international students studying abroad.[72]

Carbon Literacy courses cover the scientific underpinning of climate change and equip participants with tools, guidance, and best practices to reduce carbon emissions in our sector. Each participant develops two climate action plans to reduce the environmental impact of their work or studies that they will take forward as a result of the training: one that they can achieve largely on their own; and one that they can achieve by working with colleagues. Courses help participants to hone their skills in collaboration and communication to work with others and effect positive change.

Research and higher education institutions offer Carbon Literacy training as part of their orientation programme for new staff or integrate the training in their professional development programme. Some institutions incorporate climate actions developed through the Carbon Literacy training into their institutional sustainability plans. Sustainability working groups take the training together to share lessons learned and develop goals together. *Kate Withfield, Zeroverse, Okinawa, Japan.*

10.4.4 A Transdisciplinary Challenge-based Learning Course

The Dutch course 'Planetary Health and Resilient Health Systems' is a transdisciplinary, challenge-based programme where master's students from four universities collaborate to solve real-world problems. The course emphasises and reflects the five components of action-oriented learning.[38] Over 6–8 weeks, students work in interdisciplinary teams to design solutions for challenges posed by a local hospital's sustainability office. Through partnerships with clinicians and external stakeholders, students develop skills to address the complex issue of planetary health. Additionally, students collaborate with international students to develop an infographic to present to the class. This assignment provides students with an opportunity for not only interdisciplinary but international and intercultural collaboration to dive into the complex and persistent issue of planetary health. This course highlights the interconnectedness of human and environmental health, aiming to find sustainable solutions that support both societal needs and the planet's well-being. *Alliance of Eindhoven University of Technology, Wageningen University & Research, Utrecht University and University Medical Centre Utrecht (EWUU-alliance).*

10.4.5 A PhD Project Fully Dedicated to Greening Lab-based Research

Trained as a biochemist, Bianca Schell is now diving into laboratory sustainability for her PhD, supported by an Evonik Foundation scholarship. Her focus? To enhance lab practices by cutting down on single-use plastics, boosting lab plastic recycling that is otherwise going to incineration, and making lab operations more energy-efficient—especially when it comes to ventilation. After conducting an in-depth comparative study on the sustainability programmes LEAF and My Green Lab,[66] the research focuses on developing an open-access, tailored research-specific tool designed to help scientists implement greener lab practices. *Bianca Schell, Technical University of Darmstadt & University of Konstanz, Germany.*

10.4.6 Innovative Courses on Sustainable Research

Jeroen Dobbelaere is leading the way in integrating sustainability into scientific research through innovative courses. At the University of Vienna, the course; "Science or Science Fiction" focuses on helping master's and PhD students measure and reduce the environmental footprint of their life science research. The course covers key areas such as energy use, mobility, lab equipment, and products, while encouraging students to engage with internal and external stakeholders to catalyse change. At ISTA, Dobbelaere mentors PhD students through a cross-disciplinary course on measuring and modelling environmental impacts related to energy, mobility, and product use on campus. Students work in collaboration with faculty members, analyse current practices using data from the campus, and present their findings to drive sustainability. Additionally, he also administered a workshop format; "The Climate Café" that explores how research is impacting the environment and uses a collaborative approach to find ways to reduce CO_2 emissions in their labs, focusing on sustainable practices across both experimental and computer-based research. *Jeroen Dobbelaere, Institute of Science and Technology Austria (ISTA) and University of Vienna, Austria*

10.5 Getting Started! Transforming to Education for Sustainability and Equity in Research

Educational approaches must evolve throughout a scientist's career, from aspiring researchers to postdoctoral scholars, adapting to their changing needs. There is a role for everyone in transforming education to embed sustainability and equity in research (see Table 10.2 for inspiration). Key facilitators that accelerate the transformation of education, include role models amongst peer groups, decision makers with bold leadership, creating space for changemakers, visibility of sustainability initiatives (show and tell), the institutionalisation of a sustainability and equity culture, and joyful collaborations both within and beyond institutions. Synergy can occur when grassroots advocacy aligns with top-down support.

Across all the examples discussed, one principle stands out: the importance of starting and learning by doing. Unsure where to begin? Just take a first step! Strengthen your impact, resilience, and sense of hope by connecting with a community. If everybody does their homework on sustainability and equity well and finds ways to integrate it into their research area, we not only get good grades but can also make research fit for a just future within the planetary boundaries.

Acknowledgements

This chapter came into existence by the help of Jelle Boeve-de Pauw (Freudenthal Institute, Utrecht University, the Netherlands), Marit de Kort (Graduate School of Life Sciences, UMC Utrecht, the Netherlands; Green Labs NL), Anjali Wijnhoven (Faculty of Medicine, UMC Utrecht, the Netherlands; The CO_2-assistant), Noortje Campman (Faculty of Medicine, UMC Utrecht, the Netherlands), and Berent Prakken (Faculty of Medicine, UMC Utrecht, the Netherlands). The authors of the chapter received funding from the Dutch Ministry of Education, Culture and Science.

References

1. M. Romanello, M. Walawender, S. C. Hsu, A. Moskeland, Y. Palmeiro-Silva and D. Scamman, *et al.*, The 2024 report of the Lancet Countdown on health and climate change: facing record-breaking threats from delayed action, *Lancet*, 2024, **404**(10465), 1847–1896.
2. IPBES, Global assessment report on biodiversity and ecosystem services of the Intergovernmental Science-Policy Platform on Biodiversity and Ecosystem Services, Zenodo, 2019. Available from: https://zenodo.org/records/6417333.
3. R. S. Ulrich, View Through a Window May Influence Recovery from Surgery, *Science*, 1984, **224**(4647), 420–421.
4. World Health Organization, Green and blue spaces and mental health: new evidence and perspectives for action, Copenhagen: World Health Organization Regional Office for Europe, 2021. Available from: https://iris.who.int/bitstream/handle/10665/342931/9789289055666-eng.pdf.
5. T. Freese, N. Elzinga, M. Heinemann, M. M. Lerch and B. L. Feringa, The relevance of sustainable laboratory practices, *RSC Sustainability*, 2024, **2**(5), 1300–1336.
6. L. Lannelongue, H. E. G. Aronson, A. Bateman, E. Birney, T. Caplan and M. Juckes, *et al.*, GREENER principles for environmentally sustainable computational science, *Nat. Comput. Sci.*, 2023, **3**(6), 514–521.
7. P. Glasziou and I. Chalmers, Research waste is still a scandal—an essay by Paul Glasziou and Iain Chalmers, *BMJ*, 2018, **363**, k4645.
8. K. A. Robinson, K. Brunnhuber, D. Ciliska, C. B. Juhl, R. Christensen and H. Lund, Evidence-Based Research Series-Paper 1: What Evidence-Based Research is and why is it important?, *J. Clin. Epidemiol.*, 2021, **129**, 151–157.
9. I. Chalmers and P. Glasziou, Avoidable waste in the production and reporting of research evidence, *Lancet*, 2009, **374**(9683), 86–89.
10. H. Lund, C. B. Juhl, B. Nørgaard, E. Draborg, M. Henriksen and J. Andreasen, *et al.*, Evidence-Based Research Series-Paper 2: Using an Evidence-Based Research approach before a new study is conducted to ensure value, *J. Clin. Epidemiol.*, 2021, **129**, 158–166.
11. K. Thapa, W. J. V. Vermeulen, M. M. De Waal, P. Deutz and H. Q. Nguyên, Towards a Just Circular Economy Transition: the Case of European Plastic Waste Trade to Vietnam for Recycling, *Circ. Econ. Sustainability*, 2024, **4**(2), 851–876.
12. P. Y. Wong, Exporting Pollution, Colonizing Health, in *Plastic Waste Trade: A New Colonialist Means of Pollution Transfer*, ed. S. Gündoğdu, Springer Nature Switzerland, Cham, 2024, pp. 259–289.
13. H. Bhakuni and S. Abimbola, Epistemic injustice in academic global health, *Lancet Global Health*, 2021, **9**(10), e1465–e1470.
14. O. Dewidar, N. Elmestekawy and V. Welch, Improving equity, diversity, and inclusion in academia, *Res. Integr. Peer Rev.*, 2022, **7**(1), 4.
15. S. Ratcliffe, Nelson Mandela 1918–2013, South African Statesman, in *Oxford Essential Quotations*, Oxford University Press, 6th edn, 2018.

Available from: https://www.oxfordreference.com/display/10.1093/acref/9780191866692.001.0001/q-oro-ed6-00007046

16. M. Rieckmann, Learning to transform the world: Key competencies in Education for Sustainable Development, in *Issues and trends in education for sustainable development*, UNESCO Publishing, 2018, pp. 39–59.
17. European Commission, ResearchComp: The European Competence Framework for Researchers, European Commission, 2022. Available from: https://research-and-innovation.ec.europa.eu/document/download/7da29338-37bf-4d51-b5eb-a1571b84c7ad_en?filename=ec_rtd_research-competence-presentation.pdf.
18. European Commission, Open Science, 2024. Available from: https://research-and-innovation.ec.europa.eu/strategy/strategy-research-and-innovation/our-digital-future/open-science_en.
19. Planetary Health Alliance, What is Planetary Health?, 2025, Defining Planetary Health. Available from: https://www.planetaryhealthalliance.org/planetary-health.
20. Planetary Health Alliance, 2025, About PHA. Available from: https://www.planetaryhealthalliance.org/about-pha.
21. The Lancet Planetary Health Home Page, 2025. Available from: https://www.thelancet.com/journals/lanplh/home.
22. Planetary health: an emerging field to be developed, Royal Netherlands Academy of Arts and Sciences, Amsterdam, 2023.
23. C. A. F. Guzmán, A. A. Aguirre, B. Astle, E. Barros, B. Bayles and M. Chimbari, *et al.*, A framework to guide planetary health education, *Lancet Planet. Health*, 2021, **5**(5), e253–e255.
24. UMC Utrecht, UMC Utrecht annual report, Utrecht: UMC Utrecht, 2024. Available from: https://annualreport.umcutrecht.nl/annual-report/our-students/planetary-health.
25. Kenniscentrum Global Health, Planetary Health bril – KCGH, 2025. Available from: https://www.kcgh.nl/planetary-health-glasses.
26. K. Brundiers, M. Barth, G. Cebrián, M. Cohen, L. Diaz and S. Doucette-Remington, *et al.*, Key competencies in sustainability in higher education—toward an agreed-upon reference framework, *Sustainability Sci.*, 2021, **16**(1), 13–29.
27. A. Wiek, L. Withycombe and C. L. Redman, Key competencies in sustainability: a reference framework for academic program development, *Sustainability Sci.*, 2011, **6**(2), 203–218.
28. A. Redman and A. Wiek, Competencies for Advancing Transformations Towards Sustainability, *Front. Educ.*, 2021, **6**. Available from: https://www.frontiersin.org/journals/education/articles/10.3389/feduc.2021.785163/full.
29. F. Dekkers, Greening science: what's in it for you?, *Nature*, 2024. Available from: https://www.nature.com/articles/d41586-024-03011-2.
30. Sustainable UCL, 2022, LEAF - Laboratory Efficiency Assessment Framework. Available from: https://www.ucl.ac.uk/sustainable/take-action/staff-action/leaf-laboratory-efficiency-assessment-framework.

31. My Green Lab, 2025. My Green Lab Certification. Available from: https://www.mygreenlab.org/green-lab-certification.html.
32. Green DiSC: a Digital Sustainability Certification|Software Sustainability Institute, 2025. Available from: https://www.software.ac.uk/GreenDiSC.
33. OpenAIRE, 2025, Guide for Researchers: How to make your data FAIR. Available from: https://www.openaire.eu/how-to-make-your-data-fair.
34. *Rethinking Standards and Reference Models|Gendered Innovations in Science, Health & Medicine, Engineering, and Environment*, ed. L. Schiebinger, I. Klinge, H. Paik, I. Sánchez De Madariaga, M. Stefanick, 2011. Available from: https://genderedinnovations.stanford.edu/methods/standards.html.
35. L. W. Anderson and D. R. Krathwohl, *A taxonomy for learning, teaching, and assessing: a revision of Bloom's taxonomy of educational objectives: complete edition*, Addison Wesley Longman, Inc., 2001. Available from: https://eduq.info/xmlui/handle/11515/18824.
36. C. Loughlin, S. Lygo-Baker and Å. Lindberg-Sand, Reclaiming constructive alignment, *Eur. J. Higher Educ.*, 2021, **11**(2), 119–136.
37. G. M. Slavich and P. G. Zimbardo, Transformational Teaching: Theoretical Underpinnings, Basic Principles, and Core Methods, *Educ. Psychol. Rev.*, 2012, **24**(4), 569–608.
38. E. Sinakou, V. Donche, J. Boeve-de Pauw and P. Van Petegem, Designing Powerful Learning Environments in Education for Sustainable Development: A Conceptual Framework, *Sustainability*, 2019, **11**(21), 5994.
39. A. Cook-Sather and K. E. Matthews, Pedagogical partnership: Engaging with students as co-creators of curriculum, assessment and knowledge, in *University Teaching in Focus: A Learning Centred approach*, Routledge, 2nd edn, 2021, pp. 243–259.
40. Centre for Academic Teaching and Learning Universiteit Utrecht, Teaching and learning collection, 2025, All learning activities. Available from: https://teaching-and-learning-collection.sites.uu.nl/learning_activity/forced-debate/.
41. C. Hickman, E. Marks, P. Pihkala, S. Clayton, R. E. Lewandowski and E. E. Mayall, *et al.*, Climate anxiety in children and young people and their beliefs about government responses to climate change: a global survey, *Lancet Planet. Health*, 2021, **5**(12), e863–e873.
42. A. Levesque and R. Rocque, An innovative framework for teaching climate change: integrating emotion coping strategies and exploring a new climate emotion scale among Canadian youth, *Environ. Educ. Res.*, 2024, 1–21.
43. Green Minds MB, Climate Hope: A Made-in-Manitoba Resource ebook for Supporting the Teaching and Learning of Ecological Justice, Eco-anxiety, and Climate Action, 2023. Available from: https://docs.google.com/document/d/1VDrXxFVkKSu9PR_yPVvduKucbUf2SjEu8-pwyedMvSc/edit?usp=embed_facebook.
44. S. Walpole, D. Pearson, J. Coad and S. Barna, What do tomorrow's doctors need to learn about ecosystems? – A BEME Systematic Review: BEME Guide No. 36, *Med. Teach.*, 2015, **38**, 1–15.

45. J. Macy and C. Johnstone, *Active Hope (revised): How to Face the Mess We're in with Unexpected Resilience and Creative Power*, New World Library, 2022, p. 290.
46. Active Hope, 2025, The Authors. Available from: https://www.activehope.info/the-authors.
47. W. Finnegan and C. d'Abreu, The hope wheel: a model to enable hope-based pedagogy in Climate Change Education, *Front. Psychol.*, 2024, **15** Available from: https://www.frontiersin.org/journals/psychology/articles/10.3389/fpsyg.2024.1347392/full.
48. R. Stephen, Covey, The 7 Habits of Highly Effective People: Powerful Lessons in Personal Change, RosettaBooks LLC.
49. ActiveHope.training, Active Hope Foundations Training a free online course, 2025. Available from: https://activehope.training.
50. A. D. Groenewegen, A. M. Wijnhoven, J. Janmaat, M. A. C. de Kort and J. F. Dekkers, Guideline Socially Responsible Research (Version 1.0), 2024. Available from: https://www.kcgh.nl/handreiking-maatschappelijk-verantwoord-onderzoek-en.
51. S. Heidari, T. F. Babor, P. De Castro, S. Tort and M. Curno, Sex and Gender Equity in Research: rationale for the SAGER guidelines and recommended use, *Res. Integr. Peer Rev.*, 2016, **1**, 2.
52. Genderful Research World, 2025. Available from: https://genderfulresearchworld.com/.
53. James Lind Alliance Priority Setting Partnerships, Homepage|NIHR JLA, 2025. Available from: https://www.jla.nihr.ac.uk/.
54. American Psychological Association, APA Style, 2023. Bias-free language. Available from: https://apastyle.apa.org/style-grammar-guidelines/bias-free-language.
55. T. Rood and M. Kishna, Outline of the circular economy, PBL Netherlands Environmental Assessment Agency, 2019. Available from: https://circulareconomy.europa.eu/platform/sites/default/files/pbl-2019-outline-of-the-circular-economy-3633.pdf.
56. UK EQUATOR Centre, EQUATOR Network | Enhancing the QUAlity and Transparency Of Health Research, 2025. Available from: https://www.equator-network.org/.
57. Kenniscentrum Global Health, Elearning Kenniscentrum Global Health, 2025. Login to website|elearning.kcgh.nl. Available from: https://elearning.kcgh.nl/login/index.php.
58. University of Gloucestershire, Duurzame PABO, Frederick University, Haute École Pédagogique du canton du Vaud, Italian Association for Sustaniability Science, K. T. O. Szövetsége, *et al.* The RSP Framework – A Rounder Sense of Purpose, 2025. Available from: https://aroundersenseofpurpose.eu/framework/themodel/.
59. Framework – Inner Development Goals, 2025. Available from: https://innerdevelopmentgoals.org/framework/.
60. University of Amsterdam, Technical University of Eindhoven, Wageningen University & Research, Utrecht University, University Medical

Centre Utrecht (EWUU) alliance, Transition Makers Toolbox, 2025. A Teaching Toolbox for Societal Transitions. Available from: https://transitionmakers.nl/.
61. Waag's Co-Creation Lab, Co-Creation Navigator, 2025. Available from: https://ccn.waag.org/navigator/.
62. J. Holst, J. Grund and A. Brock, Whole Institution Approach: measurable and highly effective in empowering learners and educators for sustainability, *Sustainability Sci.*, 2024, **19**(4), 1359–1376.
63. J. Holst, Towards coherence on sustainability in education: a systematic review of Whole Institution Approaches, *Sustainability Sci.*, 2023, **18**(2), 1015–1030.
64. C. Hoadley, What is a Community of Practice and How Can We Support It?, in *Theoretical foundations of learning environments*, 2012, pp. 287–300.
65. W. Willett, J. Rockström, B. Loken, M. Springmann, T. Lang and S. Vermeulen, *et al.*, Food in the Anthropocene: the EAT–Lancet Commission on healthy diets from sustainable food systems, *Lancet*, 2019, **393**(10170), 447–492.
66. B. R. Schell and N. Bruns, Lab sustainability programs LEAF and My Green Lab®: impact, user experience & suitability, *RSC Sustainability*, 2024, **2**(11), 3383–3396.
67. M. B. Buchholz, My lab's lighter footprint, *Science*, 2024, **383**(6680), 342–342.
68. The Carbon Literacy Trust, The Carbon Literacy Project, 2025, The Carbon Literacy Project. Available from: https://carbonliteracy.com/.
69. The Carbon Literacy Trust, The Carbon Literacy Project, 2025, About Us. Available from: https://carbonliteracy.com/about-us/.
70. The Alliance for Sustainability Leadership in Education, EAUC Home| EAUC, 2025. Available from: https://www.eauc.org.uk/home.
71. Zeroverse, 2025, Zeroverse. Available from: http://zeroverse.org/.
72. Big Zero, Big Zero – Accredited Carbon Literacy Training Program, 2025. Available from: https://www.big-zero.org/.

CHAPTER 11

Technicians: The Front Line of Impact†

MATTHEW BENNET[a,b]

[a] University of East Anglia, UK; [b] University of Cambridge, UK
Email: matthew.bennett@admin.cam.ac.uk

11.1 Introduction

Technicians are vital to scientific research and innovation and make significant contributions to training the next generation of scientists. There are over 30 000 technicians working in UK universities and research institutes[1] and significant numbers working in industry. Technicians underpin, participate in, and enable scientific research, innovation, knowledge exchange and education. Despite their vital role in the scientific ecosystem, technical roles and careers have typically been less visible and recognised than many other roles and career routes. Thankfully technicians are becoming more prominent in the UK, with the contributions of these key roles being recognised, supported and increasingly visible. There are a variety of successful programmes and initiatives delivering this important focus, namely the Technician Commitment, the Research England funded Midlands Innovation (MI) TALENT programme, the UK Institute of Technical Skills and Strategy (ITSS), and the leadership and influence of the Gatsby Foundation. These provide essential insight and impact for technical careers following the 2016 Report of the Independent Panel on Technical Education which "highlighted a national shortage of technicians across all sectors and the challenges this created for raising UK productivity".[1] Furthermore, in

The Sustainability of Science: How Science Impacts the Environment, and What Can Be Done
Edited by Martin Farley

Published by the Royal Society of Chemistry, www.rsc.org

response to evidence provided by such initiatives, institutional, organisational and governmental policy is driving further focus, funding and programmes with impactful change for technical careers in the UK.

Parallel to this crucial focus on technical careers, is the necessity to address the environmental impact of scientific research; specifically, laboratory research, digital research and clinical research. Scientific research and education generate knowledge and solutions to address global societal challenges including climate change. Our reliance on science across all sectors and the unequivocal benefits of scientific advances are clear. However, we must address the significant resource requirements of scientific activities to limit the negative impacts on the environment, and align research and education with net zero challenges. Technicians are well placed and equipped to make crucial contributions to this significant challenge, as will be discussed throughout this chapter.

To understand the significance of technicians in sustainable science, we begin by exploring what technicians do; examining the definition and diversity of technical careers supporting science, with examples of UK initiatives supporting these technical careers. Expanding beyond common preconceptions and stereotypes of a technician, the breadth of roles and the skills of technical professionals will be brought to light to evidence the relevance and importance of technicians for sustainable science. We delve into the challenges for environmental sustainability in science and reveal the vital role and opportunities for the technical workforce to enable, deliver and support behavioural and operational change. In recognising their significant contributions, we shine a light on the impact technicians are making on the frontline to improve sustainability in scientific facilities.

11.2 Technicians and the Technical Community

Whilst it is useful to first define 'technicians' in order to explore their role in the context of environmental sustainability in science, this is not a simple task. Technician roles are broad, spanning across sectors and disciplines, encompassing a wide range of skills, levels, functions and job titles. Historic lack of visibility and blurred understanding of the technician role are compounded by long-standing issues with technical education systems, and the absence of clearly defined and adopted technical career pathways in the UK. Thankfully, through leadership and lobbying, a variety of focussed initiatives have been launched to develop and promote visibility and support technical careers, none more prominent than the Technician Commitment. This leading UK initiative provides a sector aligned approach for university and research institutions to address 4 key areas: visibility, recognition, career development and sustainability, for technical roles and careers. Since its launch in 2017, over 120 organisations have become signatories, creating action plans at the institutional level. The Technician Commitment is led by a steering group of sector bodies and, in addition to its signatories, supported by over 20 professional organisations from the UK Higher Education

(HE) and research sector. The initiative has resulted in a burst of activity and rise in supporting collaborations and programmes for technical staff across UK HE and research institutes. The Inception to Impact report published in 2024 provides a summary of progress and recognition of achievements.[2] The Technician Commitment first calls on signatories to identify and define their technicians. Many signatory organisations have struggled to do so, highlighting the lack of understanding of the technical workforce across much of the HE and research sector. Thankfully, defining technicians has been a feature of more recent work focussed on elevating the visibility of technicians in the UK. The Research England funded TALENT Commission Report published in February 2022, provides a valuable summary of definitions and discourse. The report finds commonality of "the emphasis on the practical element of a technician's role" within existing definitions.[1] Subsequent to this, Dr Francis Wragg authored an insightful chapter entitled 'Defining technicians'[3] in Technicians in Higher Education and Research.[4] Wragg proposes the definition below, which aims to encompass "the broad range of common and/or expected duties" without limits to discipline, sector or function.

> *"Technicians are trained and/or skilled in the techniques, tools and/or technology of their subject, providing the practical application of knowledge, and/or the expertise to best utilise this practical application through managing, teaching, and/or training others to deliver it, and/or through maintaining and developing the environment, standards, resources and/or facilities required for this practical application to be delivered"*

If we apply this definition to the breadth of scientific disciplines, it is easy to see the reliance of science in the key aspects of the technician role. We can conclude: technicians are vital for science. Furthermore, if we consider the environmental impacts resulting from science facilities and activities, we start to build a picture of the importance of technicians for the sustainability of science.

The range of job titles held by technical staff within the UK HE and research sector and the key responsibilities and contributions presented in the TALENT Commission Report[1] provide useful context for the diversity of technical roles. Job titles include Director of Technical Services, Facility Manager, Laboratory Manager, Bioinformatician, Research Technician, Teaching Technician, IT Specialist and Workshop Technician. The report highlights the range of tasks and specialisms within roles alongside a huge variety of functions delivered by technical staff over 7 key areas of responsibility: research, teaching, maintenance, health and safety, finance, infrastructure and people.

Work focused on elevating the profile and contributions of technicians has been largely focused on the UK Higher Education (HE) and research sector. This work provides value beyond this scope and is well linked to other programmes and initiatives in Further Education (FE) and industry in the UK.

One particularly useful example when exploring the diversity of technical roles is the Gatsby Foundation's Technicians We Make the Difference campaign which showcases the variety of technician roles across industries.[5] Science is dependent on many of the 100 technician roles showcased by the initiative, which also provides focus on the key skills of technicians which are of high relevance across scientific disciplines. The diversity of roles and skills of technicians clearly apply to environmental impacts of science.

Technician roles include core operational and specialist technicians in scientific labs and facilities and strategic technical leaders, Research Software Engineers (RSEs) and High Performance Computing (HPC) teams, Building Management System (BMS) technicians and technical design teams. It should be clear by now that technicians contributing to science go well beyond the stereotypical images of a technician in a lab coat bustling about in the background.

Within scientific organisations technical roles are increasingly important for the management and delivery of central facilities, operations and research platforms. Central operations and research platforms provide shared infrastructure, facilities and equipment. These vary widely, from glass washing and sterilisation facilities to cutting edge analytical equipment and digital research infrastructure. Led and operated by technical teams, they provide a good example of the diversity of skills, roles and specialisms of technical staff. Research is increasingly reliant on these central facilities as modern science moves further towards a shared model for facilities and operations with research delivered by teams. Shared facilities are well aligned with sustainable research operations, optimising space and equipment utilisation. Further opportunities for improving sustainability in science can be supported by the shared model and technicians are key enablers for this.

The future of science and technological advancement requires a growing skilled technical workforce. The lack of clearly defined technical career pathways poses a challenge to filling the UK technical skills gap. The National Technician Development Centre (NTDC) provides several resources and offerings, including The Higher Education Technical Taxonomy (HETT) Framework[6] and NTDC Survey.[7] These are designed to support universities and organisations to map technical roles by identifying the varying roles and skills across their technical workforce. An organisational approach such as this can provide insights for workforce planning and development of technical career pathways. The Institute of Technical Skills and Strategy (ITSS)[8] is also supporting the development of career frameworks, providing resources and opportunities for organisations to collaborate through the ITSS Technical Careers Pathways Lab.[9] This supports one of the recommendations of the Talent Commission, which was launched "to address the paucity of insight and knowledge about the technical skills, roles and careers seen across UK higher education and research".[1] The commission made 16 overarching recommendations to the sector, including government, funders, employers, learned societies and professional associations, which are forging change for technical careers. The commission was part of the highly impactful Research

England funded Talent programme. This programme published a number of important sector reports highlighting the key role and perspectives of technicians in a variety of areas, including Equipment Sharing, Research Culture, Health and Safety and Knowledge Exchange (KE).[10] All of which are of key relevance to technicians' role and potential for improving environmental sustainability of scientific research and innovation practice.

The UK Institute of Technical Skills and Strategy, established in 2024, is led by Director Dr. Kelly Vere MBE: pioneer, visionary leader and passionate advocate for technical careers, who began her own career as a junior technician at the University of Nottingham. Funded by Research England, part of UK Research and Innovation (UKRI), the ITSS is building on the work of the TALENT programme and is now the home to the Technician Commitment. The ITSS is accelerating and advancing technical careers, aligning and progressing the successes and insights of UK technician initiatives and programmes. At the centre of the technical community, the institute delivers connectivity across stakeholders and focussed action to support organisations, provide opportunities for technicians, create national technical networks, connect communities and conduct research to deliver insights which influence policy change. The institute presents significant opportunities for further supporting and engaging technicians, and its broader stakeholders, with sustainability in science. The ITSS and its preceding technician initiatives are funded, supported, and engage with a broad set of organisations across the UK research, education and innovation sector. These include government, funders, universities, professional associations and learned societies. Research insights, engagement and collaboration are delivering impact for technical careers and the sector. There is potential to accelerate progress towards environmental sustainability by identifying and further supporting the role of technicians within organisations' sustainability plans, programmes and strategy.

A number of UK funders and sector bodies are signatories and supporters of the Technician Commitment and are driving further positive change for technical careers.

UKRI's people and teams action plan,[11] published in 2023, builds on and brings together previous Researcher Development Concordat and Technician Commitment action plans. The single plan provides a holistic approach for people and teams across the UKRI community and actions towards shared challenges. As highlighted in the plan and the recommendations of the TALENT commission report,[1] technician and researcher roles are not always easily distinguishable as a result of the diversity of roles. Some technicians are leading experts in their research field and many others have the necessary expertise to lead and co-lead research. This is also addressed in the action plan by ensuring technicians are eligible to apply for funding opportunities where appropriate. UKRI is leading action and impactful change to ensure technical careers are supported across the sector through funding and as an employer of technicians within its strategically funded research institutions (*i.e.* Centres, Institutes and Units). Technician roles have been featured in UKRI's 101 jobs that change the world campaign,[12] which promotes visibility. Technicians' voices and expertise are increasingly included in consultation and peer review processes. Continued work to support narrative CVs[13] will also benefit those with diverse research careers such as technicians. It is of course within these narrative CVs and funding opportunities that the advancements and delivery of sustainable science are most likely to feature.

In 2023, the Engineering and Physical Sciences Research Council (EPSRC) took a further proactive step providing a £16 million funding opportunity for Research Technical Professionals (RTPs) awarded to 11 Strategic Technical Platforms, lead or co-lead by RTPs.[14,15] The range of projects signify the diversity of RTPs' focus under the EPSRC strategic remit. This investment in technician-led research and innovation is already delivering impact for the technician community and beyond. One of the projects, led by ITSS Co-Director Ian Hancox, will deliver knowledge, skills and opportunities for RTPs nationally through the ITSS Technological Specialist Network (TSN).[16] Projects such as these could benefit from sharing best practice, solutions and training on sustainability; utilising the power of national technical networks. The strategic platforms have many industry partners across the innovative projects. The Technical Platform for Innovation and Knowledge Exchange (T-PIKE) project[17] will develop innovation and Knowledge Exchange pathways for technicians, working with industry partners with a specific focus on AI and net zero. The TALENT programme published a report on the role of technicians in Knowledge Exchange,[10] showcasing a number of technician-led projects with industry partners, including a sustainable science innovation. Knowledge Exchange is another example of the broad ways technicians can use their expertise to deliver solutions for sustainable research practice which could be further supported and explored. This is also well linked to the innovation capabilities of technicians. Paul Lewis, Professor of Political Economy at King's College London (KCL), published a number of papers evidencing the importance of technical skills and knowledge for innovation. Lewis' 2019 report funded by the Gatsby Foundation, Technicians and Innovation: A Literature Review,[18] recommended "greater consideration should be given to the importance of technicians in the innovation system". Technicians' key problem-solving skills, specialist expertise and know-how for practical application may be highly valuable for providing further advances using this concept. As such, Knowledge Exchange between technicians working in scientific research, and the industry partners which develop the tools used for scientific research, such as specialist equipment, could be supported to provide solutions to accelerate progress towards sustainable science. Funded training and development opportunities for entrepreneurship, available nationally to technicians, include the Biotechnology and Biological Sciences Research Council (BBSRC) Innovation to Commercialisation of University Research (ICURe)[19] and the Nottingham University's Your Entrepreneurs Scheme (YES) programme.[20]

In order to paint a true picture of the technician landscape, this section has been intentionally heavily focussed on technician roles, the technician community and impactful technician-focused initiatives. This is in the spirit of increasing visibility of technicians and the vital importance of technical roles. Technicians are of crucial importance to the sustainability of science.

11.3 The Role of Technicians for Environmental Sustainability in Science

The growing prevalence of initiatives focused on the vital contributions of the technical workforce to science is paralleled by increasing attention on the environmental impacts of scientific activities. As organisations continue to address net zero ambitions, now is the time to connect the dots. We must recognise the potential of the technical community and the role they can play in overcoming the challenges that scientific facilities and operations pose, in order to bring about progress towards sustainable science.

The climate crisis and interlinkages with public and ecological health pose grave concern to the future of people and the planet. International, national and organisational targets set in agreements, legislation and policy aim to accelerate action to limit the impacts on global climate. Scientific research provides evidence for and solutions to the climate crisis. As such there is a growing consensus for the need to simultaneously address the environmental impacts from the scientific activities which we rely so heavily upon.

The resource requirements of science are significant and broad. Infrastructure, facilities and equipment required for science lead to high energy use and intensity. This contributes substantially to organisations scope 1 and 2 carbon footprints, particularly in comparison to their spatial footprints. Maximising space and equipment utilisation and energy efficiency are key. Beyond significant energy demands, the resources consumed by scientific activities also contribute heavily to the significant scope 3 carbon challenges organisations face. Emissions associated with the upstream supply chain being a key example. Knowledge of best practice in these areas and the challenges for sustainability in science are provided throughout this book.

Ongoing 'bottom-up' efforts have made notable advancements in the sustainability of scientific operations, as technological developments and design have concurrently delivered improvements. Technicians have contributed to both and are integral to the development of and employment of such technology. Examples of technician 'bottom up' actions will be shared later. Many further opportunities exist for widespread expansion which must be focused on, supported and led from the 'top' to achieve faster progress. Technicians are a key workforce for this change and leaders in organisations, including technical leaders, should recognise this potential. Useful 'bottom-up' initiatives for sustainable science practice must be supported and move to imperative strategic-led action. Leaders of research, innovation and education organisations must meet the need for urgent and systemic change to improve environmental sustainability, working collectively across the sector. In doing so, they should further mobilise key workforces, such as technicians and employ holistic approaches to engaging and educating the scientific workforce of today and tomorrow. Bold strategies may achieve the shared vision for net zero science to limit further damage to climate, nature and people.

The Concordat for Environmental Sustainability in Research and Innovation Practice,[21] co-developed collaboratively by the UK research and innovation sector, is an example of collective agreement and action towards environmental leadership and stewardship. Funders of scientific research are increasingly highlighting the importance of and engaging in collective action for environmentally sustainable research through statements like the Heidelberg Agreement.[22] In the UK, and further afield, funders have also developed environmental sustainability policies. Some key examples being Wellcome,[23] Cancer Research UK (CRUK)[24] and the DFG[25] in Germany. These statements and policies are 'top-down' signals from funders and research performing organisations to take ownership and leadership at all levels across the research community to address the environmental sustainability of research. In the scaling of activity there will be a heavy reliance on progress towards sustainability in functions and operations provided by the Research Technical Professionals (RTPs) and technical teams, particularly for science.

Technical staff can influence change across scientific communities and progress can be delivered as a result of inclusion, support and recognition of the technician workforce's capabilities. To drive change, it is essential to provide opportunities for technicians to further their knowledge of sustainability which they can practically apply within the varied specialisms. It is also imperative technicians are empowered and supported to deliver on these challenges with best practice and innovative solutions. The strength of technical communities, within and beyond the organisations, can maximise potential to deliver # widespread adoption of solutions to reduce environmental impacts in science operations.

There are countless examples of technical staff leading programmes of engagement and focussed activity to reduce the environmental impacts of research, particularly in laboratories, where efforts have been maturing for some years. Martin Farley, a leading expert in sustainable research, authored a chapter entitled Technicians and Environmental Sustainability in the aforementioned book Technicians in Higher Education and Research.[4] Farley's contribution includes interviews with some leaders in the technical community on how their work supports environmental outcomes. The interviews of Peter Morgan from the University of Southampton and Dr Elaine Fitzcharles from the British Antarctic Survey, provide great insight into how their roles are crucial for environmental successes.

Although there is a huge diversity of technical roles (as explored in the previous section) there is often a high proportion of technical staff working in central teams and shared centralised operations and facilities. These provide a wide array of functions and services which support and enable scientific research and teaching. There is also significant variance between and within institutions in technical staff structures. Nonetheless, the roles and functions of technical teams provide the backbone of operations in scientific facilities; unpinning research, innovation and education. Technicians also provide a spine of knowledge in facilities. Within central operational teams and facilities and in localised settings, they are so often the go-to point of contact for support and know-how for the broader staff and student communities. These are of key importance when considering sustainability in science for a number of reasons. First, is the direct impact technicians can enable within the central facilities and services they operate and manage by integrating sustainable practices. Second are the benefits which shared central operations, facilities and functions provide by improving space, equipment and resource utilisation. Finally, the less tangible impacts of the expertise, advice and influence provided by technical teams to the broader community. The latter is often more visible and formalised in the role of teaching technicians, many of whom directly teach or train students in HE, supporting the development of future scientists. Similarly, Health and Safety facing roles or aspects, such as compliance audits, training and management, are formalised in technician job descriptions. Technicians also often deliver local inductions, provide a variety of training and day-to-day support across scientific research, and are the focal point for know-how and enquiries. We can conclude that technical teams and communities provide central 'hubs', both physical operations hubs and knowledge hubs. This is of high importance for sustainability in science. The wide-reaching roles across the technician community present opportunities for technicians to embed sustainable practices in directly managed operations, and support and influence others. These central 'hubs' (technician communities) can unlock significant opportunities for sustainability and should be a focus area for engagement and training on sustainability in science. In turn, there is potential for technicians' career development and for technicians to provide more formal training provisions in sustainability for others in the future, analogous to technician-led Health & Safety training. Another common people-facing feature of the central 'hubs' technician communities provide is collaboration and connectivity with other internal groups and central teams. Through their work technicians are often conduits to finance, estate or building services, human resources and of course central sustainability teams. Externally facing, technicians are key contacts for scientific suppliers and contractors. This further demonstrates the reach of technical roles and potential this could provide for improvements in operational sustainability.

Examples of physical 'hubs' include technician-led scientific workshops, 'stores' and glass washing and sterilisation facilities. Specialist core equipment and facilities are increasingly shared centrally. All of these 'services' facilitate and support the burgeoning circular economy in scientific organisations. Consider the environmental benefits and opportunities provided by these through repair and reuse, provision of central consumables, consolidation of deliveries and equipment sharing. Furthermore, technicians often manage internal and external equipment reuse processes and supplier take-back schemes (*i.e.* pipette tip boxes). Beyond this, chemical inventories and equipment asset lists are commonly managed by technicians. Internal sharing schemes can often be simple and informal and as such tend to be 'invisible' in job descriptions (as can be much of the supporting role technicians deliver). Organisations should be conscious of the additional pressures for technicians. Individual schemes alone may not add significant time pressures but the scaling up of activities across many schemes is likely to have implications on workload. Reuse and sharing are highlighted in funders environmental policies and the Concordat for Environmental Sustainability in Research and Innovation Practice.[21] Given the key focus, and potential benefits, there may be value in establishing more formal structures and data collection to evidence, track and report progress. Moreover, more established formal services such as scientific 'stores' and workshops, could ensure environmental benefits are incorporated in business cases where required. These 'hubs' can reduce scope 3 carbon emissions associated with purchasing by extending the lifetime of equipment and consolidating consumable deliveries. Technical leaders and sustainability teams could ensure the value of these 'hubs' is backed with data on the environmental benefits through detailed assessments of the benefits. Scientific workshops, particularly those with aging technical workforces, may pose a risk to these environmental benefits in the absence of workforce succession planning. This is an example of the potential benefits of bringing together parallel focuses on technical careers and environmental sustainability.

The management of scientific equipment and infrastructure in scientific facilities are also central to many technician roles. Technicians' expertise is likely to be highly valuable when optimising the operation of high energy consuming equipment and infrastructure such as laboratory ventilation systems. Technicians should be engaged with refurbishments and new building development as their insight into design will be essential. Sustainability considerations for the management and procurement of equipment are covered in detail in other chapters of this book. These insights will be useful text for all, and I encourage technicians to take the opportunity to benefit from the expertise of the authors. Additionally, Jiteen Ahmed (Head of Technical Services at Aston University), wrote extensively on "Technicians and Procurement" in Technicians in Higher Education and Research.[4] Procurement is a key area for environmental impact and a significant challenge for upstream scope 3 carbon emissions. These emissions commonly dwarf scope 1 and 2 emissions in scientific organisations.

When we start to unravel all the components within technician roles and the relevance of these to the environmental sustainability of science, it is clear that technicians are central stakeholders and enablers for sustainable science.

'Bottom up' approaches signify that technicians across scientific organisations have engaged and embraced challenges for improving environmental performance in scientific laboratories. The laboratory management functions of technical teams are wide-ranging, among others these incorporate health and safety, management of equipment, waste streams and lab infrastructure. The adoption of "green lab certifications" such as the Laboratory Efficiency Framework (LEAF)[26] has grown significantly across the UK and European research sector in recent years. Technical teams are often the lead or are heavily involved in the work to achieve such certifications, as many of the actions are well aligned with technical roles and lab management functions. The National Technician Development Centre (NTDC) supports technicians by delivering a specialist course on sustainable laboratories through their training arm, Higher Education and Technician Educational Development (HEaTED).[27] NTDC also supported a pilot of the LEAF programme, highlighting the key role technicians play in supporting sustainable science.[28]

The activity and progress of technicians improving sustainability in laboratories has not gone unnoticed. It has been a common feature of workshops and talks at many technical community events. There are a variety of other ways the work of technicians is already being recognised, supported and shared by organisations and the sector.

The ITSS have recently added a new category for Excellence in Sustainability in the Papin Prizes which celebrate, recognise and reward technical excellence in higher education and research.[29] Technician led sustainability projects have been awarded funding by the Royal Society of Chemistry Sustainable Labs Grant.[30] The N8 research consortium awarded the technician-led Carbon Operational Net Zero (CONZ) project funding through the community building Collabor8 Fund.[31] This project specifically focuses on utilising "the expertise of technicians in laboratories and research and teaching facilities to share best practices for achieving Net Zero targets". The TALENT programme supported the formation of the MI TALENT Sustainability Knowledge Café for technicians and has hosted sustainability training through its annual online learning festival events SpringFest and WinterFest. A number of technicians[32–34] have been recognised in the Green Gown Awards,[35] the HE and research sectors leading sustainability awards programme. In numerous Higher Education Institutes (HEIs), technicians lead central lab sustainability programmes, with an increasing volume of organisations creating specific sustainability posts within sustainability or technical teams. Sustainability is an important area for technician career development and represents a new career pathway for some technicians.

The ability to draw upon the learnings and impact of these activities and share further outputs, such as guidance and case studies, could be highly valuable to scale activity and to support and recognise contributions of technicians. The strength of the technician community and network are key for this. The formation of the UK ITSS Environmental Sustainability Network

(ESN) could provide an effective mechanism to further the contributions of technicians in sustainability.

Collaborative efforts with the broader research and education community can deliver wider benefits for sustainability in science. As well as directly implementing improvements, technicians play a role in highlighting the opportunities for environmental sustainability in science, influencing and providing systems to support operational and behavioural change. However, it is important to note the success of collaboration will be determined by all groups and recognition of the importance of the key skills and expertise technicians can contribute. The ability to influence and integrate improvements may differ depending on a technician's role, institution or local management structures. Furthermore, technical teams' sphere of control to implement differs greatly across roles and organisations. Leaders within and beyond technical communities must recognise and support the role of technicians as key enablers for the progress of sustainability in science.

The roles, skills and expertise of technicians are interwoven with so many key aspects of sustainability in science. The ability to develop and practically implement solutions will rely heavily on the 'know-how' of technicians. Technicians are crucial for science and the future of science. We hope readers will take this forward in the pursuit of sustainable science.

References

1. The-TALENT-Commission-Report.pdf, 2025. Available from: https://itss.org.uk/wp-content/uploads/2024/12/The-TALENT-Commission-Report.pdf.
2. Inception to Impact – Seven Years of the Technician Commitment. 2025. Available from: https://www.techniciancommitment.org.uk/assets/technician-commitment/inception-to-impact-seven-years-of-the-technician-commitment.pdf.
3. F. Wragg, Defining technicians, in *Technicians in Higher Education and Research*, Routledge, London, 1st edn, 2024, pp. 1–19. Available from: https://www.taylorfrancis.com/books/9781003296119/chapters/10.4324/9781003296119-1.
4. *Technicians in higher education and research: an insight into technical careers, roles and contributions*, ed. K. Vere, Routledge Taylor & Francis Group, London; New York, 2025, p. 190.
5. Technicians, 2025, Technicians | We make the difference. Available from: https://www.technicians.org.uk/.
6. NTDC - HETT Framework, 2025. Available from: https://www.ntdc.ac.uk/resources/hett-framework.
7. NTDC - The Survey, 2025. Available from: https://www.ntdc.ac.uk/the-survey.
8. UK ITSS, 2025. Home. Available from: https://itss.org.uk/.
9. Technical Careers Pathway Lab. UK ITSS, 2025. Available from: https://itss.org.uk/support-for-organisations/technical-careers-pathway-lab/.

10. Resources Library. UK ITSS, 2025. Available from: https://itss.org.uk/talent-legacy/resources-library/.
11. UKRI people and teams action plan, 2023. Available from: https://www.ukri.org/publications/ukri-people-and-teams-action-plan/.
12. 101 jobs that change the world, 2025. Available from: https://www.ukri.org/news-and-events/101-jobs-that-change-the-world/.
13. Narrative CVs: what they are and why use them, 2025. Available from: https://www.ukri.org/what-we-do/supporting-healthy-research-and-innovation-culture/research-and-innovation-culture/supporting-the-community-adoption-of-r4r-like-narrative-cvs/narrative-cvs-what-they-are-and-why-use-them/.
14. Strategic Technical Platforms – Full proposals, 2023. Available from: https://www.ukri.org/opportunity/epsrc-strategic-technical-platform/.
15. Strategic Technical Platform Projects - UK ITSS, 2025. Available from: https://itss.org.uk/tsn/strategic-technical-platform-projects/.
16. Technology Specialists Network. UK ITSS, 2025. Available from: https://itss.org.uk/tsn/.
17. Technical Platform for Innovation and Knowledge Exchange (T-PIKE). Eastern Arc, 2025. Available from: https://easternarc.ac.uk/t-pike/.
18. P. A. Lewis, Technicians and Innovation: A Literature Review, *SSRN Electron. J.*, 2019, p. 6. Available from: https://www.ssrn.com/abstract=3405406.
19. BBSRC Innovation to Commercialisation of University Research (ICURe) Engage, 2025. Available from: https://www.ukri.org/what-we-do/developing-people-and-skills/bbsrc/bbsrc-innovation-to-commercialisation-of-university-research-icure-engage/.
20. YES – about, 2025. Available from: https://www.yescompetitions.co.uk/about/index.aspx.
21. Wellcome, 2025, Concordat for the Environmental Sustainability of Research and Innovation Practice. Available from: https://wellcome.org/about-us/positions-and-statements/environmental-sustainability-concordat.
22. P. M. Weber, S. Bendiscioli, G. Wallon, U. von Ahsen, A. M. de Beaufort, M. Boland, *et al.* The Heidelberg Agreement on Environmental Sustainability in Research Funding, 2024. Available from: DOI: 10.5281/zenodo.13938809.
23. Wellcome, 2025. Environmental sustainability funding policy - Funding Guidance. Available from: https://wellcome.org/research-funding/guidance/policies-grant-conditions/environmental-sustainability-funding-policy.
24. Cancer Research UK, 2024. Environmental sustainability in research. Available from: https://www.cancerresearchuk.org/funding-for-researchers/applying-for-funding/policies-that-affect-your-grant/environmental-sustainability-in-research.
25. Sustainability Guide for Research Processes, 2025. Available from: https://www.dfg.de/en/basics-topics/developments-within-the-research-system/sustainability-guide-for-research-processes.

26. UCL. Sustainable UCL, 2022. LEAF - Laboratory Efficiency Assessment Framework. Available from: https://www.ucl.ac.uk/sustainable/take-action/staff-action/leaf-laboratory-efficiency-assessment-framework.
27. HEaTED, 2025. Available from: https://www.heated.org.uk/.
28. NTDC - LEAF Reports, 2025. Available from: https://www.ntdc.ac.uk/resources/leaf-reports.
29. Papin Prizes. UK ITSS, 2025. Available from: https://itss.org.uk/papin-prizes/.
30. Royal Society of Chemistry, 2025. Sustainable Laboratories Grant. Available from: https://www.rsc.org/prizes-funding/funding/find-funding/sustainable-laboratories-grant/.
31. Partnership NR, N8 Collabor8 Fund - N8 Research Partnership %. N8 Research Partnership, 2025. Available from: https://www.n8research.org.uk/research-focus/collabor8-fund/.
32. Green Gown Awards 2016 – Research and Development: Institution – University of East Anglia - Winner | Sustainability Exchange, 2025. Available from: https://www.sustainabilityexchange.ac.uk/green_gown_awards_2016_research_and_development1.
33. Green Gown Awards 2021: Sustainability Champion–Staff- University of Exeter,Daniela Farina - Winner | Sustainability Exchange, 2025. Available from: https://www.sustainabilityexchange.ac.uk/green_gown_awards_2021_sustainability_champions5.
34. News - Sustainability Green Gown win for Nottingham pharmacy technician - University of Nottingham, 2025. Available from: https://www.nottingham.ac.uk/news/sustainability-green-gown-win-for-nottingham-pharmacy-technician.
35. Green Gown Awards UK & Ireland | Green Gown Awards, 2025. Available from: https://www.greengownawards.org/green-gown-awards-uk-ireland.

CHAPTER 12

Environmental Sustainability in Science – View from a Funder[†]

SUSAN SIMON

UKRI Sustainability, UK
Email: Susan.simon@mrc.ukri.org

Science is telling us that climate change is real, caused overwhelmingly by human activity and will have profound impact on everyday life in the future.[1] The natural world is rapidly changing. To ensure robust research results, there is the necessity for continued stable operating conditions as it ensures that the results are not due to a difference in temperature, humidity or other environmental factors. Could it be argued that therefore the basis for stable scientific research is shifting as well? Oceans, which drive the global climate, are changing – temperature and salinity has significant impact on weather patterns and models for the future of the planet's wellbeing are showing a detrimental trend.[2] As a result, facilities which house research will undergo greater stresses, challenging the status quo of operations.

And whilst science is depending on the nature that it is exploring, it is simultaneously contributing to its detriment. As funders we have powerful leverage to drive change in the sector and that establishes a responsibility in relation to environmental sustainability.

The Sustainability of Science: How Science Impacts the Environment, and What Can Be Done
Edited by Martin Farley

Published by the Royal Society of Chemistry, www.rsc.org

12.1 Role of Funders in Science

The obvious role of funders in the world of research is to fund science and drive innovation. Whilst there are some differences in which ultimate goal is their focus, the role of funders is wider than just providing the finance to enable research activities. Goals in industry would be the next successful product, while charities may seek the cure to specific diseases, and governments aim to reduce health burdens, inform policy or increase economic or military prowess.

One of the key responsibilities of research funders is to drive "excellence in science". The principle of "excellence" is a prime focus in the evaluation of research proposals, which must include aspects outside the proposed research such as ethical considerations, equality and diversity, financial costs, robustness and reproducibility of methods, as well as appropriate management, sharing of data and more.

With these principles set out, applicants (researchers and their respective institutions) are required to comply and exceed with their approach to these considerations to be successful in receiving research grants. As a consequence, funders have important leverage in driving changes in behaviour within the research community they fund.[3]

As we increasingly recognise the impact of research operations on the environment, this will necessitate that funders include a stronger focus on the reduction of the environmental impact of research. Whilst the knowledge of this impact is still developing, the urgency across the globe to reduce the contribution of science to the climate crisis is growing fast. Funders find themselves under pressure to make changes that will transform the sector.

12.2 Ethics

Research ethics were initially driven by emerging issues in medical research,[4] but apply to all scientific endeavours. They are now embedded in all considerations of research, *i.e.* developing the methodology to funding awards and conduct of the research itself. Ethics committees have been established in research institutions to ensure appropriate measures are put in place to minimise, or better eliminate, any potential harm. Critically, this has historically focussed on participants in medical research, but in recent decades the call for the inclusion of ethical research to include care for the environment has become increasingly frequent and louder.[5–7] This is not just focussed on human participants – significant change of ethical assessments for research are in place with the help of organisations, such as National Centre for the Replacement, Refinement and Reduction of Animals in Research (NC3R).

The principle of reducing the risk to individuals has been under criticism for being too narrowly focused almost entirely on participants, as the implication of environmental impacts of research on communities is a concern that should no longer be ignored.[7]

It has become clear that environmental impacts are felt more intensely in areas of the globe, which had little or no involvement in creating the problem. Activities taking place in one part of the globe are impacting on communities far removed from the originating location and the argument of the inequitable effects of climate change is being increasingly highlighted.[8,9]

The difficulty with this is that the effect of this harm is not always something that can be easily assessed or quantified. Effects from the emissions from research facilities are not necessarily felt in their respective location – but contribute to climate change globally.

Consequently, research funders must start to see the environmental impact of the research they fund through an ethical lens, failure to do so will become a reputational risk to funders in the future. This work has already begun, in part through the EU's "Do No Significant Harm" principle, which encourages researchers to consider the breadth of possible negative impacts of their research in order to mitigate them, including environmental impacts to some extent.

Similar to assessing the impact on participating individuals, research activities must consider the wider environmental impact, not just in the immediate vicinity. And this is where a change of mindset is required. Making the comparison to Health & Safety risk assessments, one must consider each element of the work – method, materials, equipment, humans, *etc.* – in terms of where the risk for harm may be. With environmental sustainability assessments the approach is similar, but the effects are much wider ranging. Carbon emissions are not necessarily only impacting locally – they do not stop at borders. Discharges of harmful substances are not always just affecting nearby geographical areas.

And this does not only refer to carbon emissions. Research activities are associated with the use of significant volumes of single-use plastics, hazardous reagents, specialist gases, high water consumption and specialist non-recyclable waste.

Simultaneously, funders must consider equality in their approach to funding awards. In particular, funders who operate internationally must make allowances for applicants who operate in developing countries, who may not be in a position to make the required changes, and the bar may therefore be set too high for these communities. Some funders have already set out environmental sustainability criteria or considerations in their funding policies, particularly in Europe, highlighting a higher standard for organisations from high-income countries.

12.3 Research Continuity and Economic Pressure

Many research activities take place in specialist research facilities – laboratories, workshops, but also museums, ships or planes (or spaceships!). All research requires some infrastructure to conduct their activities – even social sciences, which may require access to data centres to compute and perform analysis. Places of research are energy hungry as they need to power complex mechanical and electrical plants which ensures that environmental conditions, such as temperature and humidity amongst others, are maintained within very small tolerances. Many of these facilities are still using gas for boilers (used to produce heat, steam, or hot water) and utilise multitudes of gases to undertake experiments.

Recent events have demonstrated how vulnerable the research community is to spikes in the cost of utilities. This does not only increase the cost of energy, gas and water to the research facilities, it makes the goods required for operating more expensive and poses risks to a secure supply chain.

Furthermore, the increasing frequency of extreme weather conditions (heat, drought, humidity, storms) poses a risk to the continuity of research activities. Not only can they be disrupted directly by such events, but systems designed decades ago are not able to cope in maintaining the required conditions, thus potentially disrupting or invalidating results or inhibiting operations. The impact can also be more subtle in that plant is working much harder and thus has a reduced lifespan whilst using more energy. Consequently, some of the plant, such as air handling units and chillers, may need replacing more frequently with considerable cost to facility operators.

Without appropriate intervention to futureproof research facilities, the cost of research will continue to rise, resulting in financial restriction of future research calls. The challenge of addressing this remains as the process of futureproofing itself has a significant upfront cost, and increasingly tight budgets mean current *vs.* future impacts must be balanced. Another challenge is that we cannot completely predict what future facility adaptations will be needed with variable climate models. Worst case – and Covid-19 has provided an indication of these impacts – there is potential for disruption or abandonment of research. So, it is in the inherent interest of funders to drive a move towards energy-efficient buildings and research practices to protect their investments long-term.

Funders already use "value for money" criteria in selecting proposals.[10] The cost of research is tensioned against the importance and potential impact of the proposed research. Naturally, funders are keen to fund as much impactful research as possible with the available funds, and so sometimes may consider the return on investment when faced with high-cost research fields (*e.g.* particle physics). However, cost is used in monetary terms only – cost of research on the environment does not usually form part of the evaluation.

The elephant in the room is that most funders do not undertake their own research in their own facilities. Universities and research institutes, who rely on the research grants, struggle to implement the significant and costly changes to their infrastructure. And whilst most funders understand and accept that the cost of research will increase, funding the transition to net zero in research facilities is too high to be funded through research grant increases and many overhead allowances (note this varies from funder to funder).

According to the Higher Education Statistics Agency (HESA)[11] UK higher education providers have the most significant income through tuition fees, followed closely by funding through larger contracts with funding bodies and individual research grants. Additionally, universities have been very successful in securing funding from industry – mainly for undertaking research on their behalf. Concerningly, many research activities may be operating at financial loss, and are supplemented by the other income sources listed.

Today universities are under immense financial pressure. The Covid-19 pandemic, as well as inflation and national education and immigration policies, have caused huge disruptions in the coffers of higher education in the UK. As a result, transforming facilities and research activities is seen as unaffordable by some. The situation in the UK is not unusual in this either.

Is it unreasonable to demand the cost of this transformation to be borne by funders? Income of funders depends in many cases from few sources – be it government, businesses or the wider public supporting specific causes. Donations to a cause like fighting cancer or dementia are made to support research addressing these problems, and thus considering the environmental implications can conflict with the core intent of the funding by the donor.

The funding landscape for research – certainly in the UK – is complex and significant interventions can cause unintended consequences. The increasingly challenging economic environment in the UK and globally signals future restrictions in funding in all areas, including research. Charities are less likely to raise enough funds from individuals to maintain or increase research activities.

What is the solution? When resources are scarce, the things that can be done with limited resource or provide a better potential return on investment become more precious and are prioritised. Funders have their own strategies and priorities and will focus accordingly. When comparing equally good quality research proposals, the lower cost one has an advantage. And here is how the cycle closes – more efficient research methods are less costly and will provide the competitive advantage. Seemingly small actions can have huge impact. For example, simply closing fume hood sashes, or purchasing narrower units can make a big difference. Long-term, it will be important that in the costing of research activities, these impacts are integrated for all so that research as a whole can become more efficient.

12.4 Good Enough?

It is fair to say that the science community is largely aware and convinced of the climate crisis. The recognition that scientific research in itself is increasingly resource intensive drives a need to change our approach, and these changes are sometimes very difficult. In many cases no alternative methods are known that would reduce the environmental impact, as we don't fully understand yet how to achieve net zero research. Science is incredibly diverse, always evolving with new methods, materials, technologies and directions – it never stands still and trying to understand the impact of these developments is an evolving challenge. Having said that, something being a challenge or unknown has not stopped science before, in fact it is an impetus to develop new solutions.

There is little to no information that can be found on the global footprint of research activities, but it is understood that it involves energy intensive facilities and a notable level of travel, the latter mainly in form of international air travel. Both are known as being significant (air travel accounting for approximately up to 2.5% of the global carbon footprint).[12]

Science is sometimes its own enemy in that there is never a final and unequivocal answer to a question. Sarewitz[13] highlights that this is one of the key problems that science faces in the context of environmental sustainability. Any discipline of research will need to have a factual, unbiased understanding of the environmental impact of their research well beyond the direct implications (such as waste or energy consumption). But developing unbiased and unequivocal understanding is challenging, particularly when it means scientists of all disciplines may question it.

There are science disciplines that argue that their activity may produce a significant contribution to solving the climate crisis and that could in some way exempt them from the need to consider the methods. The word "may" is poignant in this – certainly in fundamental research – the nature of this research is to explore and there is no guarantee that the research will find practical solutions or when technology may be able to make use of the discovery. Yet without this research type, there are limited pipelines for future applications in practice and subsequently solutions to the world's problems.

The purpose of research justifying the environmental detriment is not an acceptable or valid argument. It would suggest there is not even an interest in trying to minimise any negative consequences. Although we understand that there may be some elements of research activities for which there is no current fully sustainable alternative, we cannot simply turn these research activities off due to their criticality for society. However, these should be the areas in which funders could focus efforts to find better alternatives.

Funders pride themselves in enabling the highest quality of research, which is expected to return results, solving current and future problems. In a corporate world the reputational implications of environmental sustainability are increasingly a part of corporate strategy as it increases competitiveness and sends a positive image to consumers[14] and investors. Similarly, funders must consider the reputational impacts of negative outcomes of research, including the environmental ones. To not do so may risk future income for funders. Conversely, funders who are seen to align with environmental outcomes may benefit from being viewed in a more positive light, though must avoid greenwashing!

Terminology and jargon around science and the competitive aspects of science always include phrases like "cutting edge", "world leading", "ground-breaking", "pioneering" and many, many more. If we subscribe to be best, the most innovative, and claim leadership, then we cannot embrace the "good enough" regarding our impact on the environment.

12.5 Culture Change

This brings me to the subject of culture change. This term is used heavily and often without clear understanding how to effect it. Everything we do in our private and professional lives has impact on the environment – positive or negative. Such impacts are not always directly or instantly felt by decision-makers, but they will be by somebody somewhere on the globe – we just do not look them directly in the eye when we make decisions. What is needed is increased consciousness of our connectedness, with the hope that this breeds more considered decision-making.

For research activities these links are very wide-ranging. Whether it is about saving energy, reducing waste, reducing travel, ensuring that research is not duplicated, or improving quality. Consideration of these angles in research will benefit all involved with research but also make it more competitive from a funder's perspective.

Environmental sustainability in the context of research is not the warm, fluffy, politically correct thing, that people in woolly jumpers and cork sandals discuss. It is critical for the continuation and validity of research now and in the future. Thousands of jobs are dependent on the "health" of the research system. Covid-19 has highlighted the extremes of research disruption and brought many organisations involved with research to the brink of collapse. The research community though has shown a capacity to adapt to culture change in these moments of disruption, and so adapting to environmental improvements has hope.

There also appears to be a mental disassociation between research estate and research activity. Alignment can be seen when research organisations look for new research activities or equipment, in that research groups must engage facility management to consider utilities and operation. However, when it comes to environmental sustainability, each element may be considered separate and seemingly without influence on the other, which in practice is clearly not the case. Part of the challenge is the lack of incentives regarding environmental sustainability. Waste or water usage is not charged to researchers and energy costs are rarely born by the users, similarly to how heavy carbon-emitters do not always feel the greatest impacts, as discussed earlier.

Not all researchers accept personal responsibility for decision-making regarding environmentally sustainable behaviour (academic travel, good lab practice, data management, need for speed, *etc.*). Equally, though many researchers, PhD students, lab technicians and senior management alike want to make the right decisions, they simply are not the experts in doing so. It should not be expected that a significant proportion of time is spent on learning how the work can be done more sustainably – that information should be available and easily accessible. By enabling scientists and people in all associated roles to do this in an easier, faster and clearer way funders can make a real difference and change the culture of research.

12.6 Travel

Travel – in particular air travel – represents a significant element of the emissions by the science community[15] and is very much an aspect in which individuals have opportunity and power to make good decisions.

Analysis of air travel activity seems to suggest that most of the emissions from air travel arise from only 10% of the researchers and with more senior professors having a 10-fold footprint compared to PhD students.[16] Furthermore, arguments by the science community tend to include the view that PhD students are the ones who need to travel to build their international networks.

The Covid-19 pandemic has shown some interesting developments from a funder's point of view regarding travel. For those funders in the medical research area, specifically, the lockdown period was one of extremely high workloads, exclusively working from home, conducting panels, conferences and meetings entirely through virtual means.[17] It has demonstrated that this is possible, and national and international travel for these activities can be significantly reduced if not completely stopped. But we have also seen detriment from the lack of social in-person interactions, blurring of the work–life balance, and loosening and breaking of networks.[18,19]

Yet, little seems to have survived from this experience. Academic travel (specifically international air travel) is increasing again towards pre-pandemic levels and whilst hybrid working has become an accepted way of working, not many organisations have yet found the right balance between virtual and in-person attendance of staff.

Funders are part of the academic world and culture, with very similar desires for attending international conferences, building relationships with other research and funding organisations, *etc.* Funders must not only drive environmental sustainability, they must be seen to do so and set exemplary standards with their own behaviour and that will include travel.

Technology has significantly improved over the last 4 years with online conferences and meetings being conducted in all areas of society – academic, commercial, public and private, and these developments should be better made use of and more often. There are some funding areas where really good leadership is shown, for example by moving toward all panel meetings taking place virtually – which also reduces the level of organisation, increases the availability of more diverse panel members, not to mention the economic saving from travel and accommodation costs.

There are also habits and cultural aspects in the academic (and probably commercial) world that are counter-productive to environmental targets in this area: in-person attendance at international conferences is still shown on CVs, with the importance of people often judged on how often and how far they fly in the world.

Again, funders may be able to affect change here – focusing on more sustainable ways of engagement could incentivise the community to follow the travel hierarchy: virtual first and air as a last resort. It is important to note that we cannot expect air travel to cease, as there are benefits to in-person engagement, but more that the volume is balanced with the resulting impacts.

12.7 Research Quality

Every funder will confirm that they aim to only fund the highest quality research, and reproducibility is a key indicator of quality.[20] Ensuring quality is assessed through a range of considerations, including applicant experience and track record, cost, and methods proposed, *etc.* Some concern within the research community is supported by studies which show that in some disciplines the replicability of research projects is as low as 30%, according to a survey by Nature,[21]although there are significant differences in the various fields.

This represents a potentially enormous amount of unnecessary effort and money, and from an environmental perspective means that many resources are likely wasted. These resources, alongside clear avenues for researchers to openly share "negative results" or null findings, represent an opportunity to improve our financial and environmental investment in funded activities.

Increased efforts are being made by funders to provide guidance to improve reproducibility and therefore integrity of research with policies and guidance. Some universities are also asking students to pre-publish results and methodologies hoping that this will improve the reproducibility.

Some of the measures include increased sample sizes, sharing data,[22] and improved sharing of methods – including code books, types of analyses, and other relevant information which facilitates the accurate replication of the experiment. Research is also working to normalise sharing of null results, through efforts to promote pre-publication of hypotheses and results. Altogether, the key is research transparency,[23] and funders are increasingly playing their role by driving open-source publication through policies and engagement.

Ultimately, this issue raises the important question of whether more preselection work is inevitably required.[24] Scarcity of funding combined with the increased cost of research itself may drive these decisions more effectively than environmental considerations.

12.8 Promoting Sharing

Some, if not most, research equipment and research facilities are expensive. Funders, who are in the position to support high-tech research equipment, want this equipment to be as well utilised as possible. Often the sources of money are charitable donations or taxpayers, and so there is an onus on ensuring that investment is as beneficial and impactful as possible. It is no longer acceptable to spend millions on very specialist and high-end equipment which is only used by few researchers or not to its full potential.

That applies of course also to funding of any research infrastructure – laboratories, data centres, *etc.* They are expensive to build and to operate and need to maximise utilisation and wider benefits.

Academic research is a very competitive and a 'hard nose' environment where advantages, such as through cutting edge equipment, may make the difference regarding career advancement and recognition. Sharing such equipment evens the playing field to some extent, and it also offers opportunities for collaborations, skills development, and 'cross fertilisation'.

Establishment of national research facilities, which make the most innovative research infrastructure accessible for researchers irrespective of which organisation they work in, are a very effective solution. They prevent multiple individual research organisations building and operating their own with insufficient utilisation or maintenance. Furthermore, they permit efficiencies and reduce the overall carbon and wider environmental impact by consolidating resources. There is also the benefit of having experts operating the facilities (including complex equipment) who can help scientists find solutions to some of the research problems. This expertise is often held by lab technicians or lab managers, who are not always recognised for their critical contributions. They are also often the drivers of the sustainability activities in their respective areas. Universities and research institutions are increasingly recognising this and have started providing career pathways for this important group in the research community.

Funders are increasingly demanding evidence of sharing equipment with the wider research community. This does not only ensure full utilisation of often very costly technology but aims to level out inequalities that exist both in terms of organisational financial standing and geographical differences (at least in the UK). More work is needed though in this area, and funders must collaborate on how to continue to improve.

12.9 Interference or Gentle Push?

With the exception of funders who own and operate research facilities, funders pay for research taking place elsewhere. Organisations have their own policies regarding travel, waste management, data and sample management, purchasing, *etc.*, though with some conformities required due to funders or national law. Scientists operating in these organisations are required to comply with those policies.

Any specific stipulations by funders on how their funds may be spent therefore has a potential to clash with the policies of the respective research organisation. Policy making is not a fast process and requires careful consideration of the entirety of the problem in order to avoid unintended consequences. But by providing the funding, funders have significant influence and responsibility to promote good practice across the research sector as their policies can reach a broad set of organisations. As such, funders must work to incorporate environmental considerations, as their reach and leverage makes them well placed to be impactful.

To take just one example, policies on sample and data management often stipulate a minimum time period for which these need to be held and managed. The easiest way of complying with this is to simply keep everything as long as possible. As a consequence, there are many freezers that are filled up with samples that are out of date (sometimes from researchers long gone) and unlikely to ever be used again. Similarly, huge volumes of data are clogging up data centres and servers. Although many funders already require appropriate data and sample management plans, the policies could be made stronger in demanding that there should not only be a minimum but also a maximum time-period for holding samples and data. These policies must be considered along with maximising impact and sharing of existing data, but if balanced, could promote more sustainable approaches to management.

The argument is often used that data and samples can (and should) be shared and by disposing of them the opportunity to share is taken away. Yet, as we are talking about time-periods of 10 years (frequently the retention period required in connection with publications in life sciences), the value of samples or data can be judged by measuring the demand from the research community for access.

We can see emerging solutions, such as in the example of Biobank UK. UK funders in this programme decided to centrally fund the creation, curation, and storage of a huge cohort of human samples, making them available only to research which meets careful considerations. This removes the burden of sample storage from the researchers themselves, whilst maximising the potential impact of them by improving comparability and increasing sample sizes.

Environmental considerations are part of every aspect of research activity – procurement, recruitment of participants, research methods, materials use, travel, *etc.* It is easy to create unwanted clashes amongst policies and/or complicate the funding landscape so much, that it is impossible for researchers and their respective organisations to be compliant to some, let alone all relevant policies. That said, we must continue to improve and drive positive action, and work with stakeholders from all angles of research activities to push a balanced approach and support long-term success.

12.10 Funding Landscape as a Problem

The landscape of research funding has various shapes in different countries and in the United Kingdom, there are a large number of charities (mainly in the biomedical area) that rely on donations from the public and altruistic contributions as well as organisations (including government departments), funded by the government, who undertake research themselves or fund such work. Within the UK there are well over 100 different research funding organisations.

Imagine now to be a researcher, who is aiming to gain funding for his/her research proposal with the opportunity to have multiple choices of potential funding bodies. Each funder has different requirements and criteria for submission of proposals but also with different policies on environmental sustainability, ethics, equality and diversity and many more relevant aspects. Proposals need to be adjusted to suit, but it is unreasonable to ask researchers to adapt their environmental credentials for each funder when ultimately, we are seeking the same goals.

Furthermore, funders increasingly join forces to fund specific areas to maximise the investments allowing much bigger research activities. But this also means that the funders must agree on the principles for the joint funding requirements, which are key elements of research integrity. And despite clear descriptions of what this entails, such as "*Care and respect for all participants in and subjects of research, including humans, animals, the environment and cultural objects*",[25] interpretations and levels of evidence required differs between funders.

From the perspective of international researcher calls or collaborations this issue is even more complex. **Simplicity is key**.

In the UK the "Concordat for Environmentally Sustainable Research & Innovation Practice" (https://wellcome.org/who-we-are/positions-and-statements/environmental-sustainability-concordat) may offer an opportunity for simple solutions – eligibility based on whether or not an organisation is a signatory or supporter or indeed evidence that organisations are committed to work on the six priorities of the concordat:

1. Leadership and system change
2. Sustainable infrastructure
3. Sustainable procurement
4. Emissions from business and academic travel
5. Collaborations and partnerships
6. Environmental impact and reporting data

Some funders have already taken the step to notify potential future applicants at what point in time being a signatory of this concordat will be a matter of being eligible for funding. This approach is particularly targeted at the institutions which house research but would expect to have impacts at group level as well.

However, this approach may not be an option for all funders. Funders of medical research, including very small charities, supported by donations from individuals, often focus on specific diseases. In other words, donors may view expending time or funds to environmental sustainability measures as a diversion from the purpose of their donation. Consequently, such charities may not feel able to make strong environmental sustainability statements or risk losing their funding base, although examples of small funders implementing strong environmental requirements can be found. Part of this challenge is due to the idea that all sustainability measures have financial costs, dismissing the fact that these measures also improve long-term financial sustainability.

The complexity of the UK funding landscape also results in very different messages being sent to the research community as well as suppliers, policy makers and the general public. A simplified, joined-up approach to environmental sustainability in research is required. Similar to how gender equality in research has become normality through the implementation of the Athena Swan Charter,[26] environmentally sustainable research must become the norm through the Concordat. We should also look beyond the UK, as research crosses borders. Funders need to work together internationally to align where possible, making the system more efficient and sustainable.

12.11 What Next?

Admittedly, it can be overwhelming to consider the long list of actions that need to take place for research to truly embrace and effectively deliver low (or no) environmental impact research. Funders expect their respective research community to be experts in their field – biology, geography, physics, chemistry, *etc.* It cannot be expected for these specialists to also be experts in understanding any materials, methods or equipment they use AND be able to identify sustainable alternatives.

Whilst there are various websites and consultants who provide some answers (not always the right ones!), there is no single tool or standard, which can be used by the researchers and funders alike to understand impact and design better research.

Funders must work closer together to establish rules or criteria that do not differ significantly for applicants. They must collate the significant pockets of information already developed by organisations to provide the necessary, validated information that researchers can use for designing their research, choosing better suppliers or indeed for funders to make better judgements of research proposals that include environmental impact assessments. These systems must equally be adaptable, as innovation allows new solutions.

But most of all, there must be some honesty from all that the ever-increasing volume of research activities is not sustainable in the way in which it is currently conducted – not in terms of the environment nor in financial aspects and uncomfortable decisions are ahead.

Our desire to do more and faster research needs to be reconsidered. Increasing calls for "slow science" are highlighting the problems with the approach to the "next big thing" in science,[27] and prioritising research in alignment with societal need, but also offering sustainable development.[28,29] This stands in direct contrast to the promises of Artificial Intelligence (AI), which allows finding solutions in a fraction of the time, saving months, years, if not decades of laborious research activity. And of course, AI alone is also not the solution to all problems and comes with its on environmental concerns – not at least in relation to the enormous capacity that is required in data centres, synonymous with high energy and water consumption.

Science is resilient and innovative and must and will find better ways of conducting its activities. Funders, like everybody else in the sector, have an important role to play, not least in leading by example. But funders alone cannot solve the problem. Everybody in this community must critically review their own behaviour and area of influence.

12.12 Doing Nothing is Not an Option; Failing is Not an Option

Change is necessary to protect research and all types of science in the future and ensuring that research remains a force for good for the whole of the globe and what is on it and in it!

NOTE: The views and opinions expressed are those of the individual and not corporate policy or statements.

References

1. Intergovernmental Panel on Climate Change, Climate Change 2021 The Physical Science Basis, IPCC, 2021.
2. Intergovernmental Panel on Climate Change, Climate Change 2022: Impacts, Adaptation and Vulnerability, IPCC, 2022.
3. A. R. Cortez, L. K. Winer, A.-F. Kassam, D. J. Hanseman, J. W. Kuethe, J. J. Sussman and R. C. Quillin, Exploring the relationship between burnout and grit during general surgery residency: A longitudinal, single-institution analysis, *Am. J. Surg.*, 2020, **219**(2), 322–327.
4. A. C. V. Armond, B. Gordijn, J. Lewis, M. Hosseini, J. K. Bodnár, S. Holm and P. Kakuk, A scoping review of the literature featuring research ethics and research integrity cases, *BMC Med. Ethics*, 2021, **22**(1), 50.
5. International Science Council, The United Nations Secretary-General's Scientific Advisory Board (SAB): Mobilizing science for a sustainable future, International Science Council, 2024. https://council.science/blog/the-united-nations-scientific-advisory-board-science-sustainable-future/.
6. S. de Saille, Innovating innovation policy: the emergence of 'Responsible Research and Innovation', *J. Responsible Innovation*, 2015, **2**(2), 152–168.
7. G. Samuel and C. Richie, Reimagining research ethics to include environmental sustainability: a principled approach, including a case study of data-driven health research, *J. Med. Ethics*, 2023, **49**(6), 428–433.
8. T. A. Deivanayagam, S. English, J. Hickel, J. Bonifacio, R. R. Guinto, K. X. Hill and D. Devakumar, Envisioning environmental equity: climate change, health, and racial justice, *Lancet*, 2023, **402**(10395), 64–78.
9. J. Oudbier, N. H. Sperna Weiland, T. Boerboom, J. H. Ravesloot, S. Peerdeman and J. Suurmond, An evidence-based roadmap to integrate planetary health education into the medical curriculum, *Med. Teach.*, 2023, **45**(3), 328–332.
10. H. W. Tuffaha, N. El Saifi, S. K. Chambers and P. A. Scuffham, Directing research funds to the right research projects: a review of criteria used by

research organisations in Australia in prioritising health research projects for funding, *BMJ Open*, 2018, **8**(12), e026207.
11. Higher Education Statistics Agency, 2024, What is the income of HE providers? HESA, in *Retrieved 25 September from Improving Reproducibility and Replicability*, National Academies Press, 2019, pp. 105–142, https://www.hesa.ac.uk/data-and-analysis/finances/income.
12. H. Ritchie, Cars, planes, trains: where do CO_2 emissions from transport come from?, 2020.
13. D. Sarewitz, How science makes environmental controversies worse, *Environ. Sci. Policy*, 2004, 7(5), 385–403.
14. M. T. Bolívar-Ramos, The impact of corporate science on environmental innovations: the role of universities and research institutions, *R & D Manage.*, 2023, **53**(3), 503–523.
15. T. Ben-Ari, G. Lefort, J. Mariette, O. Aumont, L. Jeanneau, A. Santerne and P. E. Roche, Flight quotas outperform focused mitigation strategies in reducing the carbon footprint of academic travel, *Environ. Res. Lett.*, 2024, **19**(5), 54008.
16. J. Ciers, A. Mandic, L. D. Toth and G. O. T. Veld, Carbon footprint of academic air travel: A case study in Switzerland, *Sustainability*, 2019, **11**(1), 80.
17. M. Pedaste and M. Kasemets, Challenges in Organizing Online Conferences: Lessons of the COVID-19 Era, *Educ. Technol. Soc.*, 2021, **24**(1), 92–104.
18. National Academies of Sciences, Engineering, and Medicine; Policy and Global Affairs; Committee on Women in Science, Engineering, and Medicine; Committee on Investigating the Potential Impacts of COVID-19 on the Careers of Women in Academic Science, Engineering, and Medicine. The Impact of COVID-19 on the Careers of Women in Academic Sciences, Engineering, and Medicine. Dahlberg ML, Higginbotham E, editors. Washington (DC): National Academies Press (US); 2021 Mar 9. PMID: 33705087.
19. M. L. Dahlberg and E. Higginbotham, *The Impact of COVID-19 on the Careers of Women in Academic Sciences, Engineering, and Medicine*, National Academies Press, 2021.
20. Committee on Reproducibility and Replicability in Science, *Reproducibility and Replicability in Science*, The National Academies Press, Washington, DC, 2019.
21. M. Baker, 1,500 scientists lift the lid on reproducibility, *Nature*, 2016, **533**(7604), 452–454.
22. C. Unkelbach, Increasing Replicability, *Soc. Psychol.*, 2016, **47**(1), 1–3.
23. J. B. Asendorpf, M. Conner, F. De Fruyt, J. De Houwer, J. J. A. Denissen, K. Fiedler and J. M. Wicherts, Recommendations for Increasing Replicability in Psychology, *Eur. J. Pers.*, 2013, **27**(2), 108–119.
24. T. Roberts (2019). 68 We need less research, better research, and research done for the right reasons. *BMJ evidence-based medicine*, **24**(Suppl 1), A39.
25. House of Lords Science and Technology Committee, Research Integrity, Sixth Report of Session 2017–19. H. o. Commons, 2018.

26. A. HE, Athena Swan Charter, Advance HE, 2020. Retrieved 14 September from https://www.advance-he.ac.uk/equality-charters/athena-swan-charter.
27. V. Adams, N. J. Burke and I. Whitmarsh, Slow Research: Thoughts for a Movement in Global Health, *Med. Anthropol.*, 2014, **33**(3), 179–197.
28. S. N. Lane, Slow science, the geographical expedition, and Critical Physical Geography, *Can. Geogr.*, 2017, **61**(1), 84–101.
29. I. Stengers, in *'Another Science is Possible!' A Plea for Slow Science*, Polity Press, 2018.

CHAPTER 13

Climate Activism in Science[†]

P. KNAPP,[a] A. J. PERRIN,[b] C. ACHESON,[c] V. SCAGLIOTTI*[d] AND M. VAN DER VELDEN[e]

[a] Department of Civil Engineering, Imperial College London, London, UK; [b] Department of Biology, University of York, York, UK; [c] School of Social and Political Science, University of Edinburgh, Edinburgh, UK; [d] Independent consultant at SustainLABility, Milan, Italy; [e] Department of Informatics, University of Oslo, Oslo, Norway
*Email: valeria.scagliotti@outlook.com

13.1 Introduction to the Chapter

13.1.1 Our World is in Peril

We face today a host of interlinked crises: climate breakdown, biodiversity collapse, increasing public health challenges, serious threats to democracy, and rising inequality. These crises are no longer just concepts confined to the pages of academic journals, nor are they distant threats of the future. Instead, they are stark, tangible realities that already disrupt ecosystems, communities, sectors and economies worldwide.

Indeed, for many Black, Indigenous and People of Colour in the Global South and across the world, the existential threats posed by climate and ecological collapse are not new, but rather, a continuation of historical and ongoing colonialist violence. The ascendency of the extractive industries driving climate breakdown goes hand in hand with the systemic oppression of Indigenous, racialised, and otherwise marginalised peoples.

Describing the current situation as dire seems an understatement. Earth's lands and oceans are experiencing unprecedented temperature increases,

The Sustainability of Science: How Science Impacts the Environment, and What Can Be Done
Edited by Martin Farley

Published by the Royal Society of Chemistry, www.rsc.org

shattering temperature records by wide margins. Up to two million animal and plant species worldwide are facing the threat of extinction. According to the World Health Organisation (WHO), an estimated 3.6 billion people – nearly half of the world population – are already living in areas highly susceptible to climate change.[1]

Concurrently, we are witnessing wars and conflicts all over the world, including one of the deadliest and more destructive military campaigns in recent history in Gaza and the Middle East, with catastrophic consequences for the people and the environment. Amidst these crises, economic disparities continue to widen, as shown by the fact that we might have the world's first trillionaire within the next decade, while the complete eradication of global poverty remains an elusive goal. As highlighted by the World Bank's *Pathways Out of the Polycrisis* report, global poverty reduction has nearly stalled, with nearly half of the population living in poverty (defined as living on less than $6.85 per person per day) and 8.5% of the population in extreme poverty (living on less than $2.15 per person per day).[2]

In the face of such planetary crises and injustices, what role should science play, and where does this place scientists and academic institutions?

13.1.2 Science Needs to Do More Than Making Its Practices More Sustainable

Research plays an important role in mitigation and adaptation. However, science itself comes with significant sustainability issues. Therefore, solving these urgent and interconnected crises will require a reimagined approach to scientific work, one that goes beyond simply reducing environmental footprints to encompass ethical considerations and social justice. Greening labs and campuses is no longer enough as this alone will not solve climate change nor the other environmental and social crises: it is time to put down our scientific instruments, step out of the ivory tower, and engage in collective action aimed at driving systemic change.

The authors of this chapter argue that if science wants to contribute to achieving socio-ecological justice for all, we must do much more than adjusting freezer temperatures, substituting chemicals, and reducing the use of single-use plastics. To render the future sustainable and just, we need deep structural and systemic changes, inside and outside academic and research institutions. This chapter presents some approaches and thoughts on how academics and scientists can contribute to make this happen.

13.1.3 Chapter Overview: How to Read This Chapter

This chapter consists of five short essays, each written by one (or more) of the contributors to this chapter and approved by all the other authors. Unlike the academic texts that we are used to reading and writing, the essays can be read in isolation and in any order. We encourage you to start reading the section that speaks the most to you and to take your own path through the chapter from there.

If you choose to follow the order of appearance, your exploration of this chapter will begin with a broad overview of the disconnection between rhetoric and real action in science. The first essay (Section 13.2), *Change mindsets, not just lightbulbs*, highlights how incremental shifts are insufficient in the face of the climate crisis and why a deeper transformation is needed. But what happens when scientists push for this change within their institutions? Unfortunately, despite universities' professed commitment to intellectual freedom and openness, their responses often range from passive resistance to outright hostility. Section 13.3 offers a striking example of this conflict, illustrating the challenges faced by those advocating for meaningful action.

While institutional inertia remains a major obstacle, another key issue is the overreliance on technological solutions. Section 13.4 critically examines the role of technology in addressing the climate and ecological emergency, questioning the limits of techno-optimism and technofixes. Too often, these approaches treat the symptoms rather than tackling the root causes of the crisis. However, addressing these deeper issues also requires confronting the colonialist structures that have long shaped scientific institutions. Section 13.5 explores the need for anti-colonial approaches to sustainability, shedding light on how scientific research has historically perpetuated these systems and why decolonization is crucial. This section also highlights examples of anti-colonial research methodologies and emphasizes the importance of honouring Indigenous sovereignty and knowledge within scientific and research practices.

Finally, the last essay (Section 13.6) turns to the role of activism in science. What does it mean for scientists to take an active stance, and what are the implications of doing so? Despite the risks and barriers, a growing number of scientists are stepping into activism, recognizing that their work alone is not enough. By addressing common questions and concerns, we hope to provide clarity and encouragement to those considering this path. Scientists have long been raising the alarm, frustrated that no one is listening. But if they themselves do not act as though we are in an emergency, why should anyone else?

13.1.4 Notes from the Authors

This chapter was crafted through a collaborative, semi-autonomous, and decentralised writing process, without any hierarchical structure among the authors. Our names are listed from "early career researchers" (who are most vulnerable to the long-term impacts of the climate and ecological crisis) to "more established researchers". Our guiding principle was to produce content that would neither offend nor disrespect the views and values of the other authors. To this end, we actively helped each other identify any blind spots and potentially problematic terms and perspectives. However, given our backgrounds, living, and working contexts are predominantly within northern and western European countries and institutions, we acknowledge that our perspectives and experiences do not fully represent those of scientists and institutions from other parts of the world.

We also want to acknowledge that institutionalised science is not the only form of knowledge that matters. It is not superior to the many Indigenous, subaltern, and grassroots knowledges which have been attacked or marginalised under extractive capitalism, white supremacy, and colonialism. Furthermore, we recognise that mainstream science has played a central role in legitimising and enabling the systems of harm which have brought us to the current state of emergency.

Therefore, we wish to proceed from a place of humility. While we argue that academics occupy positions of relative power and privilege under the current system, and that we have a responsibility to contribute to undoing injustices and inequalities, we also call on scientists to interrogate the structures underpinning this privilege. This means turning a critical lens back upon our own institutions, companies, universities, research, and teaching practices, and ensuring that our work does not cause further harm. It also means recognising the limitations of our work, and learning from the countless grassroots movements fighting back against oppression and injustice around the world.

May you be inspired by the perspectives that we share, and may you be motivated to take collective action – within academia or beyond – that can help bring about socioecological justice for all.

Happy reading!

13.2 Changing Mindsets, Not Just Lightbulbs

– Why Abi wrote this section –

"UK universities and research institutes have been centres of my own education and work for half my lifetime. Since 2019, I've tried to balance a scientific research career with communicating and acting on the warnings of climate and environmental scientists. In that time I have encountered significant resistance to the idea that scientists and their institutions can – and should – use our privileged positions to contribute to rapid systemic change. I often hear that something else is more important, it's too difficult, or it's somehow not our place to try. By distilling some of what I've learned as part of the scientist/activist community, I hope to counter some of those narratives and motivate others to embrace pushing for environmental and social justice as part of the role of 21st century scientists."

13.2.1 How Research Institutions and Their Scientists Can Enact Change and Empower One Another in the Era of Climate Breakdown

13.2.1.1 Matching Rhetoric with Action

Centres of scientific research – such as universities and research institutes – exist to generate and disseminate knowledge often with the stated aim of improving lives and societies. These institutions often pride themselves on being host to, and developers of, thousands of minds trained to understand and communicate complex problems and innovate to solve them. But is this what the scientific institutions of today are really prioritising or achieving? And are they responding effectively to the escalating threats of climate, ecological and societal breakdown?

Over recent years many institutions have made public declarations of a state of Climate Emergency and announced ambitions to be part of solutions. But how much evidence is there of these same institutions realigning their practices and priorities proportionate with emergency action? Working in a place like this you may well have seen stickers reminding you to turn equipment off, a reusable cup scheme, awards for sustainable practice on the walls. But has your daily reality noticeably changed? Is what you research and how you do so largely similar to how it was before environmental sustainability was part of the discourse? Have you and your colleagues received any training to help you understand the scale of the biospheric emergency we are part of, and what we can do about it? Are conversations about how our academic systems can/should/must respond in a rapidly changing world part of your experience at all?

How can the transformative potential of science and scientists be unlocked? Institutions' current modes of operation tend to stifle the changes needed to do so, but it is also within their power to facilitate and nurture cultural shifts. This needs to go far beyond reminders to turn off equipment and the elimination of single-use coffee cups; true sustainability in research requires seeing the big picture and embracing commensurate systemic change in how we work and what we value. This will mean realignment of the priority areas of research and teaching, a more holistic approach to mitigating the harms caused by our activities (explored in other sections of this book) and also looking to expand our conception of the institute's role – acting as a hub for collaboration, a resource for wider communities and a fertile ground for empowering advocates and change-makers.

13.2.1.2 Scientists' Privilege, Power and Responsibility to Accelerate Systemic Change

At its core the job of a scientist or academic is to pursue knowledge that enables us to understand specific elements of the world we are part of. Inevitably this involves evaluating and integrating information from a range of sources and disciplines, distilling what we learn into messages and principles that can be usefully shared. As such we have much of the expertise needed to comprehend and communicate the threats that all life and societies now face. When scientists publicly sound the alarm, or otherwise act to address the biospheric crises, we can be powerful messengers and advocates because of the position of trust and/or authority that is often afforded to us. We can find that our words and actions have a wider reach and a greater resonance as scientists than they would do otherwise. We might find we have greater access to platforms of communication, and that our taking action lends legitimacy to environmental movements that can suffer from unhelpful stereotyping about who it is that cares and participates.

Scientists have been warning for decades that human actions are driving irreversible and escalating climate and biodiversity threats. Could, and should, the wider scientific community be mobilising as a result, showing solidarity with the scientists who have raised alarm, and acting on the basis of the findings of our colleagues around the world? Can we expect these warnings to be listened to if we do not? Do we have greater responsibility to do so, given our privileged position to be able to understand these complex crises and our power to contribute to systemic change? What does it mean for our own role, the impact of our work and the existence of the structures that allow us to do scientific research if we do not see the kind of transformative action that is needed to protect societies as we know them from collapse? After all, there's "no research on a dead planet".[3]

13.2.1.3 Scientific Institutions as Hubs for Environmental Learning and Communication

To respond proportionately to the climate & ecological emergency, institutions must do much more than replace lightbulbs and nudge their employees to make small behavioural adjustments. By taking a more systemic view, we can see that there are many more-impactful actions open to scientific institutions (summarised in Figure 13.1). Institutions and their funders need to turn their attention towards what research areas they prioritise and whether those make sense in the context of the biospheric emergency. In early 2020, scientists across many disciplines were rapidly re-mobilised in response to the Covid-19 pandemic.[‡] Can we build on that experience to support the realignment of scientific research focus, centring the pursuit of the knowledge needed for societies to mitigate and adapt to climate and ecological breakdown? Key to achieving this will be institutional support for researchers to learn about the threats we face and – informed by that knowledge – apply our skills in new areas. Institutions, many of which are publicly funded, could also be playing a much more public role in supporting societal responses. This includes opening their doors to, supporting, learning from, collaborating, communicating, and sharing resources with local communities, under-represented groups and grassroots movements. Importantly, institutes stand to hugely increase their societal impact by cultivating conditions where their members are encouraged and empowered to be visible and effective advocates for environmental education and action.

[‡] I was one of these scientists…within weeks of the pandemic's spread in the UK my biologist colleagues and I found ourselves studying the SARS-CoV2 virus, designing and running testing pipelines, communicating with the public, health providers and politicians and more. This was supported – and often celebrated – by our employers, funders and government.

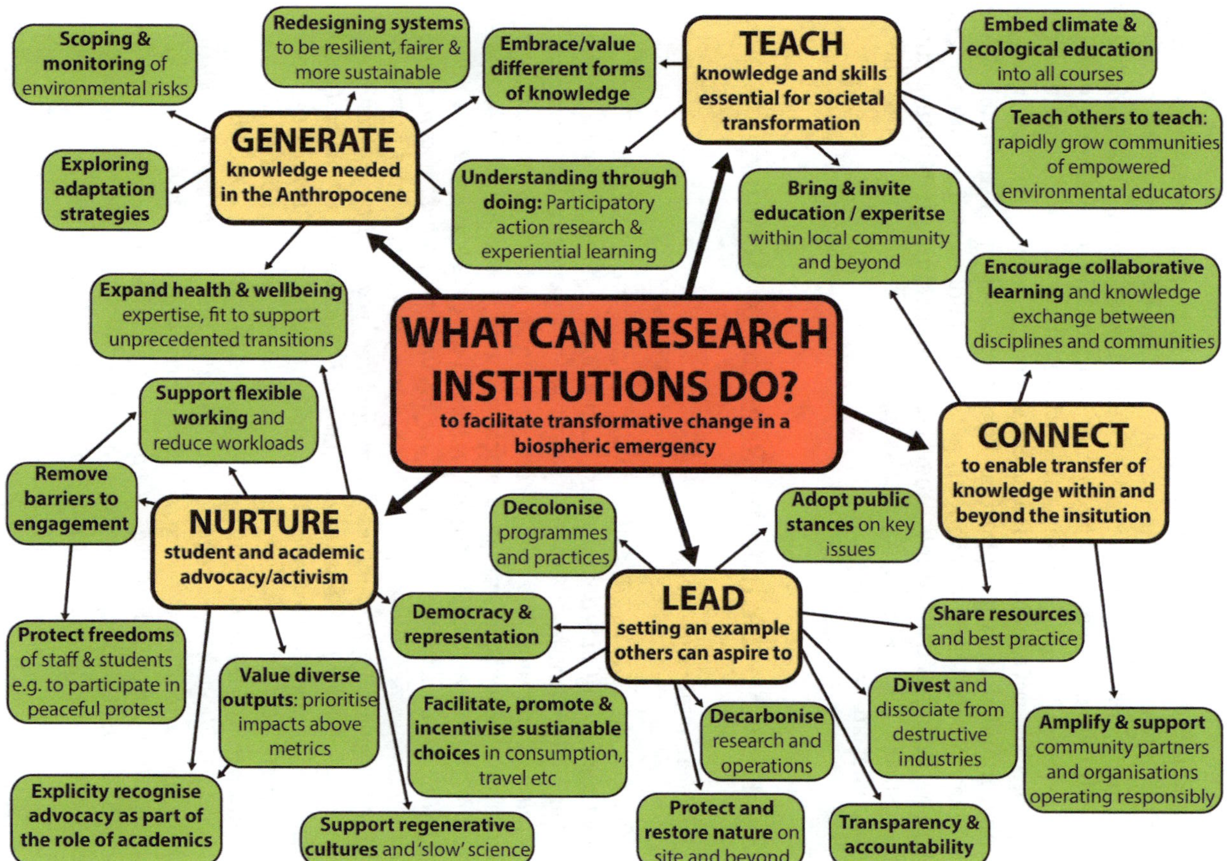

Figure 13.1 Actions scientific research institutes can take to enable transformative change in the age of climate and ecological breakdown.

13.2.1.4 Collective Action by Scientists Holds Unrealised Potential to Accelerate Systemic Change

As individuals subject to the specific pressures of academic research careers, pushing the structures we operate within to improve can often feel frustrating and futile. Whilst there is much we can do as individuals to challenge harmful practices, signpost better ones and lead by example with our own choices (see Figure 13.2), we are more effective when we organise! Joining or forming climate action groups and unions can enable us to amplify calls for systemic change whilst reducing the work required and/or risk inherent in doing so. Collective action also creates communities that are ripe for learning, connection, and support. As scientists we can play valuable roles in the wider environmental movement, providing and receiving evidence and education, acting as trusted messengers and contributing to building fairer, more resilient alternative systems. As threats to a liveable future mount rapidly, what are we waiting for?

13.3 How My University Reacted to Activism

– Why Pete wrote this section –

"I finished my PhD in 2024 and, in the years before graduating, I learnt about the scale of the climate crisis, the potential collapse of the biosphere and human society, and the urgency with which we must act. As a scientist, I felt it was my moral and ethical responsibility to act on this knowledge. It felt to me that taking direct climate action as a PhD student would be safer than it could be for staff whose jobs might be at risk. However, I was naïve in failing to consider the extent to which my university would take punitive action against me, for example by banning me from campus and restricting my employment opportunities there for five years after I graduated. I expected support from my university when I took action, but instead their response was punitive, demoralising, and intimidating. I wanted to share my experience so that it may become useful to others wishing to call out contradictions within university contexts, and to report what the university wished to hide with some considerable effort so that people may learn how these institutions may try to silence or remove those calling for uncomfortable yet critical change. I hope that you, the reader, will gain an insight not just into my experience, but of many other university students who take direct action. Although my university acted to protect their interests, I hope that my account does not dissuade you from taking action yourself, but rather that is an opportunity to see how future actions may be taken differently to avoid the response outlined. Done well, fundamental changes can be achieved, with the example of the successful Plant-based Unis campaigns."

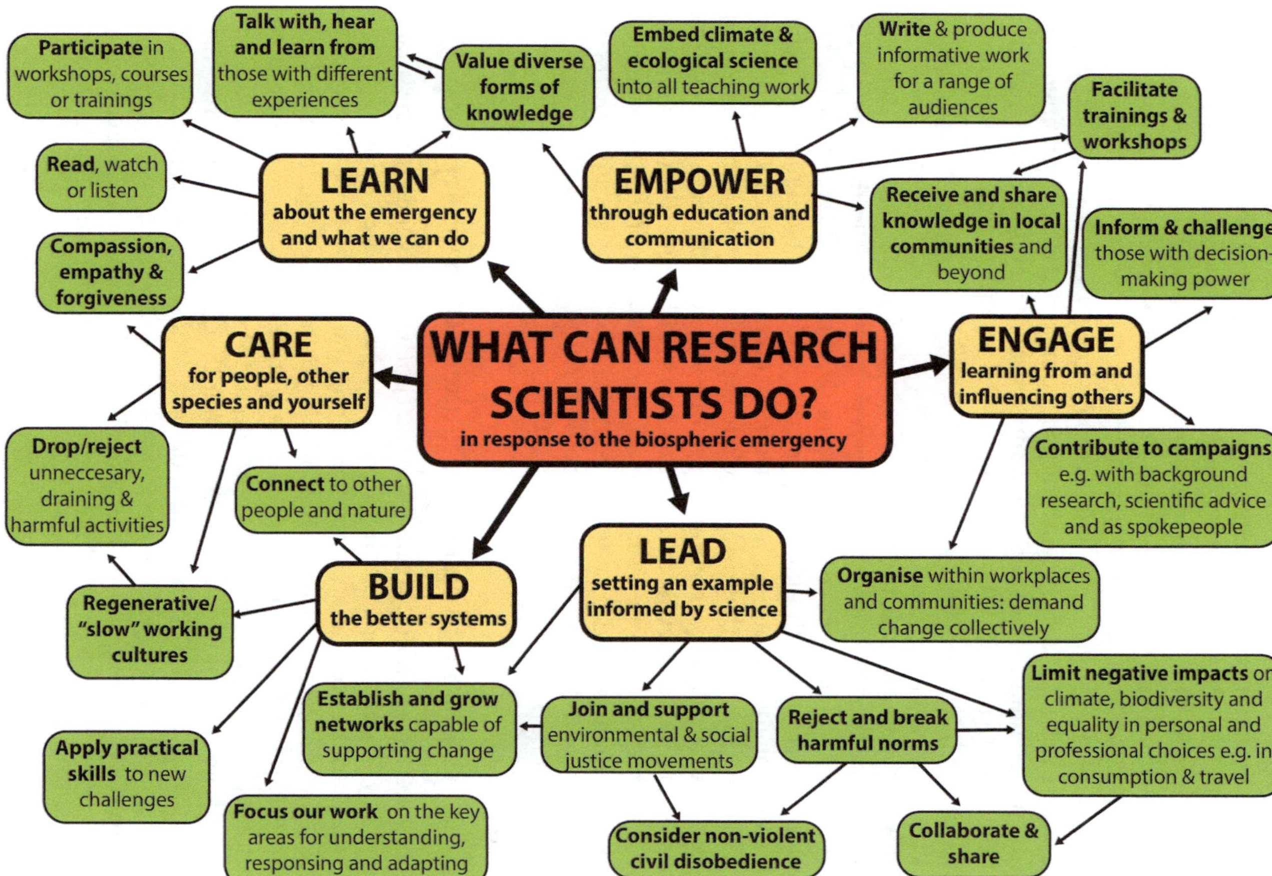

Figure 13.2 Actions research scientists can take to be part of transformative change in the age of climate and ecological breakdown.

13.3.1 I Learned About the Climate Emergency from Activists, Not Institutions

I did not learn about climate breakdown from any university course, nor from my research group's meetings, nor from an email from the university president. Neither did I learn about it from school, conversations with friends, music, film, television, nor from government. I learnt the real truth about the scale and urgency from the Extinction Rebellion protests of 2019.

At first, I did not trust the protestors. Surely, my school education, the media, and the government had a responsibility to inform me if this were the case. I believed that the government would act swiftly in an emergency. And I understood the scientific community were 'apolitical' and would act on the objective truth of both the rapid changes in climate and decline in biodiversity. I was so wrong.

Instead, I found many of those institutions and individuals I had trusted had worked closely with the fossil fuel companies who had repeatedly lied and manipulated politics, public awareness, and public discourse. At my university, I found a pervasive culture of denial, faith in techno-fixes, business-as-usual, siloed thinking, superiority, and neo-colonialism. I became frustrated at the lack of response from a community I felt I should trust. I later came to realise that disempowerment and many other barriers played a role in this culture. A common response to my early cries for help from friends, family, colleagues, and doctors was to either question the validity of my concerns, or to claim I was overreacting, lecturing others, or even that I had been manipulated by a cult. I felt alienated from society, but the validation I received from the Scientists for Extinction Rebellion made me feel human again.

13.3.2 A Hostile Response to Student Climate Activism

Once I started to properly read about the climate emergency, and my map of the many issues became more interconnected, I engaged strongly with activism. I admit that I made mistakes along the way, but my university was intolerant and punitive rather than offering opportunities to learn and grow.

In 2022, shortly after my arrest during a protest at a government building with over thirty other scientists, a photo of the arrest featured on the front cover of the student newspaper. Soon after that, I was fired from my role as a warden in the student accommodation. Although the reason given was that I had put the health and safety of the students at risk by moving a packet of wet wipes from our kitchen table to the top of the fridge, I felt that the decision was strongly linked to my involvement in this high-profile action.

A few months later, I was banned from my lab for three months. This came after borrowing lab coats, with approval and assistance from a lab technician, for a photo of the university's climate action group with fellow students, staff, and alumni.

By the end of 2023, after I had expressed my dismay at my research group's frequent flying during a group meeting, the university banned me from the entire campus for six months. A series of official complaints about my manner during this meeting had led to this escalation. For context, my research group focuses on global heating from air traffic, and they flew to locations easily reached by train to present their work. There was no sense of collective or individual responsibility to act upon the findings of their own work, or to recognise their privilege and actively pursue ways to reduce their consumption. I felt overwhelmed by the contradictions, hypocrisy, denial, and irresponsibility. During the meeting, I said: 'We need to stop talking about flying. We cannot normalise this, the planet is on fire and we have to stop flying'. There was silence. I took to Twitter that evening and vented: 'My research group works on global heating of flying but fly all the time. It's fucking terrifying'. The post received over 80 000 views and notable responses included: 'Good on you for having the courage to have the uncomfortable conversations that need to be had. We urgently need to break these taboos', and: 'Good work! Hopefully the discussion continues'. But a series of formal complaints from my research group escalated to the university banning me from all campuses for the final six months of my PhD and keeping a 5-year record of bullying and intimidation. They were most affected by my criticisms of what they regarded as personal choices, and I have since learnt to be much more careful about this. My requests for mediation to discuss our individual and collective responsibilities, and to apologise for my communication, were denied.

The university had collated a list of 'concerns from the department' to support my ban from campus. It included some bizarre references, including (i) that I had spoken with a conference caterer about improving their vegan options; (ii) confronting a keynote speaker who spoke on the topic 'Translating science into robust military capability'; and (iii) supporting a fellow activist in their efforts to stop an airport expansion. I believe in acting on my conscience, and in being a barrier against harmful actions, but these acts were held against me.

This experience added to the mental health impacts that continued after the lockdowns and isolation. The university's Faculty Dean supported me through this process, and I was even able to access funds for therapy. Fellow activists supported me, too, and I am extremely grateful for this. But the mental health impacts of being treated as a threat by an institution ignoring its own research, who thereby adds to the threat of the biosphere, were deep and painful.

It is important to reflect, of course. I have learnt that direct challenges to individuals should be avoided, and direct challenges to institutions or groups done in ways that provide anonymity could avoid you being targeted. I hope future students in similar positions will receive greater support from their Students' Union, which may need to adapt as more students will protest against universities' facilitation of current and future wars or injustices. I believe universities' opposition to challenges will ultimately lead to their undoing, contributing to the same mentality of hard-right governments defunding research and threatening the security of staff or students who do not have citizenship in that country. I found that being a member of a group of activist scientists was crucial to protect my mental health; and I can wholeheartedly recommend seeking this kind of support structure if you, the reader, ever feel alone.

13.3.3 What if Universities Took Meaningful Climate Action, Instead of Punishing Those Who Challenge Them?

I wonder how many universities actively educate their students and staff on the reality of the urgent need to reduce emissions and consumption, and how many support their students and staff in engaging in activism or other kinds of direct action? If universities provided encouragement and public backing to their students and staff, and made public statements about this support, I believe change could happen much faster. It would liberate so many people from feeling restricted in this sense. The respected voices of many academics could be amplified. If just a few universities took this step to support academic activism, the environmental movement could snowball, putting much-needed pressure on governments to listen and to act on the science. A healthier relationship with nature and with each other is possible, but I argue we need to unplug our reliance on fossil fuels and capitalism as a whole to do this. Supporting students and staff taking direct action at universities is one important part of this.

13.4 Technofixes Will Not Save the Planet

– Why Maja wrote this section –

"Years ago, I visited a local healer in a Dalit village in South-East India. We went into her garden where she grew plants and bushes. She described the medicinal powers of some of the plants. While she talked, her hands rubbed some leaves, bringing out the healing oil. She talked about the best seasons and time of day to pick leaves, and she let me smell her fingers. My research interest in meeting her and other healers was to understand what happens when Indigenous knowledges meet global technologies. I was studying a project in which local knowledge was digitalised so it could be shared globally. I followed the knowledge of the healer to the notebook of a local knowledge worker who wrote it all down; to the computer of a botanist, who 'translated' the local names of plants and bushes and trees into their Latin names; and then to a large research centre, that received project funding to 'check' the local knowledge in a lab before it was shared in the global network. I could not find her

knowledge again. It had disappeared in the technoscientific knowledge of the lab and the project's database.

The epistemic[§]*, social, and environmental violence enabled by our digital technologies has since become one of my main research interests. This section gives a very brief overview of some of the things I have learned since."*

13.4.1 Techno-optimism Rules

The scientific community generally agrees that human activity has led to unsustainable levels of greenhouse gas (GHG) emissions in the atmosphere. Still, it continues with business as usual, even if that means increasing emissions. Especially in the Global North, scientists and politicians have neglected their responsibility to reduce emissions at source. Instead, decarbonisation and mitigation technologies have been advocated, which claim to limit GHG emissions, capture CO_2 emissions directly from the air or atmosphere, or geoengineer local climates to limit the effects of global heating. These technologies still have not reached a scale and maturity that will make a difference.

There is a strong belief in the scientific community that technology will provide the needed solutions, but these solutions will come too late, if they come at all. Global fossil fuel CO_2 emissions are still rising and haven't peaked yet.[4] Latest research results show that global +2 °C will be reached by 2045, on its way to the *Point of No Return*.[5] Drastic measures to cut emissions are therefore needed right now.

Science and engineering suffer from a type of techno-optimism that is difficult to overcome, as research funding, subsidies, and scientific status depend on inventing new or improved technological solutions that will enable business to continue as usual. The promotion of carbon capture and storage (CCS) technologies enables oil and gas companies to continue with exploration and exploitation of natural resources. We see, for example, in Norway, that research and development of CCS and fossil hydrogen projects received around \$6 billion of public money – out of more than \$29.8 billion worldwide.[6] At the same time, Norway continues its aggressive oil policy worldwide, including the release of new oil exploration licences in the vulnerable Arctic area. Oil and gas companies are also actively involved in funding academic research, using profits from oil and gas production to fund academic research that focuses on making oil and gas extraction more efficient, as well as to fund transition technology research.[7–9]

[§] Epistemic violence exists when methods are used that lead to the extermination, annulment and destruction of certain knowledge and its bearers, reaching the extreme of their irrevocable loss, as *epistemicide*: https://debatesindigenas.org/en/2020/08/01/epistemic-violence-against-indigenous-peoples/

13.4.2 The Technological Fix

Developing technologies to solve problems created by humans is not new. Technology is perceived as a better or easier problem solver than people when it comes to social problems. Then, there is the widespread belief that technology is neutral. Although these two statements initially seem contradictory, they reinforce each other.

In 1967, Alvin Weinberg introduced the notion of the *Technological Fix*. He argued that social problems are more challenging to solve than technological problems. Solving social problems means changing people's behaviour and motivations towards more acceptable behaviour. He asked: "to what extent can social problems be circumvented by reducing them to technological problems?" and "to what extent can technological remedies be found for social problems without first having to remove the causes of the problem?".[10] The result is what we now call a technofix, "a technological remedy for a social problem that only involves minimal use of political power or economic incentives to change an individual's behaviour and does not require changing social norms".[11]

Fixing social problems with technology can lead to a wide range of unintended consequences that are both inevitable and unpredictable.[12] The main issue is that technofixes address symptoms rather than causes. Thus, overconsumption is addressed with 'greening' products, promises of a circular economy, and eco-efficiency or producing more with 'less'. Increases in GHG emissions due to overconsumption and ongoing oil and gas production are proposed to be managed with CCS technologies. Other technofixes to address the climate emergency are geoengineering, moving climate problems to another part of the world, or proposals for new nuclear power plants, shifting the problem of nuclear waste onto future generations.

A particular form of unintended consequences of energy-related technofixes is the *rebound effect*. The term refers to the decrease in the expected energy efficiency improvement potential of a technology because the people or businesses benefiting from the improvement will use the gains to consume more energy or other products. For example, cheaper electricity due to efficiency gains will increase energy consumption.[13] A rebound effect is also found in the field of renewable energy; the production of more renewable energy doesn't lead to a reduction of the same amount in the production of non-renewable energy.[14–16]

13.4.3 We Already Know What Technologies Work

There is currently only one group of technologies that has proven able to capture and store CO_2 on a large scale. These technologies are not developed in a lab or at a university, but are natural technologies, such as trees, wetlands, coral reefs, rainforest, and jungles. These natural climate systems are not only very vulnerable to climate changes and increased CO_2 levels in air, land, and water; they are also under attack from industrialised agriculture, mining, biofuel production, and built infrastructure. While the only thing these natural technologies need are our care and protection, we destroy them en masse. We lose 10 million hectares of forest each year.[17] We have lost 14% of our coral reefs since 2009[18] and lost 35% of our wetlands between 1970 and 2015.[19] The rate of their decline is accelerating.

13.4.4 They are Not Green Technologies

But what about so-called green technologies: technologies that produce non-fossil energy, such as solar power, batteries, and wind power? The problem is that these technologies themselves are not green. Mining the minerals for these technologies is associated with a wide variety of socio-ecological impacts, most often in the Global South. Geopolitical concerns, in terms of continuous access to these materials, are now resulting in new mining operations in Europe. Mines are proposed to be established or re-opened in areas with vulnerable ecosystems or communities and on land of Indigenous communities, resulting in widespread protests and unrest.

Land-based wind energy factories in particular have become controversial because they are often proposed or built on Indigenous land, *e.g.*, in Colombia[20] and Norway,[21] or on occupied land, such as the Golan Heights[22] and Western Sahara.[23] Renewable energy developments can endanger the livelihood of local communities, *e.g.*, in Brazil,[24] Canada,[25] and India.[26] The term "green extractivism" has become prevalent to refer to these contradictory situations in which promoted solutions to the climate and biodiversity emergencies result in the destruction of environments and livelihoods, often continuing colonial legacies.[27,28]

What is the energy produced by these so-called green technologies used for? As Dunlap and Jakobsen ask, "Is this energy providing light and heat to homes or is it fuelling socially and ecologically destructive industries?"[27] Take an example from Norway again. It is known for producing large amounts of non-fossil energy, mainly hydropower (88%) and wind power (11%). In Repparfjord, in Northern Norway, the mining company Nussir is reopening and expanding a copper mine and is announcing itself as a sustainable mine, as it claims to be the first fully electrified mine in the world. The tailings from the mining activities will be (legally) dumped in the nearby fjord. The mine will endanger marine life in the fjord and the precarious reindeer areas near the mine.[29] Another example from Norway is a Google data centre, which will be built on a 2 km^2 area that used to be a forest. It promotes itself as becoming the largest data centre in the world and will become the largest renewable energy consumer in Norway. It is widely expected that this will result in increased electricity prices for Norwegian households.[30]

13.4.5 Technology is Not Neutral

A technofix, a technology designed to solve a social problem, can hardly be called a neutral technology: a technology that has no values or interests written into its design or code. Still, the belief that technology is neutral is widespread among scientists and engineers. The idea that technologies can be used for good or evil, depending on who uses them, encourages the uncritical design and development of technologies. The responsibility for how a technology is used is delegated to the user. This seems to absolve scientists and engineers from their responsibility, both for how technology is used and for the unintended consequences of bringing these technologies into the world.

If we look at the history of some existing technologies, we can see how particular values and interests are, consciously or unconsciously, written into a technology. The most famous example is the case of the overpasses of the parkways of Long Island, New York. These are built purposely low so public transport buses cannot pass under them to reach the Long Island beaches. At the time, those beaches were used mainly by the white middle class, who could reach them by car.[31] A more recent example is facial recognition. While New York's urban planner, Rober Moses, had an apparent discriminatory intent with the low overpasses, the developers of facial recognition software did not. Still, their technology created racial bias, in particular towards black women.[32] This was the result of biased data used to train the algorithm. By uncritically accepting a data set, the developers contributed to developing racist algorithms.

Similarly, CCS is not a neutral technology. The oil and gas industry developed CCS in the 1970s to increase the amount of oil extracted from oil fields through using CO_2 for enhanced oil recovery (EOR).[33] These days, CCS is mainly perceived as a technology enabling the transition to a non-fossil fuel economy. Still, almost all large CCS projects are located in the oil and gas industry to provide CO_2 for EOR.[34]

Like many other technologies, the development of CCS technology was, and still is, funded with government subsidies, tax incentives, low-interest loans, *etc.* How public money is allocated is based on policies developed by politicians. In other words, political decisions make a technology feasible and profitable. However, this funding could also have been used for other research and development projects. In the case of CCS, we can see that government policies and subsidies are shifting from supporting EOR to promoting geological storage. While in 2023, twenty-nine of 41 operational CCS facilities were EOR, the majority of facilities now under construction are for the geological storage of CO_2.[35]

The idea that technology is neutral is part of an instrumentalist understanding of technology: namely, the belief that technologies are tools that serve the purposes of their users and have no valuative content of their own. This perspective is widespread and is based on four premises:[36]

1. Technology is indifferent to the variety of ends it can be employed to achieve
2. Technology is indifferent with respect to politics
3. Technology is rational
4. Technology furthers efficiency in whatever context it is applied

Our technologies are inherently political. The values of our socioeconomic system and culturally accepted assumptions, or rationality, permeate the designs of our technologies, inevitably supporting the dominance of the people already benefitting from the current system.[36,37] Technology designers and developers are therefore ethically responsible for what they bring into the world, be it racist overpasses, racist algorithms or technologies that lengthen the fossil energy era. However, the global mobilisation of tech workers protesting against the use of the technologies they help develop in the Gaza genocide and Israeli apartheid regime is a sign that technology designers and developers are becoming more aware of this responsibility.[38,39]

13.4.6 Another Technology is Possible

How to think differently about technology? Firstly, we need to get rid of the idea that technology alone can fix social problems. We need technologies that seek to address the damage done to our social and natural systems and to build and sustain healthy, resilient, and regenerative living systems. Such approaches put human and non-human life, not profit or economic growth, at the centre of our decision-making. We can call the outcomes of these approaches *regenerative technologies*.

Regenerative technologies are never neutral or provide technological fixes. Table 13.1 compares a regenerative perspective on technology with an instrumentalist perspective.

Technofixes can sometimes provide a temporary solution to the unwanted effect of a social problem. For example, in the current climate emergency, CCS and other climate mitigation technologies are urgently needed, but only if they are "empirically feasible, ecologically coherent, and socially just".[40] However, market-based decarbonisation and mitigation practices are often a technofix; they don't alter the way we produce and consume. To do that, we need to change the system, starting with degrowing our rich economies in the Global North and embracing radical alternatives to current systems of production and consumption.

Table 13.1 From an instrumentalist to a regenerative perspective on technology.

Instrumentalist perspective	Regenerative perspective
Technology is indifferent to the variety of ends it can be employed to achieve	Technology is biased towards the ends it is employed to achieve
Technology is indifferent with respect to politics	Technology embraces an anti-colonial, socially just, ecologically sound politics
Technology is rational	Technology is informed by alternative rationalities
Technology furthers efficiency in whatever context it is applied	Technology's efficiency and sufficiency are connected to the place in which it is put to work

13.5 The Need for Anti-colonial Stances in Sustainability Science and Research

– Why Cat wrote this section –

"As a social scientist and climate activist, my engagement with grassroots liberation movements has led me to develop an understanding of how climate breakdown is intertwined with colonialism, extractive capitalism, white supremacy, and violence. Truly effective responses to the Climate and Ecological Emergency need to dismantle these oppressive systems. Researchers and scientific institutions have a responsibility to engage deeply with these issues, and to speak out in solidarity with those who are most harmed by a status quo which is not only disastrous for nature and climate, but also sustained through violence against marginalised peoples.

A particularly potent example of where universities should be taking a stand against interconnected environmental and colonial injustices regards recent events in Palestine. The cataclysmic bombardment of the Gaza Strip by Israel, as well as attacks on the populations of the occupied West Bank and Lebanon, has resulted in the displacement of close to two million people, the near total destruction of Gaza's healthcare and civic infrastructure including schools and universities, and a catastrophic civilian death toll, the full extent of which is only beginning to be revealed (estimates based on available data put the death toll at 64 260 as of June 2024, but the true death toll is undoubtedly higher).[41] *This has been made possible by weapons exported from the USA, Germany, UK and elsewhere,*[42,43] *political support for the Israeli government by Western powers, and financing from institutions including Barclays Bank, which in addition to being the largest funder of fossil fuels in Europe, is also backing the weapons trade with Israel to the tune of over £6 billion.*[44]

In November 2023, fifteen Palestinian Higher Education institutions published a letter to the international community, calling on universities not to remain silent as our Palestinian colleagues are killed.[45] *Across Europe and North America, coalitions of students and staff have been urging their universities to heed this call for solidarity. And yet, the same Western universities which were quick to raise Ukrainian flags and unequivocally condemn the Russian invasion of Ukraine in 2022, are now turning away from the unfolding genocide in Gaza, to avoid controversy. Throughout 2024, a mass movement of students across Europe and North America constructed peaceful solidarity encampments on university campuses, to put pressure on universities and governments to divest from companies associated with Israel's military operations. The response to these protests from university administrations was extremely hostile in some cases, with police being called on protestors and disciplinary action being taken against them.*[46,47] *At the university where I did my PhD, students and staff have been organising to pressure the university to stop banking with Barclays, publicly call for a ceasefire, and to make meaningful gestures of solidarity with Palestinians, as it has done for Ukrainians since 2022. Yet, despite paying lip service to the importance of "decolonisation" in the abstract sense, and despite its declaration of a Climate Emergency in 2019, the university has repeatedly refused to call for a ceasefire in Gaza and an end to the illegal occupation of Palestinian territories, and has been slow to divest from Barclays Bank, which is actively fuelling both the climate crisis and indescribable acts of neo-colonial violence in the occupied Palestinian territories.*

Universities and research institutes often hide behind claims of political neutrality. However, they hold real power in how they choose to manage their investments and finances, and how they give or withhold social legitimacy to, or from, dominant regimes. This uncomfortable truth needs to be acknowledged. Furthermore, universities and research institutes have a responsibility to defend the democratic right of their staff and students to critique and challenge dominant structures and political decisions without fear of reprisal. One of the fundamental aims of knowledge production is to create a more thriving world. This necessitates speaking out against injustice when it occurs, and putting scientific and moral conviction ahead of appeasing the political establishment or corporate interests. It is more important than ever for universities to defend the critical voices of scientists, researchers, and students given that, in the UK where I am based, protest rights are increasingly being eroded, with non-violent climate protests, anti-racism protests, and most recently Palestine solidarity protests being criminalised.[48–50] *And yet, university executives are failing alarmingly when it comes to upholding this responsibility.*

Speaking personally, I have experienced backlash from senior colleagues for my involvement in pro-Palestine advocacy at my university, and I am certainly not alone in this. In the UK, academics who criticise the actions of the Israeli state, and speak out against settler colonialism and its impact on the human rights of Palestinians, have faced defamation and harsh penalties, putting their jobs and livelihoods at risk.[51,52] *While this came as a shock to me, it was not at all surprising to my friends who are people of colour, because they have always lived with the knowledge that the lives of the Global Majority are not valued equally in nation states built on legacies of colonialism and white supremacy.*

Climate justice and the struggle against systemic oppression around the world are interconnected. Action for a safe and liveable future needs to be for a future free from genocide, occupation, and apartheid. Universities and research institutes can make a difference by decarbonising and decolonising their investments, supply chains, and banking practices, by refusing to legitimise oppressive regimes, and by openly condemning systemic violence. But they will not do so without strong pressure from staff and students. It is our responsibility to create this pressure, and to refuse to back down when attempts are made to silence us."

13.5.1 What Does Colonialism Have to do With the Climate and Ecological Emergency?

Climate breakdown and biodiversity collapse are often described as unprecedented existential threats facing humanity. However, for many in the Global Majority [Black, Indigenous and people of colour]¶ across the world, these existential threats are nothing new, but rather, a continuation of historical and ongoing colonialist violence.

Today's global capitalist economy, which is characterised by the intensive extraction and use of fossil fuels, industrial agriculture, mineral mining, and mass-production of consumer commodities, carries echoes of the expansion of European colonial powers from the 16th century, and the extraction of resources from colonised "peripheries" to facilitate the accumulation of capital by the imperial "core".[53] As argued by scholars including Raj Patel, Jason W. Moore, and Kathryn Yusoff, the expansion of European empires into Africa, the Americas, Asia, and Oceania was propelled by an ideology which viewed nature as an exploitable commodity; separate from, and inferior to, European society. This ideology extended to view Indigenous, non-white, and non-male human beings as also part of the inferior and exploitable sphere of nature.[53,54] The seizing and enclosure of land, the rapid depletion of natural resources, and the abuse of soils, water, and air as pollution sinks to fuel capitalist production, were accompanied by the genocide of Indigenous peoples. In addition, as argued by Kehinde Andrews, the development of large-scale resource extraction, mass production, and the rapid growth of consumer markets was made possible by the subjugation and exploitation of entire populations racialised as "other" to the dominant white, colonial class – including millions of enslaved Africans.[55]

Although the world looks different today from how it did from the sixteenth to mid twentieth centuries, many of the violent and destructive power relationships established during the era of European colonial expansion, and the concomitant rise of industrial capitalism, still persist. Systems of resource extraction and capitalist accumulation, which occur through the (corporate and state-backed) appropriation of land and the devaluing and exploitation of marginalised groups, can be considered neo-colonialism.[56]

There are many examples of how neo-colonialism is present in the climate and ecological challenges we face today. Michelle Murphy argues that fossil fuel producers, for example oil refineries in Canada's "Chemical Valley", are able to carry out their highly polluting operations because the rights of the Indigenous peoples of this territory are disregarded due to: "a settler-colonial claim to the legal fiction of *terra nullius*—the land is empty and thus can be taken or anything can be done to it".[57] Far from being a thing of the past,

¶The term "Global Majority" is advocated by Black British scholar Rosemary Campbell-Stephens. It is intended to subvert the otherness and subordination implied by terms such as "BAME" [Black and Minority Ethnic"] and "People of Colour", which define racialised groups in relation to white power structures, and thus reinforce the elite status of a white minority.[93]

colonialist attitudes of entitlement to land, water, and air are very much present in how corporations exploit natural resources, cause lasting environmental damage, and disregard the impacts on local populations, who are often already marginalised on the basis of race, class, Indigeneity, or other axes of vulnerability.[58]

In addition, huge quantities of hazardous waste, which arise from industrial processes but also from consumer products including electronics and plastics, are routinely exported from high-income countries in the Global North to low-income regions of the Global South, taking advantage of the latter's weaker financial position and less stringent environmental protection laws.[59] This causes extensive harm to frontline communities, as there is a lack of effective infrastructure to process waste safely.[60,61] The dumping of toxic waste on marginalised nations and communities has been labelled "waste colonialism" – a term which speaks to the oppressive power relationships at the heart of global systems of production, consumption, and the resulting waste.[62]

Furthermore, grassroots resistance against extractive industries is met with state-sanctioned repression and violence towards Indigenous communities and other Land Defenders. Global Witness has documented the killing of 1910 Land Defenders worldwide since 2012, with the highest number of killings occurring in the Amazonian region of Latin America, where communities are fighting back against gold mining, deforestation, cattle ranching, and other destructive industries which erode Indigenous lifeways and expose the human population and wildlife to serious environmental harm.[63] The complicity of state governments in quashing grassroots resistance to extractivist industries – which can be described as a form of "internal colonialism" based on the definition provided by Eve Tuck and Wayne Yang – is a key pillar which upholds the global extractivist economy.[64]

If we want to pursue change that is genuinely transformative, then it is important that we recognise how the Climate and Ecological Emergency (CEE) is intertwined with colonialism, extractive capitalism, and white supremacy/nationalism/ethno-nationalism. These structures must be dismantled, to create futures in which all people, and more-than-human nature, can thrive. To do this, however, we must also recognise that some mainstream understandings of sustainability and climate action have been part of the problem. This needs to be interrogated before we can move forward, to ensure our work is really part of the solution.

13.5.2 Mainstream Sustainability Strategies are in Danger of Reproducing Colonialist Harms

As awareness of the CEE grows, many universities, scientists, and researchers are increasingly keen to orient their work towards sustainability. While this surge of interest should be welcomed, it is important to examine what exactly is meant by "sustainability", whom it is for, and how it interacts with structures of colonialism and extractive racial capitalism. Below are some of the problems with mainstream sustainability strategies:

- **"Downstream" sustainability innovations do not account for existing inequities.**

  Technological innovation in the name of environmental sustainability often fails to account for existing social inequities and injustices, and this can lead to new forms of harm. For example, the global plastics industry has made significant commitments to increasing the recyclability of their plastic products in recent years, as public concern over plastic pollution grows.[65] At the same time however, the supposed recyclability of plastics is also being used to justify increasing plastic production, with a particular emphasis on pushing single-use plastics into new markets in lower-income Global South countries.[66,67] This ignores the fact that these regions often lack effective infrastructure to recycle or safely process such high quantities of plastic waste, leading to a crisis of pollution, ecological damage, and negative outcomes on human health. In this way, the concept of recycling is not a solution to waste colonialism, but rather, it becomes part of the problem. This does not mean that all recycling innovations are futile. However, without upstream solutions to reduce plastic production, the social and environmental injustices of plastic pollution will only be exacerbated.
- **Sustainability narratives are being used to exert further control over land, resources and communities.**

  Decolonial critics argue that the global sustainability agenda emerged from exclusionary, top-down international policy processes, rooted in Western ideologies of progress. Failure to engage meaningfully with grassroots environmental justice movements and Indigenous communities has resulted in the prioritisation of technological efficiency and consumer behaviour change. This comes at the expense of addressing the industrial drivers of ecologically damaging land use, and of recognising, respecting and protecting the *cultural* sustainability of Indigenous peoples' lifeways. For example, Tim Frandy and B. Marcus Cederström describe how the Indigenous Lac de Flambeau Anishinaabe community in North America are prevented from practising traditional birch bark harvesting for activities such as canoe-building, because of "sustainable" forestry policies, which prioritise access for the mainstream forestry industry. This is in spite of the fact that the Anishinaabe have been sustainably maintaining forest ecosystems for generations.[68]
- **Some sustainability initiatives are funded by unethical corporations, making them tools for corporate reputation management.**

  Worryingly, the language and aesthetics of sustainability, climate change mitigation, and Net Zero are being used as a cover for new forms of colonial violence and dispossession. A notable example is BP, which in December 2023 entered a partnership deal with the British Museum to fund the museum's new strategic plan, including the development of an "energy centre" to decarbonise its operations.[69] The partnership announcement came shortly after news that BP had been granted new gas exploration licences by the state of Israel, in Palestinian territorial waters off the coast of Gaza, as Israel's mass murder and displacement of Palestinian civilians reached new heights.[70] BP's partnership with the museum is a form of greenwashing to distract from its role in climate breakdown and colonial violence. In choosing to partner with BP, the museum provides a cover to one of the most notorious climate-wrecking fossil fuel companies, undermining the museum's own efforts to become more sustainable.

- **Mainstream sustainability appropriates regenerative cultural practices of marginalised peoples.**
  Sustainability has been problematised by scholars who argue that it appropriates and erases the regenerative practices of many Global Majority communities. Many low-income, marginalised groups in the Global Majority practice forms of what Manisha Anantharaman terms "quotidian sustainability", such as waste-saving techniques and the use of non-polluting modes of transport.[71] Yet, mainstream discourses on sustainable living present many of these practices as novel innovations from a typically white or Western-educated elite.[72] Anantharaman also describes how neo-colonial ideas of "clean and green" cities are used to justify policies which target informal workers and the urban poor, displacing them from urban spaces or exposing them to new forms of exploitation, despite the fact that informal workers – for example in the waste picking and scrap dealing trades – have a long history of providing vital services to maintain cities in a sustainable way.[73]

These examples show how dominant conceptualisations of sustainability and climate action can be bound up with problematic assumptions and, at worst, violence and oppression. These power relations are often not interrogated by the institutions championing sustainability, including academia. This highlights the urgent need to pursue frameworks for sustainability and climate justice which are actively anti-colonial,[‖] and which take the lead from Indigenous and other marginalised communities.

13.5.3 Science Can Operate within Anti-colonial Frameworks

There is no universal, one-size-fits-all approach for anti-colonial research and action. Rather, as Max Liboiron and Tina Ngata argue, we must take

‖ Max Liboiron uses the terminology "anti-colonial" instead of "decolonial". This is because of concerns that "decolonisation" has been coopted within the Academy and stripped of its original meaning (Liboiron 2021).

steps to understand the specificity of where we live and work, the power relations which shape the work we do, and what forms of resistance already exist within grassroots and Indigenous communities on the ground.[74,75] That said, there are a number of guiding principles that anti-colonial and Indigenous thinkers and activists have put forward, which are useful jumping-off points in thinking about how to undertake this work.

Some instructive examples can be found in the work of the Civic Laboratory for Environmental Action Research (CLEAR), a research facility working on marine pollution at the Memorial University of Newfoundland, Canada, on the ancestral lands of the Mi'kmaq and Beothuk. Their methods include citational politics which centre Indigenous, Global Majority, and women scholars alongside non-academic sources, hiring or working collaboratively with local community members (regardless of whether they have formal academic qualifications), being accountable to the human and more-than-human relationships in whose lands they work, and prioritising collective wellbeing in the lab over expectations of individual performance.[75]

Another method advocated by CLEAR members is "community peer review". This is a process in which research must be reviewed and approved by the local community, who have the right to ask for changes to better reflect their needs, or to reject the research outright, meaning that it is not put forward for publication.[75] Respecting Indigenous peoples' right of refusal in fieldwork and research that impacts them is also an important way of subverting colonial entitlement to Indigenous knowledge and land. It honours Indigenous communities' right to determine what can be known by colonial research frameworks, and what is out of bounds.[76] Community peer review and honouring the right of refusal are significant subversions of established academic norms, in which the presumed imperative to accumulate and disseminate knowledge (to others in the academy) supersedes humility, community obligations, and justice.

13.5.4 Anti-colonial Political Action for Scientists and Researchers

There is no substitute for supporting real-world, anti-colonial struggles led by Indigenous and Global Majority movements.[64] For those of us who benefit from white privilege, and from receiving an education and credentials from institutions tied to historical and ongoing colonial power structures, this is of the utmost importance. Resistance to extractive capitalism and neo-colonialism is inseparable from support for Indigenous sovereignty and Land Back movements. In the words of Jeff Corntassel, "Land Back" means "returning to consent-based jurisdiction and rematriating self-determining authority back to Indigenous territories".[77] For settler-colonists and those who benefit from colonialism's legacies and ongoing structures, it means giving up land, power, and privilege.

From the uprising of Wet'suwet'en youth against the construction of the Coastal GasLink pipeline in Western Canada,[78] to the resistance of the Sámi against the imposition of onshore wind turbines which disrupt their traditional ways of life in Norway,[79] Indigenous groups are fighting back against extractive industries and the governments which enable them. Statements of support and solidarity, participation in on-the-ground protests, and lobbying political representatives and institutions to oppose these injustices, are critical steps if we want a genuinely transformative, justice-centred response to the climate and ecological emergency.

In addition, scientists and academics should protest against the complicity of our research institutions and universities in ongoing colonial violence. The investments, banking practices, and procurement decisions of universities are often intertwined not only with the fossil fuel industry, but also armaments, military technologies, and regimes committing acts of colonial violence. As I wrote in my personal reflection at the beginning of this essay, campaigning for divestment from the companies which fund this violence, and calling for academic boycotts of states carrying out acts of oppression against subjugated peoples, are a powerful way for scientists and academics to show solidarity with the oppressed. For example, at the time of writing, over 1100 academics and university staff in Norway have signed an open letter calling on Norwegian higher education institutions to refuse to collaborate with institutions in Israel that have ties to the ongoing occupation, apartheid, and brutal violence against Palestinians.[80]

This brief introduction, to just some of the myriad approaches to practising research and activism from an anti-colonial position, is an invitation to reflect on what being anti-colonial can look like in our own research and activism contexts. How can we identify and subvert oppressive ideologies and structures in our work? To whom are we accountable? And how can we align the work we do with grassroots struggles for dismantling colonialism, neo-colonialism, and extractive capitalism? These questions are key, if we want our work to be part of transformative ways forward in the Climate and Ecological Emergency.

13.5.5 Closing Remarks

In the face of crushing systems of violence, oppression, and extractivism, in which human lives and ecosystems are treated as disposable, it is easy to feel powerless. It is easy to be intimidated into silence, especially when facing institutional pushback. As Eve Tuck and Wayne Yang remind us, decolonisation is necessarily unsettling for centres of colonial and settler colonial power, and this includes universities.[64] Speaking out for justice is not always comfortable, and nor should it be.

In these dark times, we can take inspiration from the Global Majority movements fighting back against the destruction of their lands, bodies, and lives, for whom giving up is not an option. We can seek ways to be anti-colonial in how we conduct research, while recognising the limitations built into institutional structures. We can reject mainstream frameworks for "sustainability" which really mean sustaining violent systems. We can show up in solidarity with oppressed peoples, by speaking out against violence and oppression from our official platforms, and actively supporting grass-roots movements for justice. The work is not easy, but it is generative. Something so much better can come when we recognise how much needs to change.

13.6 The Power and Joy of Collective Action: What is It Like to Be a Scientist Activist?

– Why Abi and Valeria wrote this section –

"At the time both biomedical scientists working in London, we first met one another at a climate protest in 2019. Since then, and throughout moves to different cities and new work endeavours, we have participated in a variety of protest actions as part of scientist groups. In parallel we have also taken part in initiatives to make laboratory practices more sustainable at our institutions. However – and despite the good intentions behind these efforts – it is clear to both of us that these changes are insufficient on their own and that collective action as part of activist groups has remained where we feel we can make the biggest impact.

Whilst there are obvious reasons why being activists can be difficult, we have both hugely benefited from being involved in environmental activism and being part of a supportive and inspiring global community that surrounds it. We hope that sharing our experiences – and countering some common narratives not to act – will encourage more scientists off the side-lines and to find out what they can bring to and learn from environmental and social justice movements."

13.6.1 The Experience of Being a Scientist Activist

13.6.1.1 The Time for Action is Now…so Why Are Scientists Standing by?

Right now, scientists across disciplines are aware that we are on the verge of climate, ecological and societal collapse. We understand that the continued burning of fossil fuels and destruction of ecosystems, driven by a series of 'isms' (capitalism, colonialism, consumerism, speciesism, …) exacerbate imminent and existential threats. Scientists also know that urgent action can slow and prevent some of the worst impacts of climate and societal breakdown (discussed throughout this chapter). The pressing question, then, is: with the stakes so high, why do many scientists remain on the sidelines, seemingly more as passive observers than active participants in advocating for the urgent changes needed?

A growing number of scientists have turned their environmental concerns towards 'greening' research practices and institutions, by either founding or joining sustainability teams in their workplace. As our work is a big component of our daily life, it is natural to want to 'set things right' there – and it is important that we do. However, scientists must acknowledge the limitations of this approach in isolation and recognise that we need to go further. In recent years, we've seen the climate crisis being described as the 'biggest threat modern humans have ever faced',[81] scientific reports on the matter warning of 'code red for humanity',[82] and the classification of Earth's life support systems as 'well outside of the safe operating space for humanity'.[83] So, why isn't the scientific community reacting in proportion to its own warnings? What stops us from using our privileged positions to support the communities around the world who we know are already experiencing devastating impacts of climate breakdown first hand? And what is driving a small but growing minority of scientists towards less conventional forms of action?

The hesitation among scientists to become more actively involved in environmental advocacy and activism appears to stem from a variety of personal and institutional barriers. These include cognitive dissonance, where individuals struggle to reconcile their actions with their beliefs, fears of damaging their reputation, a lack of time, and the systemic limitations present within academic and research institutions.[3] Drawing from the published literature as well as our own experiences and those of our peers, we delve into some of the reasons scientists are starting, continuing or avoiding taking more radical action.

13.6.1.2 Climate Activism Comes in Many Flavours

Interpreted literally, "activism" can refer to any action that is taken with the intention of bringing about a social or political change. So, in the broadest sense, most of us are activists for something! More commonly "activism" is used to describe more vigorous or unconventional forms of action that go beyond more traditional methods such as writing letters. The unfortunate reality is that those letters, those warnings,[1] those petitions – and there have been many of them – have not achieved what they needed to or ought to have, often faced with powerful vested interests in sustaining a hugely destructive status quo. The failure of these methods alone and the urgent need for a transformative response to climate breakdown shows us that we need a more dynamic, multifaceted approach in this time of crisis.

The rise of global environmental movements like Fridays for Future and Extinction Rebellion has spotlighted the potential of collective climate action. These movements have been instrumental in raising awareness and motivating people to act, especially in the Global North. Importantly – though unjustly less visibly to many – Indigenous and other Global Majority communities have been battling these crises for much longer, often at great risk to their own lives and safety.[84] The Intergovernmental Panel on Climate Change (IPCC) itself, the United Nations body for assessing the science related to climate change, stated in their latest assessment from 2023 that:

> *"Mass social movements have emerged as catalysing agents in some regions, often building on prior movements including Indigenous Peoples-led movements, youth movements, human rights movements, gender activism, and climate litigation, which is raising awareness and, in some cases, has influenced the outcome and ambition of climate governance".*[1]

Social movements have used a range of different tactics to push for change. They have led legal challenges to governments and polluters, they have organised strikes and boycotts, and they have mobilised massive numbers of people to take to the streets in protest.[85,86] They have drawn on the legacy of social justice movements which have disrupted harmful or oppressive practice using non-violent forms of direct action. These methods include civil disobedience – the active withdrawal of compliance with particular rules or laws of target governments or authorities. Whilst such methods can appear vigorous or extreme to many, they can be incredibly powerful tools, especially in these desperate circumstances where more conventional methods have failed. As the United Nations Secretary General António Guterres puts it: *"We need disruption to end destruction"*.[87]

13.6.1.3 Scientists Choose Non-violent Civil Resistance Because it Works

Social science and history have shown that campaigns based on collective civil resistance are highly effective in bringing change.[88] Our societies remember and often honour movements and figures who have used non-violent civil disobedience amongst their tactics to dismantle oppressive and harmful systems. These include – but are not limited to – the American civil rights, Indian independence, South African anti-apartheid and women's suffrage movements.

As we learn from those examples, it is possible and often necessary to use disruptive, non-violent direct action to challenge and prevent the continuation of damaging and unjust practices, and to pave the way for better, fairer alternatives. Whilst the pace of change is nowhere near fast enough, collective action by 21st century environmental social movements has precipitated noticeable shifts in societal attitudes, and put political, economic, and social frameworks under increasing scrutiny.

13.6.1.4 Climate Activism is Not a Walk in the Park...

We want to be honest. Participating in direct action and/or civil disobedience as part of the climate movement can be challenging. First of all, confronting the stark realities of the ongoing climate breakdown can be daunting, because let's face it, the situation is scary. Indeed, climate scientists and activists often experience feelings of anxiety, fear, grief, dread, anger and many more. For some people these emotions can ignite action. For others it can lead to apathy and denialism. For those who do choose to engage in activism, this may well come with personal and professional repercussions. Especially when we take action publicly, or where we might interact with the police and criminal justice system, we find ourselves considering questions like: "Will I lose the respect of my family, friends, and colleagues?", "Could my job or residency status be at risk?", and "Will this affect my future employment opportunities?". An increasingly polarised media and societal discourse compounds these valid concerns about how we might be perceived and treated if we take certain actions.[89]

We acknowledge that even within our home countries (north-western Europe) – viewed as democratic despite a recent uptick in anti-protest legislation – where scientists can feel comparatively safe engaging in activism, various demographic and situational factors impact upon how straightforward it is for individual scientists to get involved. First, European scientific institutions and movements comprised of scientists are predominantly white, creating potential inclusivity issues for scientists from underrepresented backgrounds. Such barriers to equal participation could be magnified in groups/actions that will likely interact with law enforcement or risk arrest, given that people from Black and other minority backgrounds are disproportionately targeted and adversely affected by police actions and prejudice within the broader criminal justice system. Physical or mental health concerns or disabilities, immigration regulations, caring responsibilities and other interacting personal or financial circumstances, also create differences in the risks that individuals are prepared to – or can be expected to – take.

Nevertheless, people around the globe are challenging the systems that are propelling us towards climate and societal breakdown. In many instances, actions are undertaken despite the serious risks and consequences, and away from mainstream media attention. One recent example is the case of 11 Ugandan students arrested, beaten, and illegally detained for peacefully protesting against the East African Crude Oil Pipeline (EACOP),[90] a project labelled as a "mid-sized carbon bomb" by the Climate Accountability Institute and with disastrous consequences for communities and wildlife of Uganda and neighbouring countries (as well as for Earth's climate).[91] From our privileged position as scientists in the Global North, what role do we have to play in showing solidarity with and amplifying the voices of activists on these frontlines?

Acknowledging that we have a shared, but not necessarily equal, responsibility to act, and that the risks and barriers will be different for each of us, we invite you to consider what kind of actions you could be supporting or participating in. We encourage scientists to seek, shape or start groups where they feel a sense of belonging and purpose.

13.6.1.5 ...But it is Far From "All Doom and Gloom"

Most scientists we have met as part of the climate and wider environmental movement were motivated to join through a sense of duty to act, frustration with the inadequacy of more conventional approaches and/or a belief that these are the spaces where the most rapid societal change can be enacted. Most take part in full knowledge of the personal and professional challenges we may face as a result. But what so many of us – including ourselves – did not initially expect or appreciate was the positive impacts it would have on us personally.

In a world where it can be so easy to feel powerless as individuals, taking collective action can feel refreshing and empowering. This is especially true when we can see our actions changing conversations and policies. But even when we do not see these direct effects, we can often learn a huge amount in these action spaces, which are often fertile grounds for sharing resources and ideas, connecting people and making plans. Taking part in activism and being part of the communities within it can give you a sense of perspective that can liberate you from some of the less-important concerns that often occupy us in today's societies. The support and influence of these communities can also make sustainable lifestyle shifts easier and more likely. Whilst the perceived risks to our professional roles and identities are a common barrier, many scientists in our midst have, in contrast, found there to be benefits to their work life; some have found that taking a more public stance has boosted their visibility and the reach of their work, others have found new collaborations and opportunities as a result of the networks they have made through participating in activism. For both of us, the scientists activist community and the wider global movement of humans working together for a liveable future, has been a source of profound and unexpected connection, support, inspiration, friendship, joy, meaning, motivation and courage.

Not only do scientists benefit from joining social movements, but these movements also benefit from our involvement. We can put our training in understanding and communicating scientific data into practice, and harness the perception of scientists as rational, objective and trustworthy to endorse and strengthen the vital messages of movements pushing for environmental justice.

Amidst the myriad social and environmental crises unfolding worldwide, instead of pondering if scientists should engage in activism, we ask: why shouldn't we, really?[92]

We hope the table below (Table 13.2) will help scientists see activism from a different perspective and overcome common barriers to action. The time for scientists to act is now. It is valuable, it is empowering, and it is necessary.

Table 13.2 Common arguments against scientific activism and corresponding counterarguments.

	Valeria says…	Abi says…
"My actions can't change anything"	Greta Thunberg started her school strikes by sitting on her own… and we all know how that ended. So even at the individual levels, actions do have an impact. First, there are advantages when you act according to your values. As individuals, and perhaps even more as scientists, we can also inspire others and contribute to changing the current social norms. Now, imagine what could be achieved if the majority of scientists turned into activism and cooperated? For me, joining a movement of fellow scientist activists has offered a sense of community and shared purpose, providing both energy and strength. Also, I know that by acting as a collective, our actions are more effective than solitary efforts in pushing for change.	Many of the scientists I meet feel very disempowered. As individuals we often do have more power than we are generally encouraged to realise, but *collectively* we have *much* more power! Seeing people work together to challenge harmful norms and create better alternatives has felt exciting and motivating to me.
"I fear compromising my credibility as a scientist"	Being vocal doesn't mean you are being biased. By participating in activism as scientists we show our commitment to applying scientific understanding to real-world crises. How can scientists expect to convince society to change and demand systemic changes, if they themselves are still doing business-as-usual? 'Preaching water while drinking wine':[94] this is what damages scientists' credibility, not activism.	I'd flip this argument for not engaging with activism on its head! By responding with urgency and intensity to threats that are scientifically proven to be urgent and intense, are we not *more* credible?
"Science and scientists should not be political"	There was never such a thing as a-political science. The history of science is constellated by figures who have not only advanced their fields but also actively engaged with the public for critical issues, from the use of pesticides (Rachael Carson) to the use of nuclear weapons (Einstein).	Scientific evidence is not intrinsically apolitical; our findings have implications in wider society. If science shows us that particular policy directions carry serious risks or important benefits, I think we have a responsibility to make that known.

Table 13.2 (*Continued*)

	Valeria says…	Abi says…
"We should only take action via the 'proper' channels"	We all did. Scientists have been sounding the alarm for decades now, using the conventional approaches of publishing articles, informing policymakers, writing reports and signing petitions. If these methods actually worked, we would not be having this discussion now and we would not be facing climate collapse.	Pretty much every activist I know has had a long history of engaging through traditional channels (and many still do) … but the words, letters, warnings, petitions, *etc.* have too often been suppressed or dismissed. Up against some very powerful vested interests, a diversity of creative and perhaps controversial methods are necessary to push for a livable future.
"Great progress is being made" or "technology will save us"	No, technology won't do the job for us. First, there is currently no technology capable of operating at the necessary scale to fully address the environmental crisis. Second, only relying on technology will fail to tackle the environmental and social injustices that the current global economic system – promoting overexploitation and infinite growth – generates. The promotion of these so-called 'green technologies' by large corporations is therefore largely misleading, allowing them to continue profiting at the expense of millions of people who are left to already confront existential threats unjustly.	Whilst climate and sustainability are a much bigger part of our discourse than they once were, much of the progress towards decarbonisation and protecting nature has been far too slow/incremental and individualised. We need so much more ambition and urgency. We cannot rely on the promise of possible future technology, and it's absolutely not a reason to delay/dilute the actions that we know are necessary to reduce the speed and intensity of climate and ecological breakdown and its devastating impacts on life on Earth. ⟨*Refer to the section about techno-fixes*⟩
"I am too busy/stressed"	Balancing the myriad responsibilities and commitments that come with our profession is undoubtedly challenging. This, combined with the competitive nature of research fields, can often lead to burnout and mental health struggles. How on Earth could you add something as challenging as activism on top? In my case, the driving force was the realisation that I couldn't just continue with my usual research activities, fully aware of the dire state of the planet. And once you know, you can't unlearn	Our work and our lives outside of it place pressures on us that can feel very consuming and overwhelming. For me, the prospect of participating in activism initially felt like I'd be introducing an additional burden on an already overstretched person… but actually it has helped me realign my priorities and actions with what I see to be important. It has helped me understand – and benefit from – the huge power of community and collaboration… and has changed my life for the better.

	it. The silver lining is that my activities and the people I collaborate with are now more in line with my sense of purpose, and the community I have found has become an invaluable resource. However, I believe things could be much easier if academia actually provided scientists with the time and support needed to focus on sustainability and environmental crisis mitigation. Unfortunately, at present, engagement in activism is, at best, merely tolerated and often ostracised by institutions.	
"It's not my job"	I started my career because I wanted to make a positive contribution to society. So did many fellow scientists. Advocating to avert a total climate collapse is doing exactly that. Science itself is showing that humanity faces an unprecedented planetary crisis. I simply couldn't carry on doing my job as if this wasn't happening. And what is the point of doing research with the idea of potentially saving lives in the future, and then not doing what you can to mitigate the crises of the present?	Whose job is it to push for transformative action in this emergency situation? I would argue that all of us have a role to play, though some of us have more responsibility and more power to do so. Scientists are often in positions where we *can* contribute to creating fairer, more resilient systems fit for the 21st century. . .so let's do it!

Abbreviations

CEE	Climate and ecological emergency
CLEAR	Civic Laboratory for Environmental Action Research
CO_2	Carbon dioxide
CCS	Carbon capture and storage
EACOP	East African crude oil pipeline
EOR	Enhanced oil recovery
GHG	Greenhouse gas
IPCC	Intergovernmental Panel on Climate Change
WHO	World Health Organisation

Acknowledgements

We are grateful to Nikoline Borgermann, Lynn Bjerke and Janine O'Keeffe, whose actions continue to inspire us all, and who played an important role throughout this writing journey—contributing their ideas to this chapter and helping to shape its content. We hope they are pleased with the result.

References

1. IPCC, Climate Change 2023: Synthesis Report, Contribution of Working Groups I, II and III to the Sixth Assessment Report of the Intergovernmental Panel on Climate Change, IPCC, Geneva, 2023.
2. World Bank, Poverty, Prosperity, and Planet Report 2024: Pathways Out of the Polycrisis, Washington, DC, 2024.
3. A. Thierry, L. Horn, P. von Hellermann and C. J. Gardner, "No research on a dead planet": preserving the socio-ecological conditions for academia, *Front. Educ.*, 2023, **8**, 1237076.
4. P. Friedlingstein, M. O'Sullivan and M. W. Jones, *et al.*, Global Carbon Budget, *Earth Syst. Sci. Data Discuss.*, 20242024, 1–133.
5. J. E. Hansen, P. Kharecha, M. Sato, G. Tselioudis, J. Kelly, S. E. Bauer, R. Ruedy, E. Jeong, Q. Jin, E. Rignot, I. Velicogna, M. R. Schoeberl, K. von Schuckmann, J. Amponsem, J. Cao, A. Keskinen, J. Li and A. Pokela, Global Warming Has Accelerated: Are the United Nations and the Public Well-Informed?, *Environ.: Sci. Policy Sustainable Dev.*, 2025, **67**, 6–44.
6. *Oil Change International, Funding Failure: Carbon Capture and Fossil Hydrogen Subsidies Exposed*, Oil Change International, Washington, DC, US, 2024.
7. Fossil Free Research, Fossil Free Research, https://www.fossilfreeresearch.org. Accessed 9 March 2024.
8. Equinor, Basic research agreements worth NOK 315 million, https://www.equinor.com/news/archive/2019-02-01-ntnu. Accessed 7 February 2024.
9. J. Ferguson and C. Matthews, European universities accept €260 million in fossil fuel money, https://www.investigate-europe.eu/posts/european-universities-accept-260-million-euros-fossil-fuel-money. Accessed 9 March 2024.

10. A. M. Weinberg, Can Technology Replace Social Engineering?, *Am. Behav. Sci.*, 1967, **10**, 7–7.
11. H. S. Sætra and E. Selinger, The Siren Song of Technological Remedies for Social Problems: Defining, Demarcating, and Evaluating Techno-Fixes and Techno-Solutionism, *SSRN*, 2023, preprint, SSRN:4576687, https://papers.ssrn.com/abstract=4576687.
12. M. Huesemann and J. Huesemann, *Techno-Fix: Why Technology Won't Save Us Or the Environment*, New Society Publishers, 2011.
13. P. H. G. Berkhout, J. C. Muskens and J. W. Velthuijsen, Defining the rebound effect, *Energy Policy*, 2000, **28**, 425–432.
14. E. Aydın, D. Brounen and A. Ergün, The rebound effect of solar panel adoption: Evidence from Dutch households, *Energy Econ.*, 2023, **120**, 106645.
15. R. Galvin, E. Dütschke and J. Weiß, A conceptual framework for understanding rebound effects with renewable electricity: A new challenge for decarbonizing the electricity sector, *Renewable Energy*, 2021, **176**, 423–432.
16. J.-A. Vélez-Henao, C.-M. García-Mazo, J. Freire-González and D. F. Vivanco, Environmental rebound effect of energy efficiency improvements in Colombian households, *Energy Policy*, 2020, **145**, 111697.
17. FAO, Global Forest Resource Assessment 2020, http://www.fao.org/forest-resources-assessment/2020. Accessed 9 March 2024.
18. UNEP, Rising sea surface temperatures driving the loss of 14 percent of corals since 2009, http://www.unep.org/news-and-stories/press-release/rising-sea-surface-temperatures-driving-loss-14-percent-corals-2009. Accessed 9 March 2024.
19. UNFCCC, Wetlands Disappearing Three Times Faster than Forests, https://unfccc.int/news/wetlands-disappearing-three-times-faster-than-forests. Accessed 9 March 2024.
20. D. Gonzalez, Wind farms divide Indigenous communities in Colombia, https://dialogochino.net/en/climate-energy/368855-wind-farms-divide-indigenous-communities-in-colombia/. Accessed 9 March 2024.
21. IWGIA, Press Release - Sámi Activists Demand Removal of Wind Turbines in Fosen, https://www.iwgia.org/en/news/5278-press-fosen-oct2023.html. Accessed 9 March 2024.
22. M. Najib, Dozens hurt in protests against wind turbine project in Golan Heights, https://arab.news/gkeju. Accessed 9 March 2024.
23. WSRW, Western Sahara Resource Watch | Morocco plans colossal solar farm for occupied Boujdour, https://wsrw.org/en/news/morocco-plans-colossal-solar-farm-for-occupied-boujdour. Accessed 9 March 2024.
24. L. Caramel, Wind farms in Brazil are encroaching on traditional community land, https://dialogochino.net/en/climate-energy/60621-brazil-wind-energy-advances-traditional-community-land/. Accessed 9 March 2024.
25. News, Protesters in Mainland block road to wind power test site over water supply fears|CBC News, https://www.cbc.ca/news/canada/newfoundland-labrador/port-au-port-road-blockage-1.6725920. Accessed 9 March 2024.

26. S. Pal, India's Clean Energy Push, https://www.newsclick.in/india-clean-energy-push-protests-intensify-kutch-windmill-plant-forest-land. Accessed 9 March 2024.
27. A. Dunlap and J. Jakobsen, in *The Violent Technologies of Extraction: Political ecology, critical agrarian studies and the capitalist worldeater*, ed. A. Dunlap and J. Jakobsen, Springer International Publishing, Cham, 2020, pp. 43–72.
28. D. Andreucci, G. García López, I. M. Radhuber, M. Conde, D. M. Voskoboynik, J. D. Farrugia and C. Zografos, The coloniality of green extractivism: Unearthing decarbonisation by dispossession through the case of nickel, *Political Geogr.*, 2023, **107**, 102997.
29. J. Totland, Stop the Nussir mine – save Repparfjord!, https://www.nu.no/stop-the-nussir-mine-save-repparfjord/. Accessed 9 March 2024.
30. EnergiWatch, Strømanalytiker mener Google-etablering vil øke strømprisene i Sør-Norge, https://energiwatch.no/nyheter/politikk_marked/article16828794.ece. Accessed 9 March 2024.
31. L. Winner, Do Artifacts Have Politics?, *Daedalus*, 1980, **109**, 121–136.
32. J. Buolamwini and T. Gebru, in *Proceedings of the 1st Conference on Fairness, Accountability and Transparency*, PMLR, 2018, pp. 77–91.
33. J. Ma, L. Li, H. Wang, Y. Du, J. Ma, X. Zhang and Z. Wang, Carbon Capture and Storage: History and the Road Ahead, *Engineering*, 2022, **14**, 33–43.
34. J. Sekera and A. Lichtenberger, Assessing Carbon Capture: Public Policy, Science, and Societal Need, *Biophys. Econ. Sustainability*, 2020, **5**, 1–28.
35. Global CCS Institute, Global Status of CCS 2023*:* Scalig up through 20230, Global CCS Institute, Melbourne, Aus., 2023.
36. A. Feenberg, *Transforming Technology: A Critical Theory Revisited*, Oxford University Press, New York, N.Y., 2nd edn, 2002.
37. S. Costanza-Chock, *Design Justice: Towards an Intersectional Feminist Framework for Design Theory and Practice*, Social Science Research Network, Rochester, NY, 2018.
38. Tech Divestment Network, https://www.techdivestment.network. Accessed 3 February 2025.
39. No Tech For Apartheid, https://www.notechforapartheid.com. Accessed 3 February 2025.
40. J. Hickel, On Technology and Degrowth, https://monthlyreview.org/2023/07/01/on-technology-and-degrowth/. Accessed 12 July 2023.
41. UNRWA, UNRWA Situation Report #155 on the Humanitarian Crisis in the Gaza Strip and the West Bank, including East Jerusalem, https://www.unrwa.org/resources/reports/unrwa-situation-report-155-situation-gaza-strip-and-west-bank-including-east-jerusalem. Accessed 27 January 2025.
42. L. Brooke-Holland, UK Arms Exports to Israel, Research Briefing, House of Commons Library, 2025.
43. S. Kirchgaessner, 18th January 2024. 'Different rules': special policies keep US supplying weapons to Israel despite alleged abuses, https://www.theguardian.com/world/2024/jan/18/us-supply-weapons-israel-alleged-abuses-human-rights. Accessed 27 January 2025.

44. War on Want, Barclays Bank: Arming Apartheid, War on Want, 2022. Available at https://waronwant.org/news-analysis/barclays-bankrolling-genocide-apartheid?link_id=8&can_id=28f558193a30a2b93937b194de45b70e&source=email-from-the-river-to-the-sea-palestine-will-be-free&email_referrer=email_2084275&email_subject=from-the-river-to-the-sea-palestine-will-be-free. Accessed 28 January 2025.
45. Birzeit University, A unified call for justice and freedom in Palestine, https://www.birzeit.edu/en/news/unified-call-justice-and-freedom-palestine. Accessed 27 January 2025.
46. N. Kouri-Towe and S. Matthews, Debates on campus safety in response to Palestine solidarity activism show we need strategies to navigate discomfort, https://theconversation.com/debates-on-campus-safety-in-response-to-palestine-solidarity-activism-show-we-need-strategies-to-navigate-discomfort-228865. Accessed 27 January 2025.
47. M. Mirza, War on Gaza: Why are UK universities inviting police to curb student protests?, https://www.middleeasteye.net/opinion/gaza-war-uk-universities-police-crack-down-protesters-why. Accessed 27 January 2025.
48. B. Brentnall, How Britain's new laws doubly criminalise Black, Asian and Gypsy people', https://www.opendemocracy.net/en/police-bill-nationality-borders-parliament-protest-ethnic-minority-activists/. Accessed 27 January 2025.
49. E. Ng and B. Davis, More than 70 arrested at pro-Palestine rally in London ahead of Israel-Hamas ceasefire, https://www.independent.co.uk/news/uk/home-news/palestine-march-london-gaza-israel-b2682076.html. Accessed 27 January 2025.
50. C. Davies and H. Sherwood, UK professor suffered discrimination due to anti-Zionist beliefs, tribunal rules, https://www.theguardian.com/education/2024/feb/05/uk-professor-suffered-discrimination-due-to-anti-zionist-beliefs-tribunal-rules. Accessed 27 January 2025.
51. C. C. Charrett, How a Palestinian academic defeated a campaign to silence her, https://www.aljazeera.com/opinions/2022/2/10/how-a-palestinian-academic-defeated-a-campaign-to-silence-her. Accessed 27 January 2025.
52. R. Adams, UK research funding body in row with ministers over free speech in Gaza, https://www.theguardian.com/education/2023/nov/03/uk-research-funding-body-in-row-with-ministers-over-free-speech-and-gaza. Accessed 27 January 2025.
53. C. W. Chagnon, F. Durante and B. K. Gills, *et al.*, From Extractivism to Global Extractivism: the Evolution of an Organizing Concept, *J. Peasant Stud.*, 2022, **49**, 760–792.
54. R. Patel and J. W. Moore, *A History of the World in Seven Cheap Things: a Guide to Capitalism, Nature, and the Future of the Planet*, Verso, London, 2018.
55. K. Andrews, *The New Age of Empire: How Racism and Colonialism Still Rule the World*, Allen Lane, London, 2021.
56. K. Fuentes-George, The legacy of colonialism on contemporary climate governance, *Georgetown J. Int. Aff.*, 2023, **24**, 91–97.

57. M. Murphy, Some keywords toward decolonial methods: studying settler colonial histories and environmental violence from Tkaronto, *Hist. Theory*, 2020, **59**, 376–384.
58. M. Liboiron, *Pollution is Colonialism*, Duke University Press, Durham, 2021.
59. E. S. Pope, The shadowy world of hazardous waste disposal: why the Basel Convention's structure undermines its substance, *S. C. J. Int. Law Bus.*, 2017, **13**, 305–338.
60. A. Njoku, M. Agbalenyo and J. Laude, *et al.*, Environmental injustice and electronic waste in Ghana: Challenges and recommendations, *Int. J. Environ. Res. Public Health*, 2024, **21**, Article 25.
61. GAIA, Discarded: Communities on the Frontline of the Global Plastics Crisis, Report-July-12-2019-Spreads-no-marks-1.pdf (no-burn.org). Accessed 27 January 2025.
62. L. A. W. Pratt, Decreasing dirty dumping: a reevaluation of toxic waste colonialism and the global management of transboundary hazardous waste, *Tex. Environ. Law J.*, 2011, **41**, 147–180.
63. Global Witness, Standing Firm: The Land and Environmental Defenders on the frontline of the environmental crisis, https://www.globalwitness.org/en/campaigns/environmental-activists/standing-firm/. Accessed 27 January 2025.
64. E. Tuck and W. Yang, Decolonization is not a metaphor, *Decolonization: Indigeneity, Educ. Soc.*, 2012, **1**, 1-40.
65. Z. Diana, K. Reilly and R. Karasik, *et al.*, Voluntary commitments made by the world's largest companies focus on recycling and packaging over other actions to address the plastics crisis, *One Earth*, 2022, **11**, 1286–1306.
66. J. Vandenberg, Plastic politics of delay: How political corporate social responsibility discourses produce and reinforce inequality in plastic waste governance, *Global Environ. Polit.*, 2024, **24**, 122–145.
67. L. S. G. Akuoko, J. Vandenberg and J. C. Falman, *et al.*, Rethinking plastic realities in Ghana: A call for a wellbeing approach to understanding human-plastics entanglements for more equitable plastics governance, *Mar. Policy*, 2023, **158**, 105856.
68. T. Frandy and B. M. Cederström, Sustainable power: Decolonising sustainability through Anishinaabe birchbark canoe building, in *Going Beyond*, Springer, Cham, 2017, pp. 217–230.
69. V. Babbs, The British Museum has struck a new $63 million partnership deal with BP, https://news.artnet.com/art-world/british-museum-bp-partnership-2412030#:~:text=On%2520December%252019%252C%2520the%2520British,63.2%2520million)%2520deal%2520with%2520BP. Accessed 27 January 2025.
70. Energy Embargo for Palestine, Why we occupied the British Museum, https://jacobin.com/2024/02/british-museum-israel-bp-gaza. Accessed 28 January 2025.
71. M. Anantharaman, Critical sustainable consumption: a research agenda, *J. Environ. Stud. Sci.*, 2018, **8**, 53–561.

72. L. Siragusa and D. Arzyutov, Nothing goes to waste: sustainable practices of re-use among indigenous groups in the Russian North, *Curr. Opin. Environ. Sustainability*, 2020, **43**, 41–48.
73. M. Anantharaman, *Recycling Class: The Contradictions of Inclusion in Urban Sustainability*, MIT Press, Cambridge, 2024.
74. M. Liboiron, and T. Ngata, Māori Plastic Pollution Expertise and Action in Aotearoa, https://civiclaboratory.nl/2020/07/13/maori-plastic-pollution-expertise-and-action-in-aotearoa/. Accessed 02 February 2024.
75. M. Liboiron, *CLEAR Lab Book: A living manual of our values, guidelines, and protocols, Civic Laboratory for Environmental Action Research (CLEAR)*, Memorial University of Newfoundland, St. John's, v3, 2021.
76. E. Tuck and K. W. Yang, Unbecoming claims: pedagogies of refusal in qualitative research, *Quant. Enquiry*, 2014, **2**, 811–818.
77. J. Corntassel, Indigenous laws on indigenous lands: Land Back as community resurgence, *Rooted*, 2021, **1**, 15–23.
78. Amnesty International, Closing arguments heard in court case of Wet'suwet'en land defenders, https://amnesty.*ca*/human-rights-news/closing-arguments-heard-wetsuweten-court-case/. Accessed 27 January 2025.
79. M. A. Hess, Norway's treatment of Sámi Indigenous people makes a mockery of its progressive image, https://jacobin.com/2023/03/norway-sami-indigenous-people-reindeer-herding-wind-turbines-dispossession-protest. Accessed 28 January 2025.
80. M. Van der Velden and H. Hermansen, Opprop for akedemisk boikott av Israel, https://docs.google.com/forms/d/e/1FAIpQLSd5j39Wf6rd9QEEeH28v9W-ba3tOQ7l0ZmPdCVapG79O7KY1A/viewform. Accessed 28 January 2025.
81. UN, Climate Change 'Biggest Threat Modern Humans Have Ever Faced', World-Renowned Naturalist Tells Security Council, Calls for Greater Global, https://press.un.org/en/2021/sc14445.doc.htm. Accessed 30 April 2024.
82. UN News, Code red' for human driven global heating, warns UN chief, https://news.un.org/en/story/2021/08/1097362. Accessed 30 April 2024.
83. K. Richardson, W. Steffen and W. Lucht, *et al.*, Earth beyond six of nine planetary boundaries, *Sci. Adv.*, 2023, **9**, eadh2458.
84. A. Scheidel, D. D. Bene and J. Liu, *et al.*, Environmental conflicts and defenders: A global overview, *Glob. Environ. Change*, 2020, **63**, 102–104.
85. The National Archives, Protest in the 20th century, https://www.nationalarchives.gov.uk/explore-the-collection/explore-by-topic/democracy-and-protest/protest-20th-century/.
86. Easy Sociology, What is Passive Resistance? https://easysociology.com/sociology-of-social-movements/what-is-passive-resistance/#h-tactics-and-strategies.
87. UNSDG, 'Act decisively before it is too late', Guterres warns countries, laying out his priorities for 2023, https://unsdg.un.org/latest/stories/act-decisively-it-too-late-guterres-warns-countries-laying-out-his-priorities-2023. Accessed 30 April 2024.

88. M. J. Stephan and E. Chenoweth, Why civil resistance works: The strategic logic of nonviolent conflict, *Int. Secur.*, 2008, **33**, 7–42.
89. T. D. Wyatt, C. J. Gardner and A. Thierry, Actions speak louder than words: the case for responsible scientific activism in an era of planetary emergency, *R. Soc. Open Sci.*, 2024, **11**.
90. N. Lakhani, Ugandan climate activists face charges after a month in maximum security jail, https://www.theguardian.com/world/2024/feb/06/uganda-climate-pipeline-activists-students-jail-charges. Accessed 14 January 2025.
91. D. Carrington, 'Monstrous' east African oil project will emit vast amounts of carbon, data shows, https://www.theguardian.com/environment/2022/oct/27/east-african-crude-oil-pipeline-carbon. Accessed 14 January 2025.
92. T. Fossen, T, Academic Activism and the Climate Crisis: Should Scholars Protest?, *Perspect. Polit.*, 2025, 1–16.
93. R. M. Campbell-Stephens, *Educational Leadership and the Global Majority*, Palgrave Macmillan, Cham, Switzerland, 2021, pp. 1–21.
94. N. Borgermann, A. Schmidt and J. Dobbelaere, Preaching water while drinking wine: Why universitites must boost climate action now, *One Earth*, 2022, **5**, 18–21.

Subject Index